Perspectives in Control

Springer
London
Berlin
Heidelberg
New York
Barcelona
Budapest
Hong Kong
Milan
Paris
Santa Clara
Singapore
Tokyo

Dorothée Normand-Cyrot (ed.)
under the auspices of the GdR-Automatique

Perspectives in Control

Theory and Applications

a tribute to Ioan Doré Landau

Springer

Dr Dorothée Normand-Cyrot
Laboratoire des Signaux et Systèmes, CNRS - Supélec, Plateau de Moulon,
91190 Gif-sur-Yvette, France

ISBN-13:978-1-4471-1278-5

British Library Cataloguing in Publication Data
Perspectives in control : theory and applications
1.Landau, I. D. 2.Electric engineers - France 3.Electric
engineering - Congresses
I.Normand-Cyrot, D.
621.3'092
ISBN-13:978-1-4471-1278-5 e-ISBN-13:978-1-4471-1276-1
DOI: 10.1007/978-1-4471-1276-1

Library of Congress Cataloging-in-Publication Data
A catalog record for this book is available form the Library of Congress

Softcover reprint of the hardcover 1st edition 1998

Typesetting: Camera ready by editor

69/3830-543210 Printed on acid-free paper

Nothing is more practical than a good theory

L. Boltzmann

The "GdR–Automatique" decided to organize the colloquium "Perspectives in Control" to celebrate the sixtieth birthday of its founder Ioan Doré Landau. Apart from his leading scientific activities, he created and was in charge of several national and European structures which greatly contributed to the development and promotion of research in control theory.

The "GdR–Automatique", created in 1988 by I.D. Landau, is a National Co-ordinated Research Programme in Control of the CNRS and has been under his responsibility up to 1996. In 1996, under the direction of J. Descusse, the "GdR–Automatique" was officially recognized and supported by both the CNRS and the French Ministry of Research.

It is presently under the direction of D. Normand-Cyrot. The members of the Board of Directors are: H. Alla from LAG in Grenoble, A. Benveniste from IRISA in Rennes, P. Chemouil from CNET in Issy-les-Moulineaux, J.M. Coron from University of Orsay, Y. Dallery from LIP6 in Paris, J. Descusse from IRCyN in Nantes, M. Guglielmi from IRCyN in Nantes, R. Lozano from Heudiasyc in Compiègne, P. Millot from LAMIH in Valenciennes, A. Oustaloup from LAP in Bordeaux, A. Rault from PSA in Velizy, G Salut from LAAS in Toulouse, C. Samson from INRIA in Sophia Antipolis, M. Sidahmed from CETIM in Senlis, J.L. Thomas from AAR in Marcoussis.

The Colloquium received the financial support from:

CNRS – Centre National de la Recherche Scientifique
MENRT – Ministère de l'Education Nationale, de la Recherche et de la Technologie.

ACKNOWLEDGEMENTS:

Many colleagues, students and secretaries contributed to the preparation of this book and the colloquium. They were of great assistance in helping to keep the deadlines. We wish to thank all of them, with a particular mention to Nicole Labaronne and Leslie Hollett. We express our sincerest gratitude to Marianne Groen and Jochen Langer who actively contributed to the book edition.

Ioan,

quel automaticien serais-tu devenu si, un jour de septembre 1982, tu n'avais pas découvert les merveilles de la rétroaction, en croisant au large de Belle-Isle. A bord d'un fin voilier tu regardais émerveillé le κυβερνητηζ, (pilote) automatique, conserver sans faille le cap, à destination d'une réunion scientifique qui est restée dans les mémoires. Pour garder le cap, point d'électronique sophistiquée, mais un système mécanique totalement fiable. Un seul défaut qui te sauta immédiatement aux yeux : la consigne était en fait la direction du vent apparent, image du cap modulo la vitesse. Tu imaginas d'emblée la solution : adapter la consigne en fonction du vent réel, pour un cap donné ; élémentaire, pour l'expert de l'adaptatif que tu étais déjà. Mais sur ce navire tu as découvert toute la puissance de la κυβερνητικοζ, je veux dire la cybernétique ou l'art de gouverner, et ta pensée en fut profondément modifiée. Imaginons que ce jour là tu aies pris le ferry, la face de l'Automatique en eut été changée. Décidément, il y a bien trois sortes d'hommes: les vivants, les morts et ceux qui vont sur la mer. Enfin presque, car au retour de Belle-Isle, deux de nos collègues et amis, tout aussi brillants automaticiens que toi, qui m'accompagnaient, n'ont pas eu ta sagacité. Il est vrai qu'Eole ce jour là n'était pas au rendez-vous.

Bon vent Ioan.

Jacques

Ioan Doré Landau was born on July 1, 1938, in Bucharest, Romania. He received his engineering degree in electronics from the Institut Polytechnique of Bucharest in 1959, his Ph.D. degree in 1965, his degree of Docteur-es-Sciences Physiques from the University of Grenoble in 1973.

After having occupied several positions in R&D in Romania (including one year in the group of V.M. Popov), he joined the Control and Electronics Research Dept. of ALSTHOM in 1968. He was a Senior Post-doctoral Research Associate at NASA - Ames Research Center in 1971-72. From 1973 to 1976 he was an Associate Professor at the Institut National Polytechnique de Grenoble. From 1986 he is Research Director at CNRS and works at the Laboratoire d'Automatique de Grenoble (INPG/CNRS). He was Director of the LAG from 1987 to 1990. At CNRS he launched and directed the co-ordinated research programmes: Mathematical Tools and Models for Control, System Analysis and Signal Processing (1979-1982), Adaptive Systems in Control and Signal Processing (1984-1988) and Automatique (1988-1996).

His research interests encompass theory and applications in system identification, adaptive control, robust digital control and nonlinear systems. He was author and co-author of over 200 papers and several books: Adaptive Control - The Model Reference Approach (Dekker 1979) translated in Chinese, System Identification and Control Design (Hermès 1988, 1993, Prentice Hall 1990, Editura Technica 1997), Adaptive Control - Theory and Practice (with M. Tomizuka) (in Japanese - Ohm 1981) and Adaptive Control (with R. Lozano and M. M'Saad) (Springer Verlag 1997). He edited and co-edited several books in French, a series on "Mathematical Tools for Control, System Analysis and Signal Processing and Models" and on "Adaptive Control". He holds several patents and was at the origin of the transfer to industry of several software packages. He advised 29 Ph.D. students. He delivered a number of Plenary Lectures at International Conferences including ACC in Seattle in 1995 and was the keynote speaker at the ECC-97 in Brussels.

Dr. Landau received the Great Gold Medal at the Invention Exhibition Vienna (1968) for his patent on "Variable Speed Control of Induction Motors using Static Frequency Converters", the CNRS Silver Medal (1982), the "Best Review Paper Award" (1981-84) for his paper "Model Reference Adaptive Control and Stochastic Self-tuning Regulators - A Unified Approach" and the price Monpetit from the French Academy of Science. He was a Russell Severance Springer Professor at Berkeley, Dept. of Mechanical Engineering in 1992.

Dr. Landau was the General Chairman of the first European Control Conference organized in Grenoble in 1991. He was one of the founders and the first President of the European Community Control Association (ECCA). He is Editor-in-Chief of the European Journal of Control.

Doctoral Students of I. D. Landau

Name	Year	Present Affiliation
B. Courtiol	1976	ALSTHOM, France
C. Béthoux	1976	Deceased
M. España	1977	CONAE, Argentina
J.M. Dion	1977, 1983*	Lab. d'Automatique de Grenoble, France
A. Gauthier	1977	University of Bogota, Colombia
R. Bénéjean	1977	Electricity of France
H. Medeiros Silveira	1978*	University of Santa Catarina, Brazil
L. Dugard	1980, 1984*	Lab. d'Automatique de Grenoble, France
R. Lozano	1981	University of Compiègne, France
W.W. Loy	1981	National University of Singapore
H. Bourlès	1982	ENSAM, Paris, France
S. Cabrera	1984	Option, Grenoble, France
R. Ortega	1984*	LSS, SUPELEC, Gif/Yvette, France
R. Kelly	1986	CISESE, Ensenada, Mexico
M. de la Sen	1986*	University of Viscaya, Spain
M. M'Saad	1987*	University of Caen, France
C. Canudas[1]	1987	Lab. d'Automatique de Grenoble, France,
N. M'Sirdi	1988*	University Paris X, Versailles, France
M. Duque[2]	1989	University of Bogota, Colombia
M. Samaan[2]	1989	Electricity of France
A. Montano [3]	1989	Inst. Tehnologico de Torreon, Mexico
H. Abidin-Tjokronegoro	1990	University of Bandung, Indonesia
L. Schmitt[2]	1990	Schneider, Grenoble, France
B. Brogliato[4]	1991	Lab. d'Automatique de Grenoble, France,
H.N. Duong	1993	University of Ho Chi Minh, Vietnam
C. Cyrot	1993	CNRS, Marseille, France
A. Voda-Besançon	1994	Lab. d'Automatique de Grenoble, France
A. Karimi	1997	
J. Langer	1998	ITT Automotive, Frankfurt, Germany

[1] Co-advised with K.J. Åström
[2] Co-advised with M. M'Saad
[3] Co-advised with D. Normand-Cyrot
[4] Co-advised with R. Lozano
* State doctorate

Preface

This book, realized on the occasion of the Colloquium "Perspectives in Control" held at the CNRS (Paris, 29-30 June, 1998) comprises twenty-one essays of leading experts in Control chosen among the scientists, friends and students who have accompanied the career of Ioan Doré Landau and on many occasions collaborated with him. Part of these articles were presented as plenary lectures during the Colloquium.

This meeting provided a unique opportunity to solicit the views of these world-renowned authorities on some of the directions in which control discipline might evolve in various areas at the threshold of the 21st century. The variety of articles which include advanced methodological contributions, overview tutorials as well as more philosophical reflections, contributed to the richness of the book. Many of the pioneering aspects of Landau's work are dealt with in the book.

Oriented to a rather philosophical reflection is the essay of A.J. Fossard who introduces the book discussing the inherent compromise between formal fundamental knowledge and the needs of practitioners when facing control problems.

Expectations, benefits and drawbacks of adaptive control motivated several papers. Three fundamental problems, the need for a model, the control objective accuracy and the transient unstability are described by B.D.O. Anderson and M. Gevers who propose how to address these problems. A.S. Morse presents the ways of an emerging convincing computer-aided adaptive control design methodology through supervisor and switching in the framework of input-output performance theory. Identification and adaptive control of discrete-time linear systems are treated in the article by H.M. Silveira.

Several contributions are concerned with dissipativity concepts. P. Kokotovic and D. Fontaine show the wide potentiality of passivity through nonlinear stabilization in terms of dynamic backstepping and cascade passivation. The case of uncertain time-delay systems is discussed in the article of J.-M. Dion, L. Dugard, D. Ivanescu, S.I. Niculescu and V. Ionescu through a generalized Popov theory. The benchmark example of controlling the inverted pendulum is addressed by R. Lozano and I. Fantoni thanks to its passivity properties. M. Torres and R. Ortega compare the ability of three nonlinear stabilizing methods on the basis of the experimental example of a magnetically levitated ball.

New methodologies towards robust control design are proposed. K.J. Åström develops the paradox of pole placement in a robust control design and proposes a design methodology. A. Isidori revisits stabilization via output feedback and opens a totally new perspective for robustly stabilizing nonlinear unstable and non-minimum phase systems. M. Fliess, H. Sira-Ramirez and R. Marquez illustrate through concrete examples the flatness-based control synthesis, an experimental flexible structure, a dc-to-dc power converter and a PVTOL aircraft.

State concept and dissipativity meaning are at the heart of the paper by J.C. Willems whose thesis is to claim that storage is a state function. State is also emphasized by D. Auslander as a powerful tool for dealing with the complexity of modern machine control. A different way of representing discrete-time state dynamics is proposed by S. Monaco and D. Normand-Cyrot.

System identification is presented too. L. Ljung and U. Forssell answer the question of how to obtain consistent estimates of the dynamics even with an erroneous noise model in closed-loop operations. System identification and its relationship with diagnosis and model validation in the context of mathematical statistics is stressed in the overview paper by A. Benveniste, B. Delyon and M. Basseville.

Prospective applied areas are promoted. Digital and adaptive control theories are shown fundamental in mechatronics systems by M. Tomizuka. Y. Bar-Shalom, H. Wang and T. Kirubarajan develop a multi-sensor multi-target algorithm for air traffic surveillance. C.D. Mote and Jr.E.M. Mockensturm address the problem of modelling wide webs. E. Irving and A. Libaux discuss classical and advanced control for hydraulic power plants.

The editor felt that the keynote lecture given by I.D. Landau at the ECC-97 in Brussels, which enlightens his ideas on competitiveness of Control is in complete harmony with the other contributions and may constitute the conclusion of the book.

The editor wishes to thank all the authors for their outstanding work and participation in the conference so providing a remarkable collection of views on the wide area of Control.

Paris, March 1998
Dorothée Normand-Cyrot

Contents

Introduction

Part 1: Adaptive Control

its expectations, open questions and perspectives

Part 2: Passivity

a wide concept

Part 3: Control Design Methodologies

robustness issues

Part 4: State

a fundamental concept

Part 5: Identification - Supervision - Diagnosis
towards the future

Part 6: Advanced Applied Areas
theory versus applications

Conclusion

Appendix

Control: From Research to Industry, from Theory to Application

A.J. Fossard

Professeur at SUPAERO, Directeur de Recherches at l'ONERA, 10, avenue E. Belin, 31400 Toulouse, France

1. Ioan

I have been invited to participate in this Colloquium on "Perspectives in Control" organized as a tribute to our colleague Ioan D. LANDAU and, first of all, I would like to say how much I appreciate this opportunity to share this moment of friendship and reflexion about control.

According to Dorothee's announcement, the speakers were chosen among "scientists, students and post docs who marked Ioan's professional career". I should say that I don't belong to any of these categories: surely not a scientist, unfortunately no longer a student, nor a post doc, "merely" a professor in an engineering institute, specialized in Aeronautics and Space. That is the reason why my paper will deal more with reflexions about an engineer's needs and education than with science itself.

But, before going into these technical considerations, let me first recall some "souvenirs" about Ioan. It is true indeed - and it has been a privilege for me - that our paths, Ioan's and mine, crossed many a time. As far as I remember, the first time was when, at the Comite National I was the rapporteur for Ioan's candidacy report for his admission to CNRS, as "Maitre de Recherche". Some time later, while still at the Comite National, I had the honor of pushing for the RCP 567 "Models and mathematical tools for control, system analysis and signal processing". Thanks to Ioan's animation and guidance this action has been an important step in the constitution of a strong national community in Control. Since then, of course, there have been many opportunities to meet up: meetings, summer schools in LAG, conferences - with a special mention for the first European Control Conference in Grenoble - his lectures at Sup'Aero on identification and adaptative control...It would take too long to recall all of them and that is not the aim here, but I would like to remind him of just one more old point: his letter of support when, in 85, I have had to pass the habilitation!

2. A research: from the peaks or from the lacks

Let us now move on to somewhat more "technical" considerations. As said previously, however, I don't really intend to give a technical paper on a specific point - and above all on perspectives in Control - but rather to comment on some points about control which seem important to me. Indeed I think that one should be very careful when predicting the future, and other colleagues, here, are surely more competent than I am. The future is not always what was imagined decades before and the scientists in charge of keeping a close eye on any future scientific developments must, at all times, remain humble.

In 1939, at an exhibition fair in New York, there was a presentation about the future of transportation systems. There was a proposal for an airfield with runways built on circular discs, floating on mercury and rotating so that landings could be always made in the best conditions! Fortunately, this never came about, but, at a time when again a lot of discussions, reflexions and works are being led about the "Airport for the future" we can hope that all the bright researchers and engineers involved will not be so inventive.

When thinking about "advances" in Control, one should also be aware that two "streams of thought" exist and that, maybe, it is in this duality that the differences between theorists and practicists have to be found. To take again a quotation that I recently read in a review by one of our colleagues: finalized research is drawn by the weaknesses of the state of art and not pushed by its peaks, as academic research is. Most of the time, the latter alone selects its hypotheses and questions according to the opportunity that they offer with respect to a conjectured theoretical result. The search for eventual "applications" comes only after. Of course, this is legitimous but the contrary, although less frequent, is surely equally useful though to the increase in knowledge. In fact both ways, one more upstream, the other more downstream, one more rigorous, the other more pragmatic, are largely complementary.

From my experience, but, I repeat, I am not a theorist, interesting researches can be motivated, at the beginning, by real needs. For instance, in the early sixties (in 63-65) it was when looking for non linear compensators, allowing to get rid of Bode's law (by which phase and amplitude effects, for all linear networks, are conversely linked, hence limiting their capabilities) that gliding motion and its robustness were discovered. These non linear compensators, in fact called pseudo linear networks because, in terms of equivalent gain, their amplitude and phase components are only frequency dependent (like the linear networks) were basically made of two linear networks (associated with two static non linearities), one acting on the input of a relay, the output of which was modulated by the output of a redressor fed by the second linear network. In terms of state trajectories these networks led to sliding behavior with automatic amplitude chatter control. It is also in order to design "optimal" control laws for missile trajectory control, really implementable, that is function of the states and not of the time, that we have been led to work on multitime scale systems, and hence, on (forced) singular perturbations. It is the performances required by new autopilots for civil planes which led to work on eigenvector control. It is still real needs - for instance more or less disastrously high amplitude oscillations appearing in some flight conditions - and at times inputed to PIO - which led to work again on systems with saturating control and the conditions under which limit cycles could appear, function of the flight conditions, center of gravity location and closed loop desired performances. I will come later on this problem of saturated control but examples could be multiplied.

3. Towards less traditional fields of application

And now, as it seems difficult even so to completely avoid some considerations about the future of control, I would like to express, in a very few words, my feelings on the matter.

First it is clear, notwithstanding the considerable interest in control engineering over the years, since 1935, from ship rudder control to satellite attitude control, going through so many industrial plants, that support for control research is decreasing. Maybe, the problem is more general and equally acute in other fields. Maybe it is because governmental and national agencies who used to fund the theoretical and applied researches are inclined to hand over the responsibility to industry, an industry which has often been reluctant to use advanced control...as long as the classical control tools could afford a solution in terms of minimal performances or cost-performance index. This problem was raised during the 13th world IFAC congress in San Francisco in 96, during a round table organized by G.C Goodwin and J.Perkins, but I think it is still a very topical subject which can be raised again during these two days, with almost the same questions which were raised then, from "why is there reluctance on the part of industry to use advanced control?" to "are all of the remaining real world control problems too tough?"

Second I also do think that, in order to arouse a renewed interest, control should be applied to fields where it has seldom been used up to now and where its capabilities in modeling, analysis and command could be profitably used. Paradoxally enough these fields are not necessarily far from the traditional ones. I am thinking for instance about the implication, some time ago, of control people in the field of space converters where accuracy and time response are of predominant importance. Working in close cooperation with specialists in electronics they have been able to design aggregated models of these systems (of a non linear and changing structure), adapted to the understanding of the mode of operation (for instance storage effect influence) and to control,hence new control schemes known for instance as MC2 or LC3. More recently, active control of flows, in order to reduce the natural instabilities of flow which appear on aerospace vehicle surfaces or to control the position of the laminar-turbulent transition in a boundary layer, has opened fields of cooperation between control people and aerodynamicians. Other fields seem equally important. I am thinking first about biological systems. I remember that, when I was student at SupAero, my professor, Jean Charles Gille, (who was one the leading European automaticians...but, at the same time, so interested in medicine that he passed all his degrees in medicine while head of studies at SupAero) liked to draw a link between the behavior patterns of living beings and servomechanisms. Indeed analogies exist: The specific properties of control systems, their aptitude to adapt to unexpected conditions, their aptitude to operate independent of external conditions (with, of course, some limitations!), the obtention of good accuracy by operating near zero are also found in living beings, even though the words are not the same. Instead of robustness, the physiologist will talk about homeostasy, this property of living beings to maintain constant an internal milieu in spite of external modifications (thermal regulation of the human body for instance). The obtention of good accuracy by operating near zero can find its analogy in many mechanisms of physiological regulation, known as autoregulating systems, where the equilibrium is maintained through antagonistic effects (from the sugar concentration in the blood to the precise voluntary motion implying the cooperative action of agonist and antagonist muscles). I think that the experience of the control engineer can be appropriately used to understand and modelize many physiological mechanisms and hence help to control them better. Successes have been obtained already: I am thinking for instance of D.Claude's work on insuline action on diabetic people, on works about the ways to prescribe antibiotics to patients...but I am also thinking about the problems of artificial organs (about which J.F Antaki and J.E.Nostson organized a round table at the last IFAC congress) which, in order to function autonomously over a long period of time, must be able to respond to changes in physiological demands and in the level of activity of other organ systems - as do the original ones.

I am thinking too of economic and financial systems (over, of course, the classical problems of production and scheduling). To tell the truth, I no longer know how great, or low, the penetration of control methodology and theory is now in these fields. But I remember that, at the time of Forrester's books on industrial dynamics, world dynamics.., it was rather surprising to see that the economists were somehow discovering the properties of the closed loop, the effects of lags, gains...on its stability.Apparently, the situation has improved

- and for instance fuzzy control seems to be applied in these fields. However I feel that a much greater cooperation between the two worlds of economy and finances in the one hand, control on the other, should be useful, both for modeling and control (hierarchical optimal, predictive..).

4. How to transfer

This said, it is clear that the future of automatic control, whatever the field of application, will rely heavily on effective ways to transfer knowledge, from research to industry, either directly or through the young engineer's education. With different time scales, both ways are important but one should bear in mind the specific problems that they raise.

Unlike what happened twenty years ago, most of the research teams inside the university or CNRS, are now involved in "applied" studies, through contracts. Most of the time, the problem is to adapt and apply new methodologies to the control of a specific process, (for instance, H_∞ methods to the control of a missile or a turboreactor). Quite often, this work is done by young researchers, during a DEA or a doctorat and I am not sure that - with of course some exceptions - this kind of collaboration is as efficient as is often said, both for industry and the doctoral student. The reasons might be found in the fact that the "physics" of the plant and its surroundings are generally not familiar enough to the researcher, that from a practical point of view (cost performance) the methodology used might not be the best one to use for the specific application considered, while at the same time the industrial pressure is not always compatible with good research work.

In any case some kind of finalization of these cooperations over the time is important so that a common language and culture can be developed. With major industries this seems feasible - and in fact - really exists. This problem of cooperation is unfortunately more acute with small and medium enterprises (PME), at least in France. Unfortunately, because they could generally be more reactive to innovation, unfortunately because, in spite of recent efforts, they lack money to take the risks of exploratory developments. I am, for instance, thinking about the situation in parallel robotics. I am thinking of this researcher in robotics, recognized by all his peers not only as one of the best researchers in that field but as a fantastic inventor, owning a dozen of patents; His work remains almost ignored and above all unexploited, because no PME, in spite of their declared interest, ever found the money to develop his results.

Future of control, at the least in some countries is also dependant on the possibility to build real bridges between public research and the economic sector of small and medium enterprises (and not only the big ones). It is not a matter of words, but a matter of funding, specially of capital risks.

Finally I would like to say a few words about education on automatic control, rather to raise questions than to give answers. Due to the huge increase in knowledge in all the scientific fields, taking into account that the same time is alloted for the curriculum, it is clear that choices have to be made. The situation is difficult: on the one hand, it is necessary first of all, to set the foundation stones, but on the other hand, it is important to go up to the newest methodological results because it is largely through the young engineers, newly trained, that new methods can be introduced in the industry. The problem is difficult, so much the more that it might be more important for an engineer than for a researcher to have a broad spectrum of knowledge, so that he can reach in each situation the best solution.

So obviously, choices should be made. How to make them? The risk is to give a too big importance to his own center of interest or to sacrifice to ephemeral fashions (to some extent we can all plead guilty, more or less), but the risk is also to discard methods, just because they are old. (It is not because they are old that they are useless, it is not because they are not in the present style that they will not come again on fashion). Another risk is to give not enough importance to a certain topic because it seems trivial or because it is not formalized enough. The case of non linear systems and control seems to me quite significant. It has attracted considerable attention during the last years and, undoubtedly, much progress has been made. But I am not sure, contrary to what happened in the earlier times, that the transfer of these tools to Industry has been quite efficient, as it has been at the beginnings of control, in the late fifties. At that time,engineers knew how to cope with simple non linearities, due to limitations in the hardware or purposely introduced. The methods of Kochenburger, Duthil, Hamel, Cypkin, among others, were in common use in the industry. In the present days it seems, at least from an inquiry made among industrials , that the situation is more ambiguous and that the penetration of non linear methods in the industry is quite low; Many reasons, good and bad, can be given: Are non linear methods necessary? do they offer a better return in terms of performances, cost, reliability, and above all robustness, than linear ones, designed by methods for instance?

To end with this paper - still between research and education - I would like to mention the "simple" problem of linear systems subject to saturating inputs. It is one of the most frequently encountered problem in applications because actuators saturations are almost unavoidable; it is one of the most serious because such saturations can dramatically degrade the performances of closed loop systems. Paradoxally few sound and practical results exist, even in the case of the wind-up phenomena, one of the oldest phenomena recognized (along with limit cycles). Surely, many antiwind-up schemes have been designed, linear or non linear, but the practitioner feels often helpless before the selection of one of them and even more before its tuning. More generally, with respect to stability and above all performances, the situation is even worst.

As it has been said at the beginning this paper was not intended to be a technical paper. It didn't pretend either to forecast what the future of control will be. Merely, it is hoped that it will contribute to arouse some discussions and reflexions so that a future of control might exist.

Fundamental Problems in Adaptive Control

B.D.O. Anderson[1] *and M. Gevers*[2]

[1] Research School of Information Sciences and Engineering, and Cooperative Research Centre for Robust and Adaptive Systems, Australian National University, Canberra, ACT 0200, Australia, e-mail: brian.anderson@anu.edu.au

[2] CESAME, Louvain University, Bâtiment Euler, B-1348 Louvain-la-Neuve, Belgium, e-mail: gevers@csam.ucl.ac.be

Summary. [1] The paper identifies three fundamental problems in adaptive control: the need to work with models of plants which may be very accurate but are virtually never exact; the inability to know, given an unknown plant, whether a desired control objective is practical or impractical, and the possibility of transient instability, or extremely large signals occurring before convergence. A technique is advanced for addressing these problems based on controller adjustment limited by Vinnicombe metric considerations.

1. Introduction

The aim of this paper is to highlight some fundamental problems, almost at the conceptual level, with adaptive control, and indicate, at least partially, how they can be resolved.

[1] The authors acknowledge funding of activities of the Cooperative Research Centre for Robust and Adaptive Systems by the Australian Commonwealth Government under the Cooperative Research Centres Program, funding of this research by the US Army Research Office, Far East, the Office of Naval Research, Washington, and the Belgian Programme on Inter-university Poles of Attraction, initiated by the Belgian State, Prime Minister's Office for Science, Technology and Culture.

Modern adaptive control is a creation of approximately the last 20 years. The earlier results of the period, as exemplified in text books such as [5], can be encapsulated in theorems starting with words like "Suppose P is an unknown linear system that is of known degree and minimum phase ... " and ending with words like " ... all signals in the closed-loop remain bounded for all time, the adaptive controller converges, and the performance index is minimized".

There are difficulties with this sort of result.

1.1 The problem of changing experimental conditions given accurate but inexact models

It is now well known that one can find two plants whose Nyquist diagrams or impulse responses are practically indistinguishable, and a controller for the two plants for which the closed-loop behaviours are enormously different - even unstable in one instance and stable in the other [7, 1, 8]. Quoting from [8], "Modelling Principle 1: arbitarily small modelling errors can lead to arbitarily bad closed-loop performance". The higher the performance sought of the controller, the more readily this phenomenon can occur. Conversely one can attach a stabilizing controller to two plants and observe what appear to be identical closed-loop behaviours, when the open-loop behaviours of the plants are quite different [7, 8]. To quote from [8] again, "Modelling Principle 2: larger open-loop modelling errors don't necessarily lead to larger closed-loop prediction errors". (As an example of the latter phenomenon, consider a plant P_0 which is a good low frequency approximation of a plant P_1, which has a high frequency resonance not possessed by P_0. Suppose that C is a controller stabilizing P_0 and yielding a bandwidth outside of which the high frequency resonance falls. Then very similar closed-loop behaviours of the loops (C, P_0) and (C, P_1) will be observed.)

Suppose now the two plants contemplated in the above observations are a "true plant" and a model. Then one can have a marvellous open-loop model of the true plant, but the model quality collapses on introducing a controller. Alternatively, one can have a model which in conjunction with a controller very accurately captures the closed-loop behaviour of a true plant controlled the same way. However, as an open-loop model, it could be very poor.

The issue here is change of experimental conditions. Models, and the task of finding them [identification] can only have their quality evaluated for a particular set of experimental conditions. Changing from open-loop operation to closed-loop operation with a specified controller is of course one change of experimental conditions. *So is any change of a controller a change of experimental conditions*

Unless a plant model is exact, high accuracy under one set of experimental conditions does not guarantee its efficacy under changed experimental conditions. It is intuitively clear that small changes of controller should probably avoid, or ameliorate, the problem of possible loss of efficacy of a model under changed experimental conditions. But how small is small? Adaptive control theory offers no clues as to how to answer this question.

To the extent that in adaptive control (or its batch processing first cousin, iterative control and identification), one changes controller from time to time, based on an assumed model of the plant, one must reckon with the possibility that under a changed controller, the assumed model will be ineffective (if it was derived to be effective for the controller used before the change).

1.2 The problem of impractical control objective

Some years ago, the question was examined of whether the attractive robustness properties (gain and phase margin) of a linear quadratic state feedback control system would extend to the output feedback case. A celebrated paper [2] established with a simple example that there are no guaranteed margins, and presented an example of an LQG optimal system with almost zero margins.

Now contemplate an adaptive control problem, where the unknown plant (though unknown to the controller) was given exactly by that used in [2], and the desired performance index was that of [2] also. What would happen? One would expect very large transient signal values (transient instability), and great difficulty in learning the correct controller, the problems being worse the higher the noise intensity. If the unknown plant (still unknown to the controller of course) was actually given by a small perturbation of the plant of [2], the problem could be far worse.

The issue here is that the control problem posed is a very (impossibly?) demanding one in the known plant case; put another way, the closed-loop system in the known plant case is a marvellous example of a nonrobust design. Attempts to overlay adaption must cause problems.

What makes the problem especially challenging is that the very unknownness of the plant at the outset logically prevents assessment of whether the proposed control objective is or is not feasible. Of course, practitioners of adaptive control have long accepted that one should not ask for the impossible; but this begs the question of how one might know what is impossible in the face of the unknownness.

The challenge is evidently to identify in the course of operation of an adaptive control algorithm that too much is being sought, and to back off on the control objective.

As a tool for addressing the changing experimental condition problem and the impractical control objective problem, we review in the next section results of G. Vinnicombe [9] on robust stability. In Section 3., we turn to ideas for resolving the problems, in a context of iterative identification and control [6]

1.3 The transient instability problem

Recall the pseudo-quotation from a prototypical adaptive control theorem " ... all signals in the closed-loop remain bounded for all time ... ". The apparently comforting words leave a gap so large that not just a truck but an ocean liner could fit through. After all, they allow values of 10^9 volts or amps or watts (even if we would prefer more modest levels). In fact, the situation can be worse. There exist adaptive control algorithms with the following property. Let ξ_0 denote the vector of initial conditions for an adaptive control problem and $z(t)$ the vector at time t of all signals of interest. Then for arbitrary $N > 0$, there exists ξ_0 with $\| \xi_0 \| \leq 1$ and $t_1 > 0$ such that $\| z(t_1) \| > N$, while $z(.)$ is bounded; the bound is not uniform in ξ_0.

The issue is that in the course of the algorithm, controllers are connected which produce unstable behaviour - but they are removed or changed before infinite time, and by the time $t \to \infty$, the controllers have become stabilizing. We term this problem transient instability.

It is not as fundamental as problems 1.1 and 1.2, and indeed certain algorithms have been developed specifically to exclude transient instability, see eg [6].

The problem is also linked with but is not the same as the problem of changing experimental conditions.

2. Vinnicombe Metric

Consider the feedback system of Figure 1.

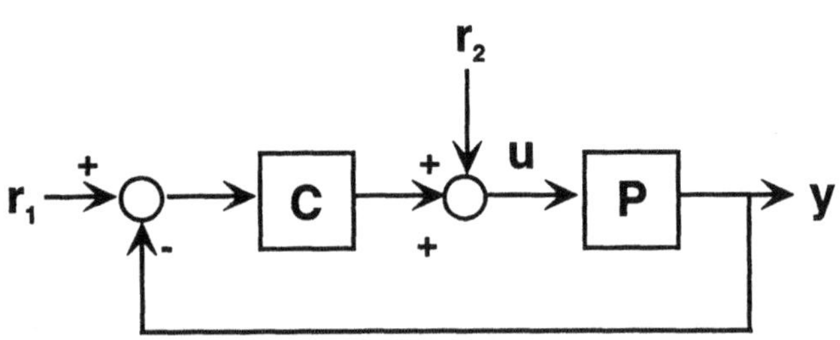

Fig. 2.1. Closed-loop system

The transfer function from $\binom{r_1}{r_2}$ to $\binom{y}{u}$ is

$$T(P,C) = \begin{bmatrix} P(I+CP)^{-1}C & P(I+CP)^{-1} \\ (I+CP)^{-1}C & (I+CP)^{-1} \end{bmatrix} \tag{2.1}$$

$$= \begin{bmatrix} P \\ I \end{bmatrix} (I+CP)^{-1} \begin{bmatrix} C & I \end{bmatrix}. \tag{2.2}$$

One can define a generalized stability margin by

$$b_{P,C} = \begin{cases} \|T(P,C)\|_{\infty}^{-1} & \text{if } (P,C) \text{ is stable,} \\ 0 & \text{otherwise.} \end{cases} \tag{2.3}$$

Here, $\|G\|_\infty = \sup_\omega \bar{\sigma}[G(j\omega)]$ and for scalar G, $\|G\|_\infty = \sup_\omega |G(j\omega)|$. Good designs correspond to $b_{P,C}$ which are well away from zero. For a scalar P and C, if $|PC(1+PC)^{-1}|$ is large at some frequency, then $|1 - PC(1+PC)^{-1}| = |(1+PC)^{-1}|$ is large at that frequency, and the closed-loop must be close to instability. If $|C(1+PC)^{-1}|$ is large at some frequency, then the plant is likely to be saturated through over-driving: perhaps a design has been attempted which seeks to achieve a wider closed-loop bandwidth than the plant open-loop bandwidth. It is also the case that if certain types of plant uncertainty are present, it is desirable for each of the entries of (1) to be small in order that the uncertainty not do too much damage. For example, it is well known that if $|PC(I+PC)^{-1}|$ is large at some frequency, one must be concerned about neglected high frequency dynamics in the plant, changing number of right half plane zeros, and output (sensor) errors. And if $|P(I+PC)^{-1}|$ is large, low frequency parameter errors and uncertainty about right half plane poles can cause problems, including output errors to input commands and disturbances, [10, 3].

A scalar measure of the difficulty of controlling a particular plant is given by

$$b_{\text{opt}}(P) = \sup_C b_{P,C} \tag{2.4}$$

and $b_{\text{opt}}(P)$ can be related by an elegant formula to the Hankel norm of a normalized coprime realization of P, [4]. This formula exposes clearly the fact that plants with right half plane poles and zeros are harder to control. The larger $b_{\text{opt}}(P)$ is, the smaller can $\| T(P,C)_\infty \|$ be made, i.e. the better one can achieve conflicting objectives of loop following, low sensitivity to noise and disturbances, avoidance of input saturation, and low sensitivity to plant modelling inaccuracies of certain types.

To understand better why $b_{P,C}$ deserves the label of stability margin, we need another concept. Vinnicombe [9] introduced a ν-gap metric defining a distance between any two linear time-invariant plants, stable or unstable, with the same input and output dimensions. For scalar plants, which will be our focus in this paper, the simplest definition is

$$\delta_\nu(P_1, P_2) = \| (I + P_2^* P_2)^{-\frac{1}{2}} (P_2 - P_1)(I + P_1^* P_1)^{-\frac{1}{2}} \|_\infty < 1, \tag{2.5}$$

if a certain winding number condition is satisfied, and $\delta_\nu(P_1, P_2) = 1$ otherwise [that $\delta_\nu(P_1, P_2)$ is in fact a metric is established in [9]; it is not *a priori* clear from (2.5).]

Now we can state some results of [9].

Proposition 2.1. *1. Given a nominal plant P_1 and a stabilizing compensator C, then (P_2, C) is stable for all plants P_2 satisfying $\delta_\nu(P_1, P_2) \leq \beta$ if and only if $b_{P_1,C} > \beta$.*

2. Given a nominal plant P and a stabilizing compensator C_1, then (P, C_2) is stable for all compensators C_2 satisfying $\delta_\nu(C_1, C_2) \leq \beta$ if and only if $b_{P,C_1} > \beta$.

Aside from the obvious point that 1 and 2 deal respectively with plant and controller adjustment, let us note several other points,

- If $b_{\text{opt}}(P_1)$ is very small (because P_1 is hard to control), then since $\beta < b_{P_1,C_1} < \sup_C b_{P_1,C} = b_{\text{opt}}(P_1)$, β must be small. So the extent of variation of P_1 to P_2 with guaranteed retention of stability with C_1 is small, as is the extent of variation of C_1 to C_2, with guaranteed retention of stability with P_1
- A more sophisticated version of the Proposition, not needed by us, can consider simultaneous variation of P_1 to P_2 and C_1 to C_2, with guarantees on closed-loop stability
- Proposition 2.1, part 1 implies that among the plants P_2 for which $\delta_\nu(P_1, P_2) = b_{P_1,C}$ one at least will not give closed-loop stability. *It does not mean* that all P_2 with $\delta_\nu(P_1, P_2) = b_{P_1,C}$ will yield an unstable closed-loop system. [Indeed, V Blondel has pointed out to us the example $P_1 = s(s+1)^{-1}, P_2 = (s+1)s^{-1}$, both of which are stabilized by $C = 1$, while also $\delta_\nu(P_1, P_2) = 1$].
- Proposition 1 does *not* connect tidily with the Youla-Kucera parametrization of all stabilizing controllers, or all plants stabilized by one controller. Exposing the details would take us too far afield here; suffice it to say that introducing a Youla-Kucera parameterization amounts to perturbing P_1 with a certain directionality and arbitrary large perturbations in a certain direction will not destroy stability, even though the perturbed P_2 may satisfy $\delta_\nu(P_1, P_2) >> b_{P_1,C}$
- Suppose C_1 stabilizes P and that C_2 replaces C_1, with the inequality $\delta_\nu(C_1, C_2) < b_{P,C_1}$ holding, so that C_2 necessarily stabilizes P. Then C_2 may result in a less robust design than C_1: In fact, it can be shown that, see [9]

$$\sin^{-1} b_{P,C_2} \geq \sin^{-1} b_{P,C_1} - \delta_\nu(C_1, C_2) \tag{2.6}$$

  (and the equality sign may be obtained)
- instead of choosing C_2 to replace C_1 and restricting $\delta_\nu(C_1, C_2)$ in order to retain stability, one can focus on limiting the change of the closed-loop transfer function matrix. In fact, it can be shown that

$$\delta_\nu(C_1, C_2) \leq \parallel T(P, C_1) - T(P, C_2) \parallel_\infty \leq \frac{\delta_\nu(C_1, C_2)}{b_{P,C_1} b_{P,C_2}} \tag{2.7}$$

and

$$\parallel T(P, C_1) - T(P, C_2 \parallel_\infty \leq \frac{\parallel T(P, C_1) \parallel_\infty^2 \delta_\nu(C_1, C_2)}{1 - \parallel T(P, C_1) \parallel_\infty \delta_\nu(C_1, C_2)}. \tag{2.8}$$

3. Iterative Control and Identification

Iterative control and identification is an approach to adaptive control which decouples the identification and controller design steps. Typically:

- with controller C_i connected to the real system P (and yielding a stable closed-loop), one identifies a model, call it $\hat{P}_j$ of P
- using $\hat{P}_j$, one redesigns the controller, to obtain C_{i+1} and a certain designed or predicted closed-loop performance. The design may in some sense be cautious (i.e. such as to ensure C_{i+1} is not greatly different from C_i). If the combined (P, C_{i+1}) true closed-loop system offers performance like that predicted using $\hat{P}_j$ and C_{i+1}, then one can redesign, each time cautiously and each time obtaining improved performance, to find C_{i+2}, C_{i+3}, until a discrepancy occurs between what is observed in the real (P, C_{i+l}) loop and the design $(\hat{P}_j, C_{i+l})$ loop. (Here, l is the first integer at which the discrepancy occurs.)
- since the loops (P, C_{i+l}) and $(\hat{P}_j, C_{i+l})$ behave differently, one re-identifies to replace $\hat{P}_j$ by $\hat{P}_{j+1}$. Of course, the identification is done in closed-loop.

Then one repeats the cycle, until the desired performance is achieved, or it is evidently unattainable.

One particular version of the above advanced for plants whose worst instability is a pole (possibly repeated) at the origin is set out in [6]. There, the successive designs $C_{i+1}, C_{i+2}, \ldots$ were obtained by the IMC (internal model control) design concept, with successive broadening of the bandwidth of the desired closed-loop system. It is evident from [6] and related works that only modest jumps in the designed closed-loop bandwidth were made at each new controller design, this being a form of caution to address the changing experimental condition problem raised in Section 1.. One can think of the approach as one which through caution limits $\parallel T(\hat{P}, C_1) - T(\hat{P}, C_2) \parallel_\infty$. By (2.7), this limits $\delta_\nu(C_1, C_2)$.

3.1 Cautious controller adjustment: what is acceptable?

Recall that (P, C_i) is the real loop, and it behaves like $(\hat{P}_j, C_i)$. We wish to adjust C_i to C_{i+1} without encountering the changed experimental condition problem of Section 1.. Can we do this in a planned way, while not knowing

P? Yes, as we shall now explain, we can, on the basis that the (P, C_i) and $(\hat{P}_j, C_i)$ loops behave similarly.

What does it mean to say the (P, C_i) and $(\hat{P}_j, C_i)$ loops behave similarly? We shall take it to mean that

Assumption 3.1. *$T(P, C_i)$ is well approximated by $T(\hat{P}_j, C_i)$.*

An immediate consequence is that

$$b_{P,C_i} \simeq b_{\hat{P}_j,C_i}. \tag{2.9}$$

By the Vinnicombe theory, we can then conclude that if C_{i+1} satisfies

$$\delta_\nu(C_i, C_{i+1}) < b_{\hat{P}_j,C_i}, \tag{2.10}$$

then C_{i+1} will be stabilizing for P. A safety play is to take C_{i+1} such that

$$\delta_\nu(C_i, C_{i+1}) < kb_{\hat{P}_j,C_i}, \tag{2.11}$$

where k is a constant in $(0, 1)$, say 0.5.

3.2 Cautious controller adjustment: what is desirable?

In the previous subsection, we have discussed what, from the point of view of stability, is an acceptable controller C_{i+1} to replace C_i. However, if our goal is to achieve a certain closed-loop performance, by minimizing a performance index say, then we need to understand which C_{i+1} we should choose in a set defined like (2.11).

To fix ideas, we shall postulate

Assumption 3.2. *For the unknown plant P, the design goal is to obtain a stabilizing compensator C to minimize a performance index $J(P, C)$.*

Suppose also that we have a model $\hat{P}_j$ of P such that $T(\hat{P}_j, C_i) \simeq T(P, C_i)$. Let $\hat{P}_j$ have a right coprime realization ND^{-1} and C_i a right coprime realization UV^{-1}. The set of all stablizing compensators of $\hat{P}_j$ is given by [10].

$$\mathcal{C} = \left\{C(Q) : C(Q) = (U - DQ)(V + NQ)^{-1}\right\}, \tag{2.12}$$

where Q (the Youla-Kucera parameter) is an arbitrary stable proper transfer function. Let us make a further postulate, that is certainly fulfilled in the H_2 and H_∞ problems.

Assumption 3.3. *The performance index $J(\hat{P}_j, C(Q))$ for $C \epsilon \mathcal{C}$ depends on Q in a convex manner.*

[Convexity is actually linearity in H_2 and H_∞ problems.] Now we can find C_{i+1} in the following way. Suppose

$$C^*_{i+1} = \arg^{\min}_{C(Q)} J(\hat{P}_j, C(Q)). \tag{2.13}$$

(When the minimum has to be replaced by an infimum, there is a minor adjustment to these calculations.) Let Q^* be such that

$$C^*_{i+1} = (U - DQ^*)(V + NQ^*)^{-1}. \tag{2.14}$$

To avoid trivialities, suppose that $Q^* \neq 0$, i.e. C_i does not minimize $J(\hat{P}_j, C)$.

If

$$\delta_\nu(C_i, C^*_{i+1}) \leq kb_{\hat{P}_j, C_i}, \tag{2.15}$$

with k the constant introduced at the end of the last section, choose

$$C_{i+1} = C^*_{i+1}. \tag{2.16}$$

Otherwise, consider the set

$$C(\alpha Q^*) = (U - \alpha DQ^*)(V + \alpha NQ^*)^{-1}, \qquad \alpha\epsilon[0, 1]. \tag{2.17}$$

Observe that $\alpha = 0$ corresponds to C_i, $\alpha = 1$ corresponds to C^*_{i+1}, and for all $\alpha\epsilon[0, 1]$, $C(\alpha Q^*)$ is stabilizing. Choose $\alpha\epsilon(0, 1)$ so that

$$\delta_\nu(C_i, C(\alpha Q^*)) = kb_{\hat{P}_j, C_i}. \tag{2.18}$$

Such an α exists, since δ_ν is a smooth function of α, taking values at $\alpha = 0$ of 0 and at $\alpha = 1$ of something in excess of $kb_{\hat{P}_j, C_i}$. Also, take

$$C_{i+1} = C(\alpha Q^*). \tag{2.19}$$

Evidently, this choice moves the controller *in the direction of* C^*_{i+1}, but not necessarily all the way; in fact, the movement is such as to retain the bound on δ_ν. Does this assist as far as the performance index is concerned? Yes, it does:

Proposition 3.1. *Suppose that the stable transfer function* Q^* *minimizes the performance index* $J(\hat{P}_j, C(Q))$ *which satisfies Assumption 3.3. Let* $\alpha\epsilon(0, 1)$. *Then*

$$J(\hat{P}_j, C(Q^*)) \leq J(\hat{P}_j, C(\alpha Q^*)) < J(\hat{P}_j, C_i),$$

(where C_i *corresponds to* $\alpha = 0$*).*

Proof. The left hand inequality follows by optimality of Q^*. For the right hand inequality observe that by the convexity property of J,

$$\begin{aligned} J(\hat{P}_j, C(\alpha Q^*)) &\leq (1 - \alpha)J(\hat{P}_j, C_i) + \alpha J(\hat{P}_j, C(Q^*)) && (2.20)\\ &< (1 - \alpha)J(\hat{P}_j, C_i) + \alpha J(\hat{P}_j, C_i) && (2.21)\\ &= J(\hat{P}_j, C_i). && (2.22) \end{aligned}$$

In [6] a performance index of the type described above was not used to determine the controller. Rather, the so called IMC design method was used, where one seeks a controller to achieve a standard closed-loop transfer function in which a single parameter, the bandwidth, appears.

This means that the controller which, in conjunction with a model $\hat{P}_j$, achieves a particular bandwidth is parameterisable by that bandwidth. It is again straightforward to compute a Vinnicombe distance between two such controllers and to set a limit on the change of bandwidth, in terms of $b_{\hat{P}_j,C_i}$.

Let us note that if C_i has been chosen to secure a closed-loop bandwidth exceeding that of the open loop plant P, the entry $C(1+PC)^{-1}$ of $T(P,C)$ will become large, in fact $O[|P^{-1}|]$ outside the plant bandwidth, and accordingly $b_{\hat{P}_j,C_i}$ will be small. This will limit the scope for further bandwidth expansion.

4. Addressing the fundamental problems

In this section, we shall explain how the design idea of Section 3. serves to address the fundamental problems raised in Section 1..

4.1 The problem of changing experimental conditions given accurate but inexact models

The key to addressing this problem is to limit the change in experimental conditions [through placing a bound on $\delta_\nu(C_i, C_{i+1})$] in such a way that the effects of the change of experimental conditions are guaranteed limited.

It is crucial that the bound, although relevant in predicting something about the interconnection of C_{i+1} with the true plant P, is computable in terms of the accurate (but inevitably inexact) model $\hat{P}_j$ - in terms of $\| T(\hat{P}_j, C_i) \|_\infty$ in fact. This is telling us about P precisely because $T(\hat{P}_j, C_i) \simeq T(P, C_i)$. The transfer function matrix $T(P, C_i)$ is not computable, and the closeness of $T(\hat{P}_j, C_i)$ to $T(P, C_i)$ can only be established on the basis of measured signals.

4.2 The problem of impractical control objective

Impractical objectives, technically, are those for which, were the plant known and the objective attained with a controller C, the quantity $\| T(P,C) \|_\infty$ would be very large.

Suppose a problem with impractical objectives is set. How will the ideas of the previous section handle this? One of several things may happen; e.g.

- the identification process, which should yield $T(\hat{P}_j, C_i) \cong T(P, C_i)$ and therefore $b_{\hat{P}_j,C_i} \cong b_{P,C_i}$, actually gives rise to errors in the transfer function estimate (the standard deviation of which may be a consequence

of the identification algorithm) of such a magnitude that the equality $b_{\hat{P}_j,C_i} \simeq b_{P_j,C_i}$ cannot be relied upon. (Examples in [6] are of this type). Equivalently, measurement times required to obtaining needed estimates of acceptable quality may simply become excessive.
To understand this better, let us explain one situation where the issue may arise. Consider the scheme of Figure 2.1, and suppose the input r_2 is zero (as is quite frequent). With measurements of r, u and y it is clear that one can obtain straightforwardly an estimate of $PC(1+PC)^{-1}$. If there is additive measurement noise, then the quality of the estimate will be poor when the output signal to noise ratio is small. Now suppose that the closed-loop bandwidth is ω_0, and we consider estimating $PC(1+PC)^{-1}$ at a frequency $10\omega_0$ with r_1 comprising white noise. Then it is conceivable that the SNR at this frequency could be 0dB. So any estimate of $P(1+PC)^{-1}$ based on computing this as C^{-1} times the estimate of $PC(1+PC)^{-1}$ will have a sizeable percentage error (as the SNR is again 0dB). If P has a resonance at $10\omega_0$, $|P(1+PC)^{-1}|$ may be larger than its estimate and in fact the dominant term in $T(P,C)$ evaluated at $10\omega_0$. This means that if $\bar{\sigma}[T(10j\omega_0)]$ approximates $\| T(P,C) \|_\infty$, our estimate of this quantity, and indeed of $b_{P,C}$ may be poor.

- $\| T(\hat{P}_j, C_i) \|_\infty$ may be so large and thus $b_{\hat{P}_j,C_i}$ so small that the scope for adjusting C_{i+1} satisfying $\delta_\nu(C_i, C_{i+1}) \leq kb_{\hat{P}_j,C_i}$ is negligible.

4.3 The problem of transient instability

Elimination of transient instability problems in the scheme of Section 3. requires firstly an additional assumption.

Assumption 4.1. *A stabilizing compensator is known for the unknown plant.*

Given this assumption, the ideas of Section 3. guarantee avoidance of transient instability. This is because all the different controllers individually are stabilizing, and they are not switched so fast as to cause instability through loss of a quasi-time-invariance assumption.

5. Conclusions

In these conclusions, we summarize the key results of the paper, and indicate directions in which further development should be both possible and beneficial.

The main messages have been these:

- any adaptive control scheme needs to embrace modest steps and/or rate of change in the controller. This statement at the qualitative level is folklore, if not a truism. By exploiting ideas of the Vinnicombe metric, one can quantify such steps or rate of change bounds.

- a desired control objective may be impractical, and in an adaptive control conext, this may be unknown; accordingly, any adaptive control scheme needs the capability of recognizing this fact, and the capability of stopping short of the objective. For a wide variety of objectives, one can rely on successive iterates achieving designs which approach the optimum.
- any adaptive control scheme needs protection against the introduction of a controller, which for a finite interval of time, is destabilizing.

What now of shortcomings and future possibilities? Let us note three:

- The Vinnicombe metric is crude, in that it associates a single number with the whole frequency axis. It is virtually certain one could refine the ideas by working with quantities such as $\bar{\sigma}\,[T(P(j\omega), C(j\omega)]$ and $\bar{\sigma}\left[(I + P_2^*(j\omega)P_2(j\omega))^{-\frac{1}{2}}(P_2(j\omega) - P_1(j\omega))(I + P_1^*(j\omega)P_1(j\omega))^{-\frac{1}{2}}\right]$ There are, after all, refinements of the theories of [9], stated in [9], which use frequency dependent quantities. One would hope that such refinements would capture the rough idea [7] that a model of a true plant can provide a basis for a high performance control design if in the vicinity of the cross-over frequency, it is highly accurate
- One area in which adaptive controller application is desired arises when plants can undergo step changes. Immediately following the step change, the previously satisfactory controller may be so unsatisfactory as to be destabilizing. None of the ideas of this paper are helpful in that regard, and quite different approaches are needed. As a general rule, in the presence of noise, hypotheses can be learnt exponentially fast, and parameters learnt at a $\frac{1}{t}$ type of rate; it follows that approaches based on hypothesis testing rather than parameter estimation probably are needed to recover stability in the quickest possible time.
- In iterative control and identification, one may well be identifying the same plant with different controllers. There is an absence of precise theory on how to characterize the performance of an identification algorithm, and even how to design one in this situation. Broadly repeating, when the second controller in considered, and one runs an identification experiment, one does not want to throw away all the information obtained when the first controller was connected. More precisely, how best should one use it? Continuously-varying identification algorithms appear to handle this problem better, focussing as they do on recursive parameter identification; this however has clear disadvantages, for example if there is a risk of initially unexcited high frequency resonances in the plant, and even undermodelling of the plant order.

References

1. R. L. Dailey and M. S. Lukich, "Recent results in identification and control of a flexible truss structure", *Proc Amer Control Conf*, Georgia, 1988, pp 1468-1473.
2. J. C. Doyle, "Guaranteed margins in LQG Regulators", *IEEE Trans Auto Control*, Vol AC-23, 1978, pp 664-665.
3. J. C. Doyle, J. E. Wall and G. Stein, "Performance and robustness analysis for structured uncertainty", *Proc IEEE Conf on Decision and Control*, Orlando, 1982, pp 629-634.
4. K. Glover and D. McFarlane, "Robust stabilization of normalized coprime factor plant descriptions with H_∞ - bounded uncertainty", *IEEE Trans Auto Control*, Vol 34, 1989, pp 821-830.
5. G. C. Goodwin and K. S. Sin, *Adaptive Filtering, Prediction and Control*, Prentice Hall, New Jersey, 1984
6. W. S. Lee, B.D.O. Anderson, I. M. Y. Mareels and R. L. Kosut, "On some key issues in the windsurfer approach to adaptive robust control", *Automatica*, Vol 31, 1995, pp 1619-1636.
7. R. Schrama, *Approximate Identification and Control Design*, PhD Thesis, Delft University of Technology, 1992.
8. R. E. Skelton, "Model Error Concepts in Control Design", *Int Journal Control*, Vol 49, 1989, pp 1725-1753.
9. G. Vinnicombe, "Frequency domain uncertainty and the graph topology", *IEEE Trans Auto Control*, Vol 38, 1993, pp 1571-1383.
10. K. Zhou, J. C. Doyle and K. Glover, *Robust and Optimal Control*, Prentice Hall, New Jersey, 1996.

A Bound for the Disturbance - to - Tracking - Error Gain of a Supervised Set-Point Control System

A. S. Morse

Center for Computational Vision and Control, Department of Electrical Engineering, Yale University New Haven, Connecticut, 06520-208267, USA, e-mail: morse@euler.me.berkeley.edu

Dedicated to Ioan D. Landau on the Occasion of His Sixtieth Birthday

Summary. [1] The aim of this paper is to provide a simple analysis of the dynamical behavior of a set-point control system consisting of a poorly modelled process, an integrator and a multi-controller supervised by an estimator-based algorithm employing dwell-time switching. For a slowly switched multi-controller implementation of a finite family of linear controllers, explicit upper bounds are derived for the normed-value of the process's allowable unmodelled dynamics as well as for the system's disturbance-to-tracking error gain.

1. Introduction

Much has happened in adaptive control since Ioan Landau published his pioneering monograph in 1979 [8]. The solution to the classical model reference problem is by now well understood. Provably correct algorithms exist which, at least in theory, are capable of dealing with unmodelled dynamics, noise, right-half-plane zeros, and even certain types of nonlinearities – and a number of excellent texts and monographs have been written covering many of these advances [3, 1, 14, 11, 13, 5, 6].

[1] This research was supported by the National Science Foundation, the Army Research Office, and the Air Force Office of Scientific Research

However despite the impressive gains made since 1979, there remain many important, unanswered questions: Why, for example, is it still so difficult to explain to a novice why a particular algorithm is able to functions correctly in the face of unmodelled process dynamics and $\mathcal{L}^{\infty}$ bounded noise? How much unmodelled dynamics can a given algorithm tolerate before loop-stability is lost? How do we choose an adaptive control algorithm's many design parameters to achieve good disturbance rejection, transient response, etc.?

It is our view that eventually there will be satisfactory answers to all of these questions, that adaptive control will become much more accessible to non-specialists, that we will be able to much more clearly and concisely quantify unmodelled dynamics norm bounds, disturbance-to-controlled output gains, and so on and that because of this we will see the emergence of a bona fide computer-aided adaptive control design methodology which relies much more on design principals then on trial and error techniques. It is with these ends in mind, that this paper has been written.

In the sequel we provide a relatively uncluttered analysis of the dynamical behavior of a set-point control system consisting of a poorly modelled process, an integrator and a multi-controller supervised by an estimator-based algorithm employing dwell-time switching. The system has been considered previously in [9]. It has been analyzed in one form or another in [7, 2, 10, 12] and elsewhere under various assumptions. It has been shown in [10] that the system's supervisor can successfully orchestrate the switching of a sequence of candidate set-point controllers into feedback with the system's imprecisely modeled siso process so as (i) to cause the output of the process to approach and track a constant reference input despite norm-bounded unmodelled dynamics, and constant process disturbances and (ii) to insure that none of the signals within the overall system can grow without bound in response to bounded disturbance, be they constant or not. The objective of this paper is to re-derive these same results in a much more straight forward manner. This will be done for a supervisory control system in which the number of candidate controllers is finite, and the switching between candidate controllers is constrained to be "slow" in a sense to be made precise in the sequel. These restrictions not only greatly simplify the analysis in comparison with that given in [10], but also make it possible to derive reasonably explicit upper bounds for the process's allowable unmodelled dynamics as well as for the system's disturbance-to-tracking error gain.

The overall supervisory control system to be considered is described in §2. The main theorem characterizing the system's behavior is re-stated in §3. A simple, informal proof of the theorem is carried out in §4. Explicit bounds for the process's allowable unmodelled dynamics as well as for the system's disturbance-to-tracking error gain appear (3.22) and (3.24) respectively.

2. The Overall System

The aim of this section is to describe the structure of the supervisory control system to be considered in this paper. We begin with a description of the process.

2.1 The Process

The overall problem of interest is to construct a control system capable of driving to and holding at a prescribed set-point r, the output of a process modeled by a dynamical system with large scale uncertainty. The process is presumed to admit the model of a siso linear system Σ_P whose transfer function from control input u to measured output y is a member of a known class of admissible transfer functions of the form

$$\mathcal{C}_P = \bigcup_{p \in \mathcal{P}} \{\nu_p + \delta : |\delta| \leq \epsilon_p\}$$

where $\mathcal{P}$ is a finite set of indices,

$$\nu_p \triangleq \frac{\alpha_p}{\beta_p}$$

is a prespecified, strictly proper, *nominal transfer function*, ϵ_p is a real non-negative number, δ is a proper stable transfer function whose poles all have real parts less than the negative of a prespecified *stability margin* $\lambda > 0$, and $|\cdot|$ is the shifted infinity norm

$$|\delta| = \sup_{\omega \in \mathbb{R}} |\delta(j\omega - \lambda)|$$

It is assumed for each $p \in \mathcal{P}$, that β_p is monic and that α_p and β_p are coprime. All transfer functions in $\mathcal{C}_P$ are thus proper, but not necessarily stable rational functions. Prompted by the requirements of set-point control, it is further assumed that the numerator of each transfer function in $\mathcal{C}_P$ is nonzero at $s = 0$. The specific model of the process to be controlled is shown in Figure 3.1.

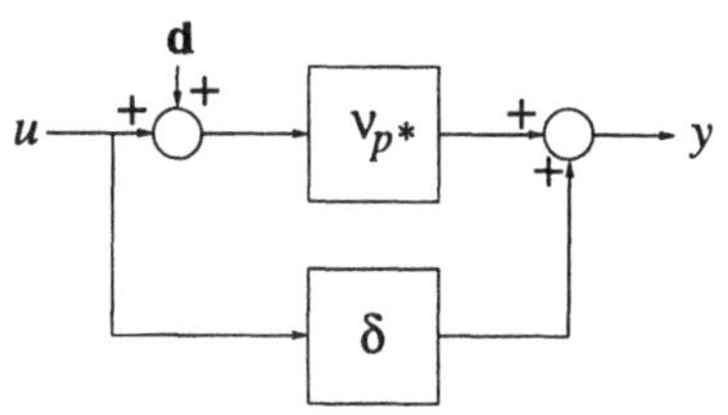

Fig. 3.1. Process Model

Here y is the process's measured output and **d** is a disturbance.

2.2 The System to Be Supervised

Presumed given is an indexed family of "off-the-shelf" loop controller transfer functions $\mathcal{K} \stackrel{\Delta}{=} \{\kappa_p : p \in \mathcal{P}\}$ with at least the following property:

Stability Margin Property: *For each $p \in \mathcal{P}$, $-\lambda$ is greater than the real parts of all of the closed-loop poles*[2] *of the feedback interconnection*

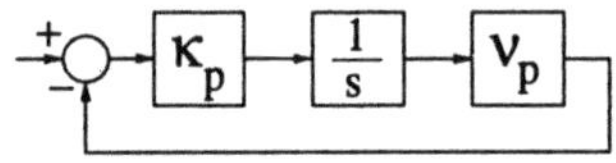

Fig. 3.2. Feedback Interconnection

Also presumed given is an integer $n_C \geq 0$ and a family of n_C-dimensional realizations $\{A_p, b_p, f_p, g_p\}$, one for each $\kappa_p \in \mathcal{K}$. These realizations are required to be chosen so that for each $p \in \mathcal{P}$, $(c_p, \lambda I + A_p)$ is detectable and $(\lambda I + A_p, b_p)$ is stabilizable. As noted in [9], there are a great many different ways to construct such realizations, once one has in hand an upper bound n_κ on the McMillan Degrees of the κ_p. Given such a family of realizations, the sub-system to be supervised is thus of the form shown in Figure 3.3

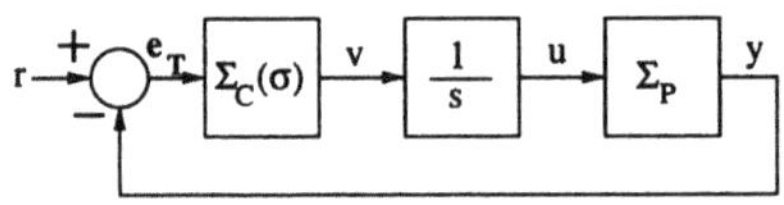

Fig. 3.3. Supervised Sub-System

where $\Sigma_C(\sigma)$ is the n_C-dimensional "state-shared" dynamical system

$$\dot{x}_C = A_\sigma x_C + b_\sigma \mathbf{e_T} \qquad v = f_\sigma x_C + g_\sigma \mathbf{e_T}, \tag{3.1}$$

called a *multi-controller*, v is the input to the *integrator*

$$\dot{u} = v, \tag{3.2}$$

$\mathbf{e_T}$ is the *tracking error*

$$\mathbf{e_T} \stackrel{\Delta}{=} r - y, \tag{3.3}$$

and σ is a piecewise constant *switching signal* taking values in $\mathcal{P}$.

[2] By the closed-loop poles are meant the zeros of the polynomial $s\rho_p\beta_p + \gamma_p\alpha_p$, where $\frac{\alpha_p}{\beta_p}$ and $\frac{\gamma_p}{\rho_p}$ are the reduced transfer functions ν_p and κ_p respectively.

2.3 The Supervisor

The problem of interest is to construct a provably correct "supervisor" which is capable of generating σ so as to achieve 1. *global boundedness* {of all system signals} in the face of an arbitrary but bounded disturbance input and 2. *set-point regulation* {i.e., $\mathbf{e_T} \to 0$} in the event that the disturbance signal is constant. The functioning of the supervisor to be considered can be explained informally in terms of the "multi-estimator" architecture shown in Figure 3.4. Here each y_p is a suitably defined estimate of y which would be asymptotically

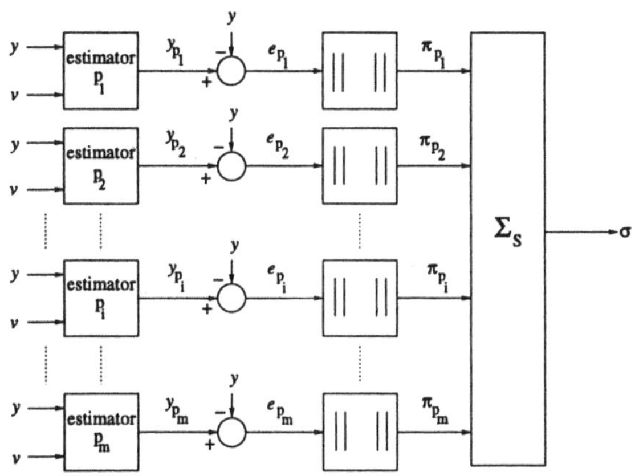

Fig. 3.4. Estimator-Based Supervisor

correct if ν_p were the process model's transfer function and there were no noise or disturbances. For each $p \in \mathcal{P}$, $e_p = y_p - y$ denotes the pth output estimation error and π_p is a "normed" value of e_p or a "performance signal" which is used by the supervisor assess the potential performance of controller p. Σ_S is a switching logic whose function is to determine σ on the basis of the current values of the π_p. The underlying decision making strategy used by such a supervisor is basically this: From time to time select for σ, that candidate control index q whose corresponding performance signal π_q is the smallest among the π_p, $p \in \mathcal{P}$. Motivation for this idea is obvious: the nominal process model whose associated performance signal is the smallest, "best" approximates what the process is and thus the candidate controller designed on the basis of that model ought to be able to do the best job of controlling the process.

The specific supervisor considered below admits a slightly different realization than that shown in Figure 3.4, even though the two supervisors are input-output equivalent. Internally the supervisor we want to discuss consists of three subsystems: a *multi-estimator dynamic* Σ_E, a *performance-weight generator* Σ_W, and a *dwell-time switching logic* Σ_D.

Σ_E is a n_E-dimensional linear dynamical system of the form

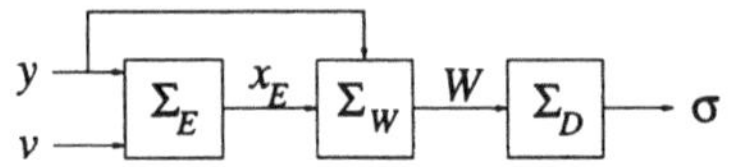

Fig. 3.5. Estimator-Based Supervisor

$$\dot{x}_E = \begin{bmatrix} A_E & 0 \\ 0 & A_E \end{bmatrix} x_E + \begin{bmatrix} b_E \\ 0 \end{bmatrix} y + \begin{bmatrix} 0 \\ b_E \end{bmatrix} v \tag{3.4}$$

where $n_E \triangleq 2(n_\nu + 1)$ and (A_E, b_E) is a parameter-independent, $n_\nu + 1$-dimensional siso, controllable pair with $\lambda I + A_E$ stable. Here n_ν is an upper bound on the McMillan Degrees of the ν_p, $p \in \mathcal{P}$. In [9] it is explained how to construct a function $p \longmapsto c_p$ so that for each $p \in \mathcal{P}$,

$$\left\{ \begin{bmatrix} A_E & 0 \\ 0 & A_E \end{bmatrix} + \begin{bmatrix} b_E \\ 0 \end{bmatrix} c_p, \begin{bmatrix} 0 \\ b_E \end{bmatrix}, c_p \right\}$$

is a stabilizable realization of $\frac{1}{s}\nu_p$ whose uncontrollable eigenvalues have real parts less than $-\lambda$. The c_p are used in the definition of Σ_W which will be given in a moment. The c_p also enable us to define *output estimation errors*

$$e_p \triangleq c_p x_E - y, \quad p \in \mathcal{P} \tag{3.5}$$

While these error signals are not actually generated by the supervisor, they play an important role in explaining how the supervisor functions.

The supervisor's second subsystem, Σ_W, is a causal dynamical system whose inputs are x_E and y and whose state and output W is a "weighting matrix" which takes values in a linear space $\mathcal{W}$. W together with a suitably defined *performance function* $\Pi : \mathcal{W} \times \mathcal{P} \to \mathbb{R}$ determine a scalar-valued *performance signal* of the form

$$\pi_p \triangleq \Pi(W, p) \tag{3.6}$$

which is viewed by the supervisor as a measure of the expected performance of controller p. Σ_W and Π are defined by

$$\dot{W} = -2\lambda W + \begin{bmatrix} x_E \\ y \end{bmatrix} \begin{bmatrix} x_E \\ y \end{bmatrix}' \tag{3.7}$$

and

$$\Pi(W, p) = \begin{bmatrix} c_p & -1 \end{bmatrix} W \begin{bmatrix} c_p & -1 \end{bmatrix}' \tag{3.8}$$

respectively. The definitions of Σ_W and Π are prompted by the observation that if π_p are given by (3.6), then

$$\dot{\pi}_p = -2\lambda \pi_p + e_p^2, \quad p \in \mathcal{P} \tag{3.9}$$

because of (3.5), (3.7) and (3.8).

The supervisor's third subsystem, called a *dwell-time switching logic* Σ_D, is a hybrid dynamical system whose input and output are W and σ respectively, and whose state is the ordered triple $\{X, \tau, \sigma\}$. Here X is a discrete-time matrix which takes on sampled values of W, and τ is a continuous-time variable called a *timing signal.* τ takes values in the closed interval $[0, \tau_D]$, where τ_D is a prespecified positive number called a *dwell time.* Also assumed prespecified is a *computation time* $\tau_C \leq \tau_D$ which bounds from above for any $X \in \mathcal{W}$, the time it would take a supervisor to compute a value $p = p_X \in \mathcal{P}$ which minimizes $\Pi(X, p)$. Between "event times," τ is generated by a reset integrator according to the rule $\dot{\tau} = 1$. Event times occur when the value of τ reaches either $\tau_D - \tau_C$ or τ_D; at such times τ is reset to either 0 or $\tau_D - \tau_C$ depending on the value of Σ_D's state. Σ_D's internal logic is defined by the computer diagram shown in Figure 3.6 where p_X denotes a value of $p \in \mathcal{P}$ which minimizes $\Pi(X, p)$.

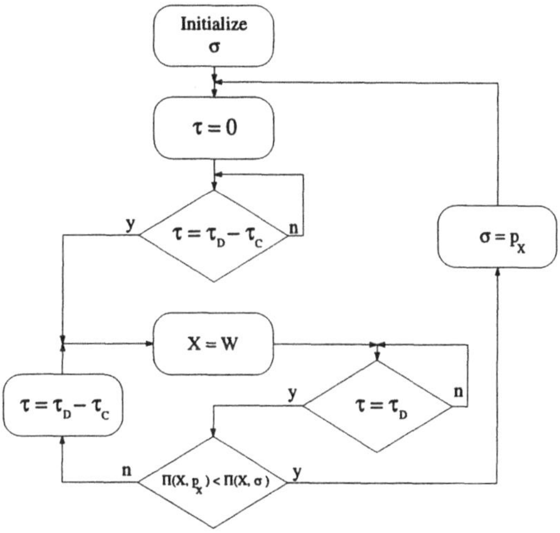

Fig. 3.6. Computer Diagram of Σ_D

In the sequel we call a piecewise-constant signal $\bar{\sigma} : [0, \infty) \to \mathcal{P}$ *admissible* if it either switches values at most once, or if it switches more than once and the set of time differences between each two successive switching times is bounded below by τ_D. We write $\mathbb{S}$ for the set of all admissible switching signals. Because of the definition of Σ_D, it is clear its output σ will be admissible. This means that switching cannot occur infinitely fast and thus that existence and uniqueness of solutions to the differential equations involved is not an issue.

3. Discussion

The overall system just described, admits a block diagram description of the form

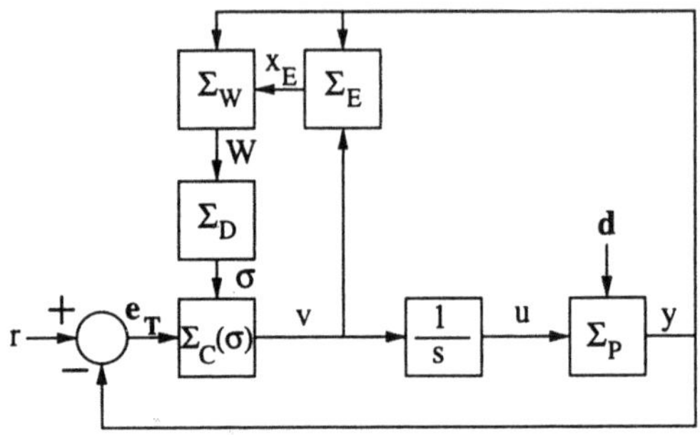

Fig. 3.7. Supervisory Control System

The following theorem is proved in [10]:

Theorem 3.1. *Let $\tau_C \geq 0$ be fixed. Let τ_D be any positive number no smaller than τ_C. There are positive numbers ϵ_p, $p \in \mathcal{P}$, for which the following statements are true provided Σ_P has a transfer function in $\mathcal{C}_P$.*

1. **Global Boundedness:** *For each constant set-point value r, each bounded piecewise-continuous disturbance input* **d**, *and each system initialization, u, x_C, x_E, W, and X are bounded responses.*
2. **Tracking and Disturbance Rejection:** *For each constant set-point value r, each constant disturbance* **d**, *and each system initialization, y tends to r and u, x_C, x_E, W, and X tend to finite limits, all as fast as $e^{-\lambda t}$.*

The theorem implies that the overall supervisory control system shown in Figure 3.7 has the basic properties one would expect of a non-adaptive set-point control system.

4. Analysis

The aim of this section is to re-derive Theorem 3.1 in a much more straight forward manner than in [10]. This will be done for a supervisory control system in which the switching between candidate controllers is constrained to be "slow" in a sense to be made precise in §4.1. This restrictions not only greatly simplifies the analysis, but also make it possible to derive reasonably explicit bounds for the process's allowable unmodelled dynamics as well as for the system's disturbance-to-tracking-error gain.

In the sequel we will invariably ignore initial condition dependent terms which decay to zero as fast as $e^{-\lambda t}$, as this will make things much easier to follow. A more thorough analysis which would take these terms into account can carried out in essentially the same manner.

4.1 Slow Switching

Assume that r is a constant and let x denote the composite state

$$x = \begin{bmatrix} \bar{x}_E \\ x_C \end{bmatrix} \tag{3.10}$$

where $\bar{x}_E$ is the shifted state

$$\bar{x}_E = x_E + \begin{bmatrix} A_E^{-1} b_E \\ 0 \end{bmatrix} r \tag{3.11}$$

It is then possible to show in a straightforward manner, that for any $q \in \mathcal{P}$ and any given piecewise constant switching signal $\sigma : [0, \infty) \to \mathcal{P}$, whether generated by Σ_D or not, the relationships between e_q, e_σ, v, and $\mathbf{e_T}$ determined by (3.1)-(3.5) are given by a system of equations of the form

$$\left.\begin{array}{rcl} e_\sigma & = & c_{\sigma q} x + e_q \\ \dot{x} & = & A_{\sigma\sigma} x + h_\sigma e_\sigma \\ v & = & f_{\sigma\sigma} x + g_\sigma e_\sigma \\ \mathbf{e_T} & = & e_\sigma - \bar{c}_\sigma x \end{array}\right\} \tag{3.12}$$

where $d \triangleq \text{column}\{-b_E,\ 0,\ 0,\ b_C\}$, $b \triangleq \text{column}\{0,\ b_E,\ b_C,\ 0\}$, and for all $p, l \in \mathcal{P}$,

$$f_{pl} \triangleq \begin{bmatrix} -g_p c_l & f_p \end{bmatrix} \qquad h_p \triangleq b g_p + d \qquad c_{pl} \triangleq \begin{bmatrix} c_p - c_l & 0 \end{bmatrix} \qquad \bar{c}_l \triangleq \begin{bmatrix} c_l & 0 \end{bmatrix}$$

and

$$A_{pl} \triangleq \text{block diagonal}\,\{A_E,\ A_E,\ A_C,\ A_C\} - d \begin{bmatrix} c_l & 0 \end{bmatrix} + b f_{pl}$$

One readily verifiable and important property of the matrices defined above is that for each $p, l \in \mathcal{P}$, $(c_{pl}, \lambda I + A_{pl})$ is a detectable matrix pair [9]. This is a consequence of certainty equivalence, the Stability Margin Property, the requirements that $\lambda I + A_E$ be a stability matrix and that for $p \in \mathcal{P}$, $\{\lambda I + A_p,\ b_p,\ f_p,\ g_p\}$ be a stabilizable and detectable system. Since $c_{pp} = 0$, $p \in \mathcal{P}$, this means that for each such p, $\lambda I + A_{pp}$ must be a stability matrix [9]. In the sequel we will assume the following.

Slow Switching Assumption: *The dwell time τ_D is large enough so that for each admissible switching signal $\sigma : [0, \infty) \to \mathcal{P}$, $\lambda I + A_{\sigma\sigma}$ is an exponentially stable matrix.*

It is possible to compute an explicit lower bound for τ_D for which this assumption holds [9].

4.2 Norms

It is especially useful to introduce the following. For any piecewise-continuous function $z : [0, \infty) \to \mathbb{R}^n$, and any times $t_2 > t_1 \geq 0$, let us write $||z||_{\{t_1,t_2\}}$ for the exponentially weighted 2-norm

$$||z||_{\{t_1,t_2\}} \triangleq \sqrt{\int_{t_1}^{t_2} e^{2\lambda t}|z(t)|^2 dt}$$

Note that $e^{2\lambda T}\pi_p(T) = ||e_p||^2_{\{0,T\}}$, $T \geq 0$, $p \in \mathcal{P}$, because of (3.9) and the assumption that $W(0) = 0$.

For any time-varying SISO linear system Σ of the form $y = c(t)x + d(t)u$, $\dot{x} = A(t)x + b(t)u$ we write

$$\left\| \begin{matrix} A & b \\ c & d \end{matrix} \right\|$$

for the induced norm $\sup\{||y_u||_{\{0,\,\infty\}} : u \in \mathcal{U}\}$ where y_u is Σ's zero initial state, output response to u and $\mathcal{U}$ is the space of all piecewise continuous signals u such that $||u||_{\{0,\,\infty\}} = 1$. The induced norm of Σ is finite whenever $\lambda I + A(t)$ is {uniformly} exponentially stable.

We note the following easily derived facts. If $e^{-\lambda t}||u||_{\{0,t\}}$ is bounded on $[0, \infty)$ {in the $\mathcal{L}^\infty$ sense}, then so is y_u provided $d = 0$ and $\lambda I - A(t)$ is exponentially stable. If u is bounded on $[0, \infty)$ in the $\mathcal{L}^\infty$ sense, then so is $e^{-\lambda t}||u||_{\{0,t\}}$. If $u \to 0$ as $t \to \infty$, then so does $e^{-\lambda t}||u||_{\{0,t\}}$.

4.3 Block Diagrams

Using the diagram of Σ_P in Figure 3.1 together with (3.1)-(3.5) and (3.12), it is not difficult to verify that, up to initial condition dependent terms decaying to zero as fast at $e^{-\lambda t}$, the relationships between $\mathbf{d}, e_\sigma, v$, and $\mathbf{e_T}$ are as shown in the block diagram in Figure 3.8 where ω_E is the characteristic polynomial of the estimator matrix A_E [9]. In developing this diagram we've represented the system defined by (3.12) as two separate subsystems, namely

$$\begin{aligned} \dot{x}_1 &= A_{\sigma\sigma}x_1 + h_\sigma e_\sigma \\ v &= f_{\sigma\sigma}x_1 + g_\sigma e_\sigma \end{aligned} \qquad \begin{aligned} \dot{x}_2 &= A_{\sigma\sigma}x_2 + h_\sigma e_\sigma \\ \mathbf{e_T} &= e_\sigma - \bar{c}_\sigma x_2 \end{aligned}$$

where $x_1 = x_2 = x$. Note that the signal in the block diagram labeled $\mathbf{b}$, will tend to zero if $\mathbf{d}$ is constant because of the zero at $s = 0$ in the numerator of the transfer function in the block driven by $\mathbf{d}$.

Let us note that each of the five blocks in Figure 3.8 represents an exponentially stable linear system. It is convenient at this point to introduce certain "system gains" associated with these blocks. In particular, let us define for $p \in \mathcal{P}$

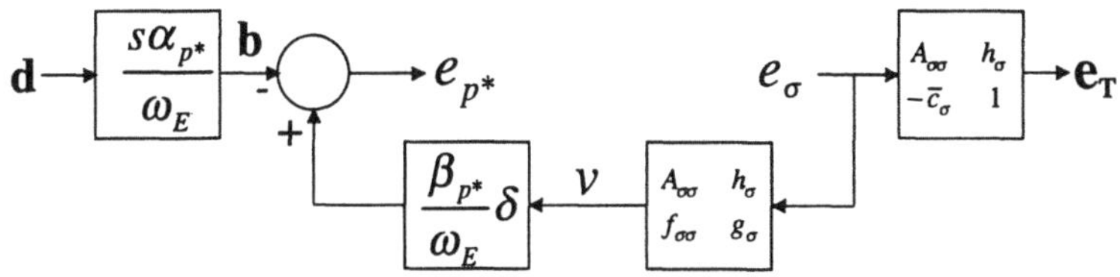

Fig. 3.8. Block Diagram I

$$\mathfrak{a}_p \triangleq \left\| \frac{s\alpha_p}{\omega_E} \right\|, \quad \mathfrak{b}_p \triangleq \sqrt{2}\left\| \frac{\beta_p}{\omega_E} \right\| \left\{ \sup_{\sigma \in \mathbb{S}} \left\| \begin{matrix} A_{\sigma\sigma} & h_\sigma \\ f_{\sigma\sigma} & g_\sigma \end{matrix} \right\| \right\}, \quad \mathfrak{c} \triangleq \sqrt{2} \sup_{\sigma \in \mathbb{S}} \left\| \begin{matrix} A_{\sigma\sigma} & h_\sigma \\ -\bar{c}_\sigma & 1 \end{matrix} \right\|$$

where, as defined earlier, $\| \cdot \|$ is the shifted infinity norm and $\mathbb{S}$ is the set of all admissible switching signals. In the light of Figure 3.8, it is easy to see that

$$||\mathbf{e_T}||_{\{0,\, t\}} \leq \frac{\mathfrak{c}}{\sqrt{2}}||e_\sigma||_{\{0,\, t\}}, \quad t \geq 0, \tag{3.13}$$

$$||e_{p^*}||_{\{0,\, t\}} \leq \epsilon_{p^*}\frac{\mathfrak{b}_{p^*}}{\sqrt{2}}||e_\sigma||_{\{0,\, t\}} + ||\mathbf{b}||_{\{0,\, t\}}, \quad t \geq 0, \tag{3.14}$$

and

$$||\mathbf{b}||_{\{0,\, t\}} \leq \mathfrak{a}_{p^*}||\mathbf{d}||_{\{0,\, t\}} \quad t \geq 0 \tag{3.15}$$

where ϵ_{p^*} is the norm bound on δ. The inequality in (3.13) bounds the norm of $\mathbf{e_T}$ in terms of the norm of e_σ whereas (3.14) and (3.15) bound the norm of e_{p^*} in terms of the norms of e_σ and $\mathbf{d}$. To develop a bound for the norm of $\mathbf{e_T}$ in terms of the norms of $\mathbf{d}$, it would therefore be enough to establish a bound for the norm of e_σ in terms of the norm of e_{p^*}. As a first step toward this end, we shall make use of the following result which is a direct consequence of dwell time switching.

4.4 Dwell-Time Switching

Lemma 4.1. *Suppose that $\mathcal{P}$ contains $m > 0$ elements, that W is generated by (3.7), that the π_p, $p \in \mathcal{P}$, are defined by (3.6) and (3.8), that $W(0) = 0$, and that σ is the response of Σ_D to W. For each fixed time $T > 0$, there exists a piecewise-constant function $\psi_T : [0, \infty) \to \{0, 1\}$ such that*

$$\int_0^\infty \psi_T(t)dt \leq m(\tau_D + \tau_C) \tag{3.16}$$

and

$$||(1 - \psi_T)e_\sigma + \psi_T e_q||_{\{0,T\}} \leq \sqrt{m}||e_q||_{\{0,T\}}, \quad q \in \mathcal{P} \tag{3.17}$$

This lemma is proved in the appendix.

4.5 Block Diagram II

Let us fix $T > 0$. In view of (3.17) there is a piecewise constant signal $\psi_T : [0, \infty) \to \{0, 1\}$ satisfying (3.16) such that

$$||(1 - \psi_T)e_\sigma + \psi_T e_{p^*}||_{\{0,T\}} \leq \sqrt{m}||e_{p^*}||_{\{0,T\}} \tag{3.18}$$

But as noted before, what we need is a bound for the norm of e_σ, not $(1 - \psi_T)e_\sigma + \psi_T e_{p^*}$. To get around this, we first note from (3.12) that for $q \in \mathcal{P}$

$$e_\sigma - e_q = c_{\sigma q} x \qquad \dot{x} = A_{\sigma\sigma} x + h_\sigma e_\sigma \tag{3.19}$$

Next consider the block diagram in Figure 3.9 which depicts this sub-system with $q \triangleq p^*$ together with an additional block and summing junctions representing the formulas

$$e_\sigma = \psi_T e_\sigma + (1 - \psi_T)e_\sigma \qquad \text{and} \qquad e_\sigma = e_{p^*} + e_\sigma - e_{p^*}$$

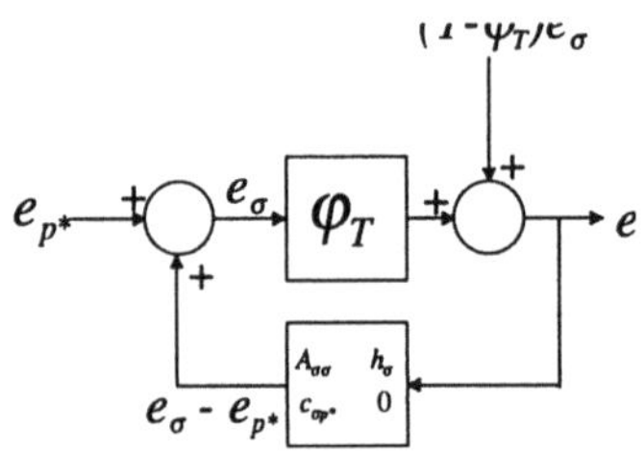

Fig. 3.9. Block Diagram II

Let us define for $q \in \mathcal{P}$

$$\boxed{\mathfrak{v}_q \triangleq \sup_{\sigma \in \mathcal{S}} \sup_{t \geq 0} \int_0^t |w_q(t, \tau)e^{\lambda(t-\tau)}|^2 d\tau}$$

where $w_q(t, \tau) \triangleq c_{\sigma(t)q} \Phi(t, \tau) h_{\sigma(\tau)}$ and $\Phi(t, \tau)$ is the state transition matrix of $A_{\sigma\sigma}$. Note that each $\mathfrak{v}_q$ is finite because of the Slow Switching Assumption. Using Cauchy-Schwartz it can easily be shown with $\mathfrak{v}_q$ so defined that

$$||\psi_T(w_q \circ e_\sigma)||_{\{0,\, t\}} \leq \sqrt{\mathfrak{v}_q \int_0^t \psi_T^2 ||e_\sigma||^2_{\{0,\, \mu\}} d\mu}, \qquad t \geq 0 \tag{3.20}$$

where $w_q \circ e_\sigma$ is the zero initial state output response of (3.19).

From Figure 3.9 it is clear that

$$e_\sigma = \psi_T(e_{p^*} + w_{p^*} \circ e_\sigma) + (1 - \psi_T)e_\sigma$$

Rearranging terms and taking norms we thus obtain

$$||e_\sigma||_{\{0,\, t\}} \leq ||(1 - \psi_T)e_\sigma + \psi_T e_{p^*}||_{\{0,\, t\}} + ||\psi_T(w_{p^*} \circ e_\sigma)||_{\{0,\, t\}}, \quad t \geq 0$$

Moreover $||(1-\psi_T)e_\sigma + e_{p^*}||_{\{0,t\}} \leq ||(1-\psi_T)e_\sigma + e_{p^*}||_{\{0,T\}}$, $t \in [0,\ T]$. Using (3.18) we thus get

$$||e_\sigma||_{\{0,\, t\}} \leq \sqrt{m}||e_{p^*}||_{\{0,T\}} + ||\psi_T(w_{p^*} \circ e_\sigma)||_{\{0,\, t\}}, \quad t \geq 0$$

Taking squares

$$||e_\sigma||^2_{\{0,\, t\}} \leq 2m||e_{p^*}||^2_{\{0,\, T\}} + 2||\psi_T(w_{p^*} \circ e_\sigma)||^2_{\{0,\, t\}}, \quad 0 \leq t \leq T$$

Using (3.20) with $q = p^*$

$$||e_\sigma||^2_{\{0,\, t\}} \leq 2m||e_{p^*}||^2_{\{0,\, T\}} + 2\mathfrak{v}_{p^*} \int_0^t \psi_T^2 ||e_\sigma)||^2_{\{0,\, \mu\}} d\mu, \quad 0 \leq t \leq T$$

Hence by the Bellman-Gronwall Lemma

$$||e_\sigma||^2_{\{0,\, T\}} \leq 2m||e_{p^*}||^2_{\{0,\, T\}} e^{2\mathfrak{v}_{p^*} \int_0^T \psi_T^2 dt}$$

From this, (3.16), and the fact that $\psi_T^2 = \psi_T$, we arrive finally at an expression for the norm of e_σ in terms of e_{p^*}, namely

$$||e_\sigma||_{\{0,\, T\}} \leq \sqrt{2m} e^{\mathfrak{v}_{p^*} m(\tau_D + \tau_C)} ||e_{p^*}||_{\{0,\, T\}}, \quad T \geq 0 \tag{3.21}$$

4.6 Stability Margin

Examination of (3.21) and (3.14) reveals that if ϵ_{p^*} satisfies the small gain condition

$$\boxed{\epsilon_{p^*} < \frac{e^{-\mathfrak{v}_{p^*} m(\tau_D + \tau_C)}}{\mathfrak{b}_{p^*}\sqrt{m}}} \tag{3.22}$$

then (3.21) and (3.14) can be combined to give

$$||e_\sigma||_{\{0,\, T\}} \leq \frac{\sqrt{2}}{\frac{e^{-\mathfrak{v}_{p^*} m(\tau_D + \tau_C)}}{\sqrt{m}} - \epsilon_{p^*}\mathfrak{b}_{p^*}} ||\mathbf{b}||_{\{0,\, T\}}, \quad T \geq 0 \tag{3.23}$$

The inequality in (3.22) provides an explicit bound for the allowable process dynamics.

4.7 Global Boundedness

The global boundedness condition of Theorem 3.1 can now easily be justified as follows. Suppose $\mathbf{d}$ is bounded on $[0, \infty)$. Then so must be $e^{-\lambda t}||\mathbf{b}||_{\{0,\, t\}}$. Hence by (3.23), $e^{-\lambda t}||e_\sigma||_{\{0,\, t\}}$ must be bounded on $[0, \infty)$ as well. This, the differential equation for x in (3.12), and the exponential stability of $\lambda I + A_{\sigma\sigma}$ then imply that x is also bounded on $[0, \infty)$. In view of (3.10) and (3.11), x_E and x_C must also be bounded. Next recall that the zeros of ω_E {i.e., the eigenvalues of A_E} have negative real parts less than $-\lambda$, and that the transfer function $\frac{\beta_{p^*}}{\omega_E}\delta$ in Figure 3.8 is strictly proper. From these observations and the block diagram in Figure 3.8 one readily concludes that e_{p^*} is bounded on $[0, \infty)$. Hence from the formulas in (3.12) for e_σ, v and $\mathbf{e_T}$ one concludes that these signals are also bounded. In view of (3.3), y must be bounded. Thus W must be bounded because of (3.7). Finally note that u must be bounded because of the boundedness of $y - \mathbf{n}$ and v and because of the observability of the cascade interconnection of (3.2) with any minimal realization of Σ_P. This, in essence, proves Claim 1 of Theorem 3.1.

4.8 Convergence

Now suppose that $\mathbf{d}$ is a constant. Examination of Figure 3.8 reveals that $\mathbf{b}$ must tend to zero as fast as $e^{-\lambda t}$ because of the zero at $s = 0$ in the numerator of the transfer function from $\mathbf{d}$ to $\mathbf{b}$. This implies that $||\mathbf{b}||_{\{0,\, \infty\}} < \infty$. Therefore $||e_\sigma||_{\{0,\, \infty\}} < \infty$ because of (3.23). Hence e_σ must tend to zero as fast as $e^{-\lambda t}$. So therefore must x because of the differential equation for x in (3.12). In view of (3.10) and (3.11) $\bar{x}_E$ and x_C must tend to zero as well. From Block Diagram I in Figure 3.8 it now can be seen that e_{p^*} tends to zero. Hence from the formulas in (3.12) for e_σ, v and $\mathbf{e_T}$ one concludes that these signals must tend to zero as well. In view of (3.3), y must tent to r . Thus W must approach a finite limit because of (3.7). Finally note that u tend to a finite limit because y and v do and because of the observability of the cascade interconnection of (3.2) with any minimal realization of Σ_P. This, in essence, proves Claim 2 of Theorem 3.1.

4.9 A Bound on the Disturbance - to - Tracking - Error Gain

By combining the inequalities in (3.13), (3.15) and (3.23) we obtain an inequality of the form

$$||\mathbf{e_T}||_{\{0,\, T\}} \leq \mathfrak{g}_{p^*}||\mathbf{d}||_{\{0,\, T\}}, \quad T \geq 0$$

where

$$\boxed{\mathfrak{g}_{p^*} = \frac{\mathfrak{c}\mathfrak{a}_{p^*}}{\frac{e^{-\mathfrak{v}_{p^*} m(\tau_D + \tau_C)}}{\sqrt{m}} - \epsilon_{p^*}\mathfrak{b}_{p^*}}} \tag{3.24}$$

Thus $\mathfrak{g}_{p^*}$ bounds from above the overall system's disturbance - to - tracking - error gain.

5. Concluding Remarks

The formula for $\mathfrak{g}_{p^*}$ in (3.24) and the stability margin bound in (3.22) are probably the most explicit discovered so far for an estimator-based adaptive control system with the properties outlined in Theorem 3.1. We believe that even simpler expressions than these can be found for the system under consideration. For example, it is likely that instead of (3.22), it will suffice to bound ϵ_{p^*} by $\frac{1}{\mathfrak{b}_{p^*}\sqrt{m}}$. Under certain conditions, it also possible to derive useful relationships between system gains and the the shifted infinity norms of the transfer functions of the constant linear systems being switched. For example, for τ_D sufficiently large, any strict upper bound on the family $\{\sqrt{2}|1 - \bar{c}_p(sI - A_{pp})^{-1}h_p| \;:\; p \in \mathcal{P}\}$ is an upper bound on $\mathfrak{c}$ [4]. Results such as these suggest that a bona fide, input-output performance theory for adaptive control may be within our reach.

6. Appendix

In the sequel, σ is a fixed switching signal, $t_0 \stackrel{\Delta}{=} 0$, t_i denotes the ith time at which σ switches and p_i is the value of σ on $[t_{i-1}, t_i)$; if σ switches at most $n < \infty$ times then $t_{n+1} \stackrel{\Delta}{=} \infty$ and p_{n+1} denotes σ's value on $[t_n, \infty)$. Any time X takes on the current value of W is called a *sample time.* We use the notation $\lfloor t \rfloor$ to denote the sample time just preceding time t, if $t > \tau_D - \tau_C$, and the number zero otherwise. Thus, for example, $\lfloor t_0 \rfloor = 0$ and $\lfloor t_i \rfloor = t_i - \tau_C$, $i > 0$.

To prove Lemma 4.1 will need the following algebraic fact

Lemma 6.1. *For all* $\mu_i \in [0, 1]$, $i \in \{1, 2, \ldots, m\}$

$$\sum_{i=1}^{m}(1 - \mu_i) \leq (m-1) + \prod_{i=1}^{m}(1 - \mu_i) \tag{3.25}$$

Proof of Lemma 6.1: Set $x_i = 1 - \mu_i$, $i \in \{1, 2, \ldots, \}$. It is enough to show that for $x_i \in [0, 1]$, $i \in \{1, 2, \ldots, \}$

$$\sum_{i=1}^{j} x_i \leq (j-1) + \prod_{i=1}^{j} x_j \tag{3.26}$$

for $j \in \{1, 2, \ldots, m\}$. Clearly (3.26) is true if $j = 1$. Suppose therefore that for some $k > 0$, (3.26) holds for $j \in \{1, 2, \ldots, k\}$. Then

$$\begin{aligned}\sum_{i=1}^{k+1} x_i &= x_{k+1} + \sum_{i=1}^{k} x_i \\ &\leq x_{k+1} + (k-1) + \prod_{i=1}^{k} x_i \\ &\leq (1 - x_{k+1})\left(1 - \prod_{i=1}^{k} x_i\right) + x_{k+1} + (k-1) + \prod_{i=1}^{k} x_i \\ &= k + \prod_{i=1}^{k+1} x_i\end{aligned}$$

By induction, (3.26) thus holds for $j \in \{1, 2, \dots, m\}$. ■

Proof of Lemma 4.1: Let $\mathcal{P}_T$ be the image of $[0, T]$ under σ. Let k be that integer for which $T \in [t_{k-1},\ t_k)$. For each $p \in \mathcal{P}_T$, let $\mathcal{I}_p$ denote the set of nonnegative integers i such that $\sigma = p$ for $t \in [t_{i-1}, t_i)$. Let j_p denote the largest integer in $\mathcal{I}_p$. Note that $j_{p_k} = k$.

For each $i \in \{1, 2, \dots, k\}$ define

$$\bar{t}_i = \begin{cases} t_i & \text{if } i < k \\ T & \text{if } i = k \end{cases}$$

The definition of dwell-time switching then implies that for $p \in \mathcal{P}_T$,

$$\begin{aligned}\pi_p(\lfloor t_{i-1} \rfloor) &\leq \pi_q(\lfloor t_{i-1} \rfloor), && \forall q \in \mathcal{P}, i \in \mathcal{I}_p \\ \pi_p(\lfloor \bar{t}_i \rfloor - \tau_C) &\leq \pi_q(\lfloor \bar{t}_i \rfloor - \tau_C), && \forall q \in \mathcal{P}, i \in \mathcal{I}_p \text{ if } \bar{t}_i - t_{i-1} > \tau_D\end{aligned}$$

Setting $i \stackrel{\Delta}{=} j_p$ and using the fact that $e^{2\lambda t}\pi_p(t) = ||e_p||^2_{\{0,t\}}$, $p \in \mathcal{P},\ t \geq 0$, we obtain the expressions

$$\left.\begin{aligned}||e_p||^2_{\{0,\ \lfloor t_{j_p-1} \rfloor\}} &\leq ||e_q||^2_{\{0,\ \lfloor t_{j_p-1} \rfloor\}}, && \forall q \in \mathcal{P} \\ ||e_p||^2_{\{0,\ \lfloor \bar{t}_{j_p} \rfloor - \tau_C\}} &\leq ||e_q||^2_{\{0,\ \lfloor \bar{t}_{j_p} \rfloor - \tau_C\}}, && \forall q \in \mathcal{P}, \text{ if } \bar{t}_{j_p} - t_{j_p-1} > \tau_D\end{aligned}\right\} \tag{3.27}$$

For each $p \in \mathcal{P}_T$, let $\phi_p : [0, \infty) \to \{0, 1\}$ be that piecewise-constant signal which is zero everywhere except on the interval

$$[\lfloor t_{j_p-1} \rfloor,\ \bar{t}_{j_p}), \qquad \text{if } \bar{t}_{j_p} - t_{j_p-1} \leq \tau_D$$

or

$$[\lfloor \bar{t}_{j_p} \rfloor - \tau_C,\ \bar{t}_{j_p}), \text{ if } \bar{t}_{j_p} - t_{j_p-1} > \tau_D$$

In either case ϕ_p has support no greater than $\tau_D + \tau_C$ and is idempotent {i.e., $\phi_p^2 = \phi_p$}. It follows that if

$$\psi_T \triangleq 1 - \prod_{p \in \mathcal{P}_T} (1 - \phi_p), \tag{3.28}$$

then ψ_T is also idempotent and has support no greater than $m(\tau_D + \tau_C)$. In view of the latter property, (3.16) must be true.

The definitions of the ϕ_p imply that for $p \in \mathcal{P}_T$ and $l \in \mathcal{P}$

$$||(1-\phi_p)e_l||^2_{\{0,\bar{t}_{j_p}\}} = \begin{cases} ||(1-\phi_p)e_l||^2_{\{0,\ \lfloor t_{j_p-1} \rfloor\}} & \text{if } \bar{t}_{j_p} - t_{j_p-1} \leq \tau_D \\ ||(1-\phi_p)e_l||^2_{\{0,\ \lfloor \bar{t}_{j_p} \rfloor - \tau_C\}} & \text{if } \bar{t}_{j_p} - t_{j_p-1} > \tau_D \end{cases}$$

From this and (3.27) we obtain for all $q \in \mathcal{P}$

$$||(1-\phi_p)e_p||^2_{\{0,\bar{t}_{j_p}\}} = ||e_p||^2_{\{0,\ \lfloor t_{j_p-1} \rfloor\}} \leq ||e_q||^2_{\{0,\ \lfloor t_{j_p-1} \rfloor\}} = ||(1-\phi_p)e_q||^2_{\{0,\bar{t}_{j_p}\}}$$

if $\bar{t}_{j_p} - t_{j_p-1} \leq \tau_D$ and

$$\begin{aligned} &||(1-\phi_p)e_p||^2_{\{0,\bar{t}_{j_p}\}} = \\ &= \ ||e_p||^2_{\{0,\ \lfloor \bar{t}_{j_p} - \tau_C \rfloor\}} \leq ||e_q||^2_{\{0,\ \lfloor \bar{t}_{j_p} - \tau_C \rfloor\}} = ||(1-\phi_p)e_q||^2_{\{0,\bar{t}_{j_p}\}} \end{aligned}$$

if $\bar{t}_{j_p} - t_{j_p-1} > \tau_D$. From this and the fact that

$$||(1-\phi_p)e_q||^2_{\{0,\bar{t}_{j_p}\}} \leq ||(1-\phi_p)e_q||^2_{\{0,T\}},\ q \in \mathcal{P},\ p \in \mathcal{P}_P$$

there follows

$$||(1-\phi_p)e_p||^2_{\{0,\bar{t}_{j_p}\}} \leq ||(1-\phi_p)e_q||^2_{\{0,T\}}, \quad \forall p \in \mathcal{P}_T,\ \ q \in \mathcal{P} \tag{3.29}$$

Now

$$\begin{aligned} &||(1-\psi_T)e_\sigma||^2_{\{0,T\}} = \\ &= \ \sum_{p \in \mathcal{P}_T} \sum_{i \in \mathcal{I}_p} ||(1-\psi_T)e_p||^2_{\{t_{i-1},\bar{t}_i\}} \leq \sum_{p \in \mathcal{P}_T} ||(1-\psi_T)e_p||^2_{\{0,\bar{t}_{j_p}\}} \end{aligned} \tag{3.30}$$

In view of (3.28) we can write

$$\sum_{p \in \mathcal{P}_T} ||(1-\psi_T)e_p||^2_{\{0,\bar{t}_{j_p}\}} = \sum_{p \in \mathcal{P}_T} \left|\left| \left\{ \prod_{l \in \mathcal{P}_T} (1-\phi_l) \right\} e_p \right|\right|^2_{\{0,\bar{t}_{j_p}\}} \tag{3.31}$$

But

$$\sum_{p \in \mathcal{P}_T} \left|\left| \left\{ \prod_{l \in \mathcal{P}_T} (1-\phi_l) \right\} e_p \right|\right|^2_{\{0,\bar{t}_{j_p}\}} \leq \sum_{p \in \mathcal{P}_T} ||(1-\phi_p)e_p||^2_{\{0,\bar{t}_{j_p}\}}$$

From this, (3.29), (3.30), and (3.31) it follows that

$$||(1-\psi_T)e_\sigma||^2_{\{0,T\}} \leq \sum_{p\in\mathcal{P}_T} ||(1-\phi_p)e_q||^2_{\{0,T\}}, \quad \forall\ q \in \mathcal{P}$$

Thus for $q \in \mathcal{P}$

$$\begin{aligned}
||(1-\psi_T)e_\sigma||^2_{\{0,T\}} &\leq \sum_{p\in\mathcal{P}_T} \int_0^T \{e_q e^{\lambda t}\}^2 (1-\phi_p)^2 dt \\
&= \int_0^T \{e_q e^{\lambda t}\}^2 \left\{ \sum_{p\in\mathcal{P}_T} (1-\phi_p)^2 \right\} dt \\
&= \int_0^T \{e_q e^{\lambda t}\}^2 \left\{ \sum_{p\in\mathcal{P}_T} (1-\phi_p) \right\} dt
\end{aligned}$$

This, Lemma 6.1 and (3.28) imply that

$$\begin{aligned}
||(1-\psi_T)e_\sigma||^2_{\{0,T\}} &\leq \int_0^T \{e_q e^{\lambda t}\}^2 \left\{ m - 1 + \prod_{p\in\mathcal{P}_T} (1-\phi_p) \right\} dt \\
&= \int_0^T \{e_q e^{\lambda t}\}^2 \{m - \psi_T\} dt \\
&= \int_0^T \{e_q e^{\lambda t}\}^2 \{m - \psi_T^2\} dt
\end{aligned}$$

Hence

$$||(1-\psi_T)e_\sigma||^2_{\{0,T\}} \leq m||e_q||^2_{\{0,T\}} - ||\psi_T e_q||^2_{\{0,T\}} \tag{3.32}$$

Now

$$||(1-\psi_T)e_\sigma||^2_{\{0,T\}} + ||\psi_T e_q||^2_{\{0,T\}} = ||(1-\psi_T)e_\sigma + \psi_T e_q||^2_{\{0,T\}}$$

because $\psi_T(1-\psi_T) = 0$. From this and (3.32) it follows that

$$||(1-\psi_T)e_\sigma + \psi_T e_q||^2_{\{0,T\}} \leq m||e_q||^2_{\{0,T\}}$$

and thus that (3.17) is true. ■

References

1. K. J. Åström and B. Wittenmark. *Adaptive Control.* Addison-Wesley, 1994.
2. D. Borrelli, A. S. Morse, and E. Mosca. Discrete-time supervisory control of families of 2-degree of freedom linear set-point controllers. *IEEE Transactions on Automatic Control*, 1997. to appear.

3. G. C. Goodwin and K. S. Sin. *Adaptive Filtering Prediction and Control.* Prentice Hall, 1984.
4. J. P. Hespanha. A bound for the induced norm of a switched linear system in terms of a bound on the infinity norms of the transfer functions of the constant linear systems being switched. Technical report, Yale University, March 1998.
5. P. A. Ioannou and J. Sun. *Robust Adaptive Control.* Prentice Hall, 1996.
6. Miroslav Krstić, Ioannis Kanellakopoulos, and Petar Kokotović. *Nonlinear and Adaptive Control Design.* Adaptive and Learning Systems for Signal Processing, Communications, and Control. John Wiley & Sons, New York, 1995.
7. S. R. Kulkarni and P. J. Ramadge. Model and controller selection policies based on output prediction errors. *IEEE Transactions on Automatic Control*, 41:1594–1604, 1996.
8. I. D. Landau. *Adaptive Control: The Model Reference Approach.* Marcel Dekker, New York, 1979.
9. A. S. Morse. Supervisory control of families of linear set-point controllers - part 1: Exact matching. *IEEE Transactions on Automatic Control*, pages 1413–1431, oct 1996.
10. A. S. Morse. Supervisory control of families of linear set-point controllers - part 2: Robustness. *IEEE Transactions on Automatic Control*, 42:1500–1515, nov 1997.
11. K. S. Narendra and A. M. Annaswamy. *Stable Adaptive Systems.* Prentice Hall, 1989.
12. K. S. Narendra and J. Balakrishnan. Adaptive control using multiple models. *IEEE Transactions on Automatic Control*, pages 171 –187, feb 1997.
13. J. W. Polderman and I. Mareels. *Adaptive Systems - An Introduction.* Birkhäuser, 1996.
14. Shankar Sastry and Mark Bodson. *Adaptive Control: Stability, Convergence and Robustness.* Prentice Hall, Englewood Cliffs, New Jersey, 1989.

A Discrete-Time Algorithm for Identification and Adaptive Control

H. M. Silveira

Departamento de Engenharia Elétrica - CTC, Universidade Federal de Santa Catarina UFSC, C.P. 476, Trindade - Florianópolis - Santa Catarina 88040-900 Brésil, e-mail: silveira@eletrosul.gov.br

Summary. This paper establishes a global convergent algorithm independent of bounded signals restriction for parameter identification and adaptive control of discrete-time monovariable deterministic linear systems. An application of the algorithm to identification and adaptive control is presented.

1. Introduction

In this paper we shall be concerned with identification and adaptive control of a linear time-invariant system. The decreasing gain algorithm [1, 2] is a powerful algorithm for identification and adaptive control. However, this approach needs an additional proof to establish the conditions of parameter identification.

The main purpose of this paper is to propose a modification in this classical algorithm which allows us to prove the results we need independently of bounded signals. Section 2. presents the main theorem which defines the proposed algorithm, Section 3. and Section 4. present an application of the algorithm in identification and in adaptive control, respectively.

2. The Adaptation Algorithm

The analysis of identification and adaptive control will be based on the following technical result:

Theorem 2.1. *Given*

$$y(k) = v^T(k)p \tag{4.1}$$

where $p, v(k)$ *are an unknown* $[m \times 1]$ *parameter vector and an* $[m \times 1]$ *known vector sequence and* $y(k)$ *is a known scalar sequence, respectively.*

Let $V(k) = [v(0)|v(1)|\ldots|v(k)]$ be an $[m \times (k+1)]$ matrix.

If the adaptation algorithm is

$$\theta(-2) = \theta(-1), \tag{4.2}$$

$$F(-1) = I, \tag{4.3}$$

where I is the unit matrix of order m,

$$\begin{aligned} F(k) = F(k-1) - \\ -F(k-1)v(k)[1 + v^T(k)F(k-1)v(k)]^{-1}v^T(k)F(k-1), \end{aligned} \tag{4.4}$$

$$\begin{aligned} \theta(k) = \theta(k-1) + F(k)[\theta(k-1) - \theta(k-2)] + \\ +F(k-1)[1 + v^T(k)F(k-1)v(k)]^{-1}v(k)[y(k) - v^T(k)\theta(k-1)], \end{aligned} \tag{4.5}$$

where $F(k)$ is an $[m \times m]$ matrix and $\theta(k)$ is an $[m \times 1]$ parameter estimate vector, then

1. $\lim_{k\to\infty} [\theta(k) - \theta(k-1)] = 0$,
2. $\lim_{k\to\infty} [y(k) - v^T(k)\theta(k)] = 0$,
3. If, in addition, $V(i)$ has rank m for a certain $i < \infty$, then $\theta(k)$ converges to p.

Proof. From Eq. (4.4) and the matrix inversion lemma [2], it follows that

$$F(k)^{-1} = F(k-1)^{-1} + v(k)v(k)^T. \tag{4.6}$$

Considering Eq. (4.3), the solution of Eq. (4.6) is

$$F(k)^{-1} = I + v(0)v(0)^T + v(1)v(1)^T + \ldots + v(k)v(k)^T.$$

But, $V(k)V(k)^T = v(0)v(0)^T + v(1)v(1)^T + \ldots + v(k)v(k)^T$, so

$$F(k)^{-1} = I + V(k)V(k)^T. \tag{4.7}$$

Since

$$\theta(k-1) + F(k-1)[1 + v^T(k)F(k-1)v(k)]^{-1}v(k)[y(k) - v^T(k)\theta(k-1)]$$

is equal to

$$F(k)[F^{-1}(k-1)\theta(k-1) + v(k)y(k)],$$

Eq. (4.5) can be rewritten

$$F^{-1}(k)\theta(k) - \theta(k-1) = F^{-1}(k-1)\theta(k-1) - \theta(k-2) + v(k)y(k). \quad (4.8)$$

Considering Eq. (4.2) and Eq. (4.3), the solution of Eq. (4.8) is

$$F^{-1}(k)\theta(k) - \theta(k-1) = v(0)y(0) + v(1)y(1) + \ldots + v(k)y(k).$$

So,

$$F^{-1}(k)\theta(k) - \theta(k-1) = V(k)Y(k) \quad (4.9)$$

where $Y(k)^T = [y(0), y(1), \ldots, y(k)]$.

Therefore, $V(k)Y(k) = V(k)V(k)^T p$. Considering Eq. (4.7), Eq. (4.9) can be written

$$[I + V(k)V(k)^T]\theta(k) = \theta(k-1) + V(k)V(k)^T p \quad (4.10)$$

From Eq. (4.10), it follows that

$$[I + V(k)V(k)^T](\theta(k) - p) = (\theta(k-1) - p). \quad (4.11)$$

Along the Eq. (4.11),

$$\begin{aligned}(\theta(k) - p)^T(\theta(k) - p) - (\theta(k-1) - p)^T(\theta(k-1) - p) = \\ -2(\theta(k) - p)^T V(k)V(k)^T(\theta(k) - p) \\ -(\theta(k) - p)^T V(k)V(k)^T V(k)V(k)^T(\theta(k) - p).\end{aligned} \quad (4.12)$$

Since $(\theta(k)-p)^T(\theta(k)-p)$ is a bounded, from zero, non-increasing sequence, it converges. So, independently of bounded signal conditions, we find that

$$\lim_{k\to\infty} (\theta(k) - p)^T V(k)V(k)^T(\theta(k) - p) = 0$$

and

$$\lim_{k\to\infty} (\theta(k) - p)^T V(k)V(k)^T V(k)V(k)^T(\theta(k) - p) = 0.$$

Then

$$\lim_{k\to\infty} V(k)^T(\theta(k) - p) = 0, \quad (4.13)$$

and

$$\lim_{k\to\infty} V(k)V(k)^T(\theta(k) - p) = 0. \quad (4.14)$$

From Eq. (4.14) and Eq. (4.11), we conclude that

$$\lim_{k\to\infty} [\theta(k) - \theta(k-1)] = 0,$$

and the first part of the theorem is proved.

Using Eq. (4.13), we have

$$\lim_{k\to\infty} V(k)^T(\theta(k) - p) = \lim_{k\to\infty} V(k)^T\theta(k) - Y(k) = 0.$$

Hence,

$$\lim_{k\to\infty} [y(k) - v^T(k)\theta(k)] = 0$$

and the second part of the theorem is proved.

Since

$$\lim_{k\to\infty} V(k)^T\theta(k) - p) = 0 \tag{4.15}$$

and the rank of $V(i)$ is m for a certain $i < \infty$, we conclude that the rank of $V(k)$ is m for $k > i$. Then, from Eq. (4.15),

$$\lim_{k\to\infty} (\theta(k) - p) = 0$$

and we conclude that $\theta(k)$ converges to p. This completes the proof.

3. Identification

For process identification purposes, Theorem 2.1 can be used. This Section will show an application of this algorithm in this type of problem.

Consider the single-input single-output (SISO) discrete system, controllable and observable,

$$\begin{aligned} y(k) &= -\alpha_1 y(k-1) - \alpha_2 y(k-2) - \ldots - \alpha_n y(k-n) + \\ &+ \beta_1 u(k-1) + \beta_2 u(k-2) + \ldots + \beta_n u(k-n), \end{aligned} \tag{4.16}$$

where $y(k)$ is the output, $u(k)$ is the input, α_i and β_i, $(i = 1, \ldots, n)$, are the unknown parameters.

Defining

$$p^T = [\alpha_1, \alpha_2, \ldots, \alpha_n, \beta_1, \beta_2, \ldots, \beta_n] \tag{4.17}$$

and

$$\begin{aligned} v^T(k) &= [-y(k-1), -y(k-2), \ldots, -y(k-n), \\ &\quad u(k-1), u(k-2), \ldots, u(k-n)], \end{aligned} \tag{4.18}$$

the algorithm defined by Eq. (4.2), (4.3), (4.4) and (4.5) can be applied. Assuming that $V(i)$ has rank $2n+1$ for a certain $i < \infty$, $\theta(k)$ will converge to p, where $V(k) = [v(0)|v(1)| \ldots |v(k)]$.

4. Adaptive Zero Regulator

For adaptive control, Theorem 2.1 can also be used. This Section will show an example of such an application.

Consider the single-input single-output (SISO) discrete system defined by Eq. (4.16). The adaptive controller is

$$\begin{aligned} u(k) &= -q_1(k)u(k-1) - q_2(k)u(k-2) - \ldots - q_n(k)u(k-n) + \\ &\quad +p_1(k)y(k-1) + p_2(k)y(k-2) + \ldots + p_n(k)y(k-n). \end{aligned} \tag{4.19}$$

The adaptation algorithm is

$$\theta_p(-2) = \theta_p(-1), \tag{4.20}$$

$$F_p(-) = I, \tag{4.21}$$

where I is the unit matrix of order $2n+1$,

$$\begin{aligned} F_p(k) &= F_p(k-1) - \\ &\quad -F_p(k-1)v_p(k)[1 + v_p^T(k)F_p(k-1)v_p(k)]^{-1}v_p^T(k)F_p(k-1), \end{aligned} \tag{4.22}$$

$$\begin{aligned} \theta_p(k) &= \theta_p(k-1) + F_p(k)[\theta_p(k-1) - \theta_p(k-2)] + \\ &\quad +F_p(k-1)[1 + v_p^T(k)F_p(k-1)v_p(k)]^{-1} \\ &\quad v_p(k)[y(k-1) - v_p^T(k)\theta_p(k-1)], \end{aligned} \tag{4.23}$$

where $F(k)$ is a $[(2n+1)\times(2n+1)]$ matrix, $\theta_p(k)$ is a $[(2n+1)\times 1]$ parameter estimate vector and

$$\begin{aligned} v_p^T(k) = [&-y(k-2), -y(k-3), \ldots, \\ &-y(k-n-1), u(k-1), u(k-2), \ldots, u(k-n-1)], \end{aligned} \tag{4.24}$$

$$\theta_p(k) = [a_1(k), a_2(k), \ldots, a_n(k), b_0(k), b_1(k), \ldots, b_n(k)], \tag{4.25}$$

$$\theta_c(-2) = \theta_c(-1), \tag{4.26}$$

$$F_c(-1) = I, \tag{4.27}$$

where I is the unit matrix of order $2n$,

$$\begin{aligned} F_c(k) &= F_c(k-1) - \\ &\quad -F_c(k-1)v_c(k)[1 + v_c^T(k)F_c(k-1)v_c(k)]^{-1}v_c^T(k)F_c(k-1), \end{aligned} \tag{4.28}$$

$$\begin{aligned} \theta_c(k) &= \theta_c(k-1) + F_c(k)[\theta_c(k-1) - \theta_c(k-2)] + \\ &\quad +F_c(k-1)[1 + v_c^T(k)F_c(k-1)v_c(k)]^{-1} \\ &\quad v_c(k)[s(k) - v_c^T(k)\theta_c(k-1)], \end{aligned} \tag{4.29}$$

$$\begin{aligned} s(k) &= [(r_1 - a_1)q^{-1} + \ldots + \\ &\quad (r_n - a_n)q^{-n} + r_{n+1}q^{-n-1} + \ldots + r_{2n}q^{-2n}]y(k), \end{aligned} \tag{4.30}$$

where q^{-1} is the delay operator, the polynomial

$$1 + r_1 q^{-1} + \ldots + r_{2n} q^{-2n} \tag{4.31}$$

is known and it has all roots inside the unit circle, $F_c(k)$ is a $[2n \times 2n]$ matrix and

$$\begin{aligned} \theta_c(k) &= [q_1(k), q_2(k), \ldots, q_n(k), p_1(k), \ldots, p_n(k)], & (4.32)\\ v_c^T(k) &= [A(k)y(k-1), \ldots, A(k)y(k-n), \\ &\qquad -B(k)y(k-1), \ldots, -B(k)y(k-n)], & (4.33)\\ A(k) &= 1 + a_1(k)q^{-1} + \ldots + a_n(k)q^{-n}, & (4.34)\\ B(k) &= b_1(k)q^{-1} + \ldots + b_n(k)q^{-n} & (4.35) \end{aligned}$$

Theorem 4.1. *If the adaptive regulator defined by Eqs. (4.19) through (4.35) is applied on the system defined by (4.16), then* $\lim_{k \to \infty} y(k) = 0$.

Proof. For a sufficiently large k, the following conclusions are valid.

Using Eqs. (4.20) through (4.25) and Theorem 2.1, it follows that

$$A(k)y(k) = B(k)u(k). \tag{4.36}$$

Defining

$$\begin{aligned} Q(k) &= 1 + q_1(k)q^{-1} + \ldots + q_n(k)q^{-n}, & (4.37)\\ P(k) &= p_1(k)q^{-1} + \ldots + p_n(k)q^{-n}, & (4.38) \end{aligned}$$

Eq. (4.19) can be rewritten

$$Q(k)u(k) = P(k)y(k). \tag{4.39}$$

From Eqs. (4.36) and (4.39), and considering that $\theta_p(k) = \theta_p(k-1)$ and $\theta_c(k) = \theta_c(k-1)$ we have

$$[Q(k)A(k) - P(k)B(k)]y(k) = 0. \tag{4.40}$$

Using Eqs. (4.26) through (4.35) and Theorem 2.1, it follows that

$$[Q(k)A(k) - P(k)B(k)]y(k) = [1 + r_1 q^{-1} + \ldots + r_{2n} q^{-2n}]y(k). \tag{4.41}$$

From Eqs. (4.40) and (4.41),

$$[1 + r_1 q^{-1} + \ldots + r_{2n} q^{-2n}]y(k) = 0. \tag{4.42}$$

Since this system is stable, it yields $\lim_{k \to \infty} y(k) = 0$. The proof is completed.

Eq. (4.42) shows that the poles of the feedback system were arbitrarily defined. The identity defined by (4.40) is classical. The proof that there are polynomials P and Q such that the identity exists is also classical.

This algorithm was tested in a second-order process with poles and zeros outside the unit circle and gave good results. This algorithm can be used in the synthesis of adaptive robust control, by transforming it into an adaptive zero regulator.

References

1. Landau I.D., Silveira H.M., "A Stability Theorem with Applications to Adaptive Control", IEEE Trans. on Automat. Control, 1979.
2. Landau I.D., Lozano R., M'Saad M., "Adaptive Control", Springer Verlag, 1997.

Backstepping and Passivation in the Presence of an Integrator with Throughput

D. Fontaine and P. Kokotović

Center for Control Engineering, University of California, Santa Barbara, CA 93106, USA, e-mail: dan@seidel.ece.ucsb.edu, petar@ece.ucsb.edu

Dedicated to Ioan Landau for his 60th Birthday

Summary. [1] This paper presents two controllers that globally stabilize a nonlinear system augmented with an integrator and throughput. Neither affine in control nor of triangular structure, the system is not in the class to which backstepping or forwarding designs are applicable.

1. Introduction

The longevity of passivity as a design tool is remarkable. The senior author of this note remembers when back in the late 1950's, passivity was used as a network synthesis concept, without ever being related to feedback control. Then came the seminal work of Popov, and its impact on adaptive control through the work of Parks [11], Anderson [1], and above all, Landau [6, 7, 8], who inspired the use of passivity ("hyperstability") as a framework for adaptive control design.

Passivity and dissipativity, generalized by Willems [14], Hill [3] and others, were introduced to nonlinear control studies in the late 1980's and are very much alive today [2, 12]. Again Landau helped, this time as a catalyst in insisting that backstepping is but a recursive way of avoiding obstacles to passivity. With his coworkers he backed his claim in [10].

[1] Supported under NSF grant ECS-9528370 and AFOSR grant F49620-95-1-0409.

2. Problem Formulation

Several methodologies are available for the stabilization of nonlinear systems with triangular structure affine in the control variable. Backstepping [5] is used for systems in feedback form (lower triangular), while feedforward (upper triangular) systems are stabilized with forwarding [13]. When systems do not have either of the two triangular structures, or are not affine in the control variable, controllers are more difficult to design. This note considers the system

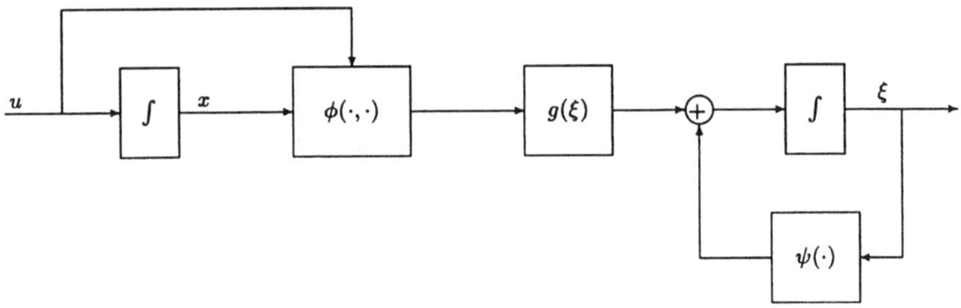

Fig. 5.1. System augmented with integrator and throughput

$$\dot{\xi} = \psi(\xi) + g(\xi)\phi(x,u) \qquad \xi \in \mathbb{R}^n, \quad \psi(0) = 0 \tag{5.1}$$
$$\dot{x} = u \qquad x, u \in \mathbb{R} \tag{5.2}$$

in which the control u appears in a non-affine way through the function $\phi(x,u)$. This system can be viewed as the ξ-subsystem (5.1) augmented by the integrator (5.2) with the throughput $\phi(x,u)$. Because of the throughput, integrator backstepping is not applicable. We allow $\dot{\xi} = \psi(\xi)$ to be unstable which means that neither forwarding [13] nor low gain [9] designs can be applied. Instead, we present two different passivating designs that globally asymptotically stabilize the equilibrium $\xi = 0$, $x = 0$. Of the two designs, *dynamic backstepping* is new, while *cascade passivation* is an application of a result in [13].

3. Dynamic Backstepping

3.1 Design

This design can be seen as an extension of backstepping to a non-affine system which is not in feedback form. If the non-affine term $\phi(x,u)$ were simply x, we would use it as a virtual control and proceed with integrator backstepping in the standard way. We extend this idea by assigning to $\phi(x,u)$ a virtual control law $\alpha(\xi)$ that renders the ξ-subsystem globally asymptotically stable (GAS). If we could enforce $\phi(x,u)=\alpha(\xi)$, differentiation would give $\dot{\alpha}=\frac{\partial\phi}{\partial x}u+\frac{\partial\phi}{\partial u}\dot{u}$. We proceed with this design, under the following restrictions.

Assumption 3.1.

1. *There exists a C^1 function $\alpha(\xi)$ such that the system $\dot{\xi}=\psi(\xi)+g(\xi)\alpha(\xi)$ is GAS.*
2. $\left(\frac{\partial\phi}{\partial u}\right)^{-1}\left(\frac{\partial\phi}{\partial x}\right)>0$, and $\left|\left(\frac{\partial\phi}{\partial x}\right)\right|\geq c>0\ \forall u,x$

By assigning $\phi(x,u)=\alpha(\xi)$ and differentiating, we have in fact performed backstepping of α through the parallel connection of an integrator and the throughput $\phi(x,u)$. In doing so we have constructed the dynamic compensator

$$\dot{u}=\left(\frac{\partial\phi}{\partial u}\right)^{-1}\left(-\frac{\partial\phi}{\partial x}u+\dot{\alpha}\right). \tag{5.3}$$

To account for the fact that $\phi(x,u)=\alpha(\xi)$ can not be enforced, we define the error variable $y=\phi(x,u)-\alpha(\xi)$ and obtain

$$\begin{aligned}\dot{\xi} &= \psi(\xi)+g(\xi)(\alpha(\xi)+y) && (5.4)\\ \dot{y} &= \frac{\partial\phi}{\partial x}u+\frac{\partial\phi}{\partial u}\dot{u}-\dot{\alpha}(\xi,y), && (5.5)\end{aligned}$$

where $\dot{\alpha}(\xi,y)=\frac{\partial\alpha}{\partial\xi}\left(\psi(\xi)+g(\xi)(\alpha(\xi)+y)\right)$. Treating $\dot{y}=\omega$ as the new control variable we rewrite the system (5.1-5.2) with the dynamic compensator as

$$\begin{aligned}\dot{y} &= \omega && (5.6)\\ \dot{\xi} &= \psi(\xi)+g(\xi)(\alpha(\xi)+y) && (5.7)\\ \dot{u} &= \left(\frac{\partial\phi}{\partial u}\right)^{-1}\left(-\frac{\partial\phi}{\partial x}u+\dot{\alpha}(\xi,y)+\omega\right). && (5.8)\end{aligned}$$

With the storage function $S=\frac{1}{2}y^2$, the derivative $\dot{S}=\omega y$ verifies passivity from input ω to output y. This passivity property, verified with a positive *semidefinite* storage function, is not sufficient for asymptotic stability. In addition we need the following detectability property.

Definition 3.1. *For a system $\dot{x}=f(x,u)$, $y=h(x,u)$ with zero input, let $Z\subset\mathbb{R}^n$ denote its largest positively invariant set contained in $\{x\in\mathbb{R}^n \mid y=h(x,0)=0\}$. If $x=0$ is asymptotically stable conditionally[2] to Z, then the system is Zero Sate Detectable (ZSD).*

[2] A solution is asymptotically stable conditionally to Z if it is asymptotically stable in the sense of Lyapunov for all initial conditions $x_0\in Z$, see [4, 13].

We see that for the system (5.6-5.8) with $y = 0$ the set $Z = \{\xi, 0, u\}$ is invariant. To establish ZSD, we analyze this system confined to Z:

$$\begin{aligned}\dot{\xi} &= \psi(\xi) + g(\xi)\alpha(\xi) &(5.9)\\ \dot{u} &= \left(\frac{\partial\phi}{\partial u}\right)^{-1}\left(-\frac{\partial\phi}{\partial x}u + \dot{\alpha}(\xi,0)\right) &(5.10)\end{aligned}$$

Its global asymptotic stability is immediate from the following boundedness property of the u-subsystem (5.10).

Lemma 3.1. *If Assumption 3.1 is satisfied, the system*

$$\dot{u} = \left(\frac{\partial\phi}{\partial u}\right)^{-1}\left(-\frac{\partial\phi}{\partial x}u + \dot{\alpha}(\xi,0)\right)$$

is Input to State Stable (ISS) from input ξ to state u.

Proof: The derivative of $W = \frac{1}{2}u^2$ is

$$\begin{aligned}\dot{W} &= \left(\frac{\partial\phi}{\partial u}\right)^{-1}\left(-\frac{\partial\phi}{\partial x}u^2 + u\dot{\alpha}(\xi,0))\right)\\ \dot{W} &\leq -\left(\frac{\partial\phi}{\partial u}\right)^{-1}\left(+\frac{cu^2}{2} + \frac{1}{2}|u|(c|u| - 2|\dot{\alpha}(\xi,0)|)\right)\end{aligned}$$

When $|u| > \frac{2|\dot{\alpha}(\xi,0)|}{c}$, $\dot{W}$ is negative. The function $\dot{\alpha}(\xi,0)$ is C^1 and $\alpha(0,0) = 0$. Therefore u is bounded if ξ is, and moreover, u goes to zero if ξ does. □

Proposition 3.1. *Under Assumption 3.1, the system (5.6-5.8) is rendered GAS by the output feedback $\omega = -y$.*

Proof: Inside the set $Z = \{\xi, 0, u\}$, the ξ-subsystem 5.9 is GAS by Assumption 3.1, so ξ is bounded and converges to zero. Lemma 3.1 and the resulting global asymptotic stability of the system (5.9-5.10) verify that the full system (5.6-5.8) is ZSD. By Theorem 2.28 from [13], passivity with respect to a positive semidefinite storage function and the ZSD property are sufficient to globally asymptotically stabilize the system with the feedback $\omega = -y$. □

3.2 Discussion and Examples

The existence of a stabilizing $\alpha(\xi)$ is necessary. This is equivalent to stabilizability of the ξ-subsystem and, hence, is not restrictive. To see that the second part of Assumption 3.1 is restrictive, let $\phi(\xi, u) = \phi_1(u) + \phi_2(x)$. It follows that both derivatives $\frac{\partial\phi_1}{\partial u}$ and $\frac{\partial\phi_2}{\partial x}$ must be of the same sign and not equal to zero. Examples of such functions are $\sigma + \sigma^3$ and $e^\sigma - e^{-\sigma}$, where σ is u or x. Also, while $\tan^{-1}(u)$ is acceptable for $\phi_1(u)$, $\phi_2(x) = \tan^{-1}(x)$ is not allowed, because Lemma 3.1 requires $\frac{\partial\phi_2}{\partial x}$ to be bounded away from zero.

Example 3.1. We apply dynamic backstepping to the system

$$\dot{\xi} = \xi^2 + x + u + u^3 \tag{5.11}$$

$$\dot{x} = u \tag{5.12}$$

where $g(\xi) = 1$, $\phi_1(u) = u + u^3$, $\phi_2(x) = x$, and $\psi(\xi) = \xi^2$. We select $\alpha(\xi) = -\xi^2 - \xi$ as a stabilizing virtual control for the ξ-subsystem. With the error variable

$$y = \phi_1(u) + \phi_2(x) - \alpha(\xi) = u + u^3 + x + \xi^2 + \xi, \tag{5.13}$$

the ξ-subsystem becomes $\dot{\xi} = -\xi + y$. Assumption 3.1 is satisfied so we can use the dynamic compensator (5.8)

$$\dot{u} = \left(\frac{\partial \phi_1}{\partial u}\right)^{-1} \left(-\frac{\partial \phi_2}{\partial x} u + \dot{\alpha}(\xi, y) + \omega\right) \tag{5.14}$$

$$= \left(\frac{\partial}{\partial u}\left(u + u^3\right)\right)^{-1} \left(-u + \dot{\xi}\frac{\partial \alpha}{\partial \xi} + \omega\right) \tag{5.15}$$

$$= \frac{1}{1 + 3u^2}\left(-u - (2\xi + 1)(-\xi + y) + \omega\right). \tag{5.16}$$

Now, by Proposition 3.1 the system (5.11-5.12) is rendered GAS with $\omega = -y$.

□

Remark 3.1. For $\phi_1(u) = u + u^3$ the inverse $\phi^{-1}(u)$ exists for all u and $\phi(u)u > 0$ for all $u \neq 0$. Therefore the system (5.11-5.12) with the output ξ, is globally input/output linearizable with ISS zero dynamics. We stress that this would not be the case if instead of $\phi_1(u) = u + u^3$ we had $\phi_1(u) = \tan^{-1}(u)$ because input/output linearization design is impossible beyond $|u| < \frac{\pi}{2}$. However, dynamic backstepping still yields a global result.

4. Cascade Passivation

4.1 Design

For our second design we interpret the system (5.1-5.2) as the cascade connection of two subsystems with input feedforward and output feedback passivity properties defined in [13]. Assuming $g(\xi) \neq 0$ for all ξ, we let the two subsystems Σ_1 and Σ_2 be

$$\Sigma_1 \left\{ \begin{array}{rcl} \dot{\xi} & = & \psi(\xi) + g(\xi)u_1 \\ y_1 & = & h(\xi) \end{array} \right. \tag{5.17}$$

$$\Sigma_2 \left\{ \begin{array}{rcl} \dot{x} & = & u_2 \\ y_2 & = & \phi(x, u_2) \end{array} \right. \tag{5.18}$$

Our first task is to choose an output function $h(\xi)$. If S_1 is a storage function for Σ_1, then its derivative is $\dot{S}_1 = \mathcal{L}_\psi S_1 + \mathcal{L}_g S_1 u_1$. Therefore we take $h(\xi) = \mathcal{L}_g S_1$ as the output y_1.

Now we make an assumption about the passivity properties of each subsystem. We assume that, in the terminology of [13], Σ_1 is Output Feedback Passive (OFP) and Σ_2 is Input Feedforward Passive (IFP). These properties quantify excess or shortage of passivity using scalar functions $\rho(\cdot)$ and $\nu(\cdot)$.

Assumption 4.1.
There exist C^1 functions $\rho(\cdot), \nu(\cdot)$ and positive definite, radially unbounded C^1 functions S_1, S_2 such that

1. *$\rho(y_1)y_1$ and $\nu(u_2)u_2$ are sign definite*
2. *$\dot{S}_1 \leq -\rho(y_1)y_1 + u_1 y_1$*
3. *$\dot{S}_2 \leq -\nu(u_2)u_2 + u_2 y_2$*

We now analyze when the feedback $u_2 = -y_1$ will render the interconnection GAS. The derivative of the cascade storage function $S = S_1 + S_2$ is $\dot{S} \leq -\rho(y_1)y_1 - \nu(y_1)y_1$. Hence, $\dot{S}(y_1)$ can be either positive or negative. Even when $\dot{S}(y_1)$ is non-positive, it is still only *semidefinite.* Thus we require that both Σ_1 and Σ_2 be zero state detectable. The ZSD property of Σ_1 is equivalent to stabilizability with respect to the storage function S_1 as a control Lyapunov function. For Σ_2, the ZSD property is equivalent to $\phi(x, 0) = 0 \Rightarrow x = 0$.

Proposition 4.1. *Under Assumption 4.1, let S_1 be a control Lyapunov function for Σ_1, and $\phi(x, 0) = 0 \Rightarrow x = 0$. Then if*

$$\rho(\sigma)\sigma + \nu(\sigma)\sigma > 0 \quad \text{for all} \quad \sigma \neq 0, \tag{5.19}$$

the system (5.1-5.2) is rendered GAS with the feedback control $u_2 = -y_1 = -\mathcal{L}_g S_1$.

Proof: $\dot{S} \leq 0 \Rightarrow y_1 \to 0$, thus ZSD of Σ_1 implies $\xi = 0$. This implies that $g(0)\phi(x, 0) = 0$ and $\phi(x, 0) = 0$. ZSD of Σ_1 and $\phi(x, 0) = 0$ imply that $x = 0$. □

4.2 Discussion and Examples

For a comparison with the backstepping design we again let $\phi(\xi, u) = \phi_1(u) + \phi_2(x)$. First we assess the passivity of a scalar Σ_1 subsystem with $g(\xi) = 1$. Using the storage function $S_1 = \frac{1}{2}\xi^2$, the output for Σ_1 is $y_1 = \mathcal{L}_g S_1 = \xi$. The derivative of S_1 along the solutions of (5.17) is

$$\begin{aligned}\dot{S}_1 &= \xi\psi(\xi) + u_1\xi && (5.20)\\ &= y_1\psi(y_1) + u_1 y_1 && (5.21)\end{aligned}$$

Thus, if $\psi(y_1)y_1$ is sign definite, Σ_1 is OFP($\rho(y_1)$) with $\rho(y_1) = -\psi(y_1)$.

Next we determine the passivity of the Σ_2 subsystem (5.18) which consists of $\dot{x} = u_2$ and the output $y_2 = \phi_1(u_2) + \phi_2(x)$. Let us assume that $\phi_2(x)x > 0$ for all $x \neq 0$. Then a storage function for the Σ_2 is $S_2 = \int_0^x \phi_2(x)dx$. Its derivative is,

$$\begin{aligned}\dot{S}_2 &= \phi_2(x)u_2 && (5.22)\\ &= -\phi_1(u_2)u_2 + u_2 y_2 && (5.23)\end{aligned}$$

where we have used the fact that $\phi_2(x) = y_2 - \phi_1(u)$. If $\phi_1(u_2)u_2$ is sign definite, then Σ_2 is IFP($\nu(u_2)$) with $\nu(u_2) = \phi_1(u_2)$.

Both Σ_1 and Σ_2 are ZSD, so the only remaining condition to be satisfied is $\rho(\sigma)\sigma + \nu(\sigma)\sigma > 0$ for all $\sigma \neq 0$.

Example 4.1. The same system in Example 3.1 can be seen as the cascade of two subsystems. The Σ_1 subsystem is

$$\Sigma_1 \left\{ \begin{aligned} \dot{\xi} &= \xi^2 + u_1 \\ y_1 &= \xi \end{aligned} \right. . \qquad (5.24)$$

The derivative of its storage function $S_1 = \frac{1}{2}\xi^2$ is

$$\begin{aligned}\dot{S}_1 &= \xi(\xi^2) + \xi u_1 && (5.25)\\ &= (y_1^2)y_1 + u_1 y_1 && (5.26)\\ &\leq \frac{1}{2}(y_1 + y_1^3)y_1 + u_1 y_1 && (5.27)\end{aligned}$$

Thus, Σ_1 is OFP($\rho(y_1)$) with $\rho(y_1) = -\frac{1}{2}(y_1 + y_1^3)$. We proceed to Σ_2,

$$\Sigma_2 \left\{ \begin{aligned} \dot{x} &= u_2 \\ y_2 &= x + u_2 + u_2^3 \end{aligned} \right. \qquad (5.28)$$

and use the storage function $S_2 = \int_0^x (x)dx = \frac{1}{2}x^2$. Its derivative is

$$\begin{aligned}\dot{S}_2 &= xu_2 && (5.29)\\ &= (y_2 - u_2 - u_2^3)u_2 && (5.30)\\ &= -(u_2 + u_2^3)u_2 + u_2 y_2 && (5.31)\end{aligned}$$

which means that Σ_2 is IFP($\nu(u_2)$) with $\nu(u_2) = -(u_2 + u_2^3)$. Assigning $u_2 = -y_1$, we obtain $\rho(y_1)y_1 + \nu(y_1)y_1 = \frac{1}{2}(y_1^2 + y_1^4) > 0$ for all $y_1 \neq 0$. By Proposition 4.1 the cascade system Σ_1, Σ_2 with feedback $u_2 = -y_1$ is GAS.
□

5. Conclusions

Although the example system (5.11-5.12) admits both designs presented here as well as input/output linearization, more common are the systems to which only one of the three designs is applicable. If $\psi(\xi) = \xi^4$ instead of ξ^2, Σ_2 no longer has enough excess passivity, and either dynamic backstepping or I/O linearization must be used. When $\phi_1(u) = \tan^{-1}(u)$, the excess of passivity in Σ_2 is again insufficient and dynamic backstepping is the only option. On the other hand, in systems where Σ_1 has an excess of passivity, the cascade passivation design admits functions $\phi_1(u)$ to which neither dynamic backstepping nor I/O lineariztion apply. For example, if $\psi(\xi) = -\xi_1 - \xi_1^3$ and $\phi_1(u) = -u^2$, cascade passivation is the only applicable design. In this sense, the two designs presented here, along with I/O linearization, offer a complementary set of three design options.

Acknowledgment: The authors acknowledge the helpful comments of Rodolphe Sepulchre.

References

1. B. D. O. Anderson. A system theory criterion for positive real transfer matrices. *Siam Journal of Control*, 5:171–182, 1967.
2. C. Byrnes, A. Isidori, and J. Willems. Passivity, feedback equivalence, and the global stabilization of minimum phase nonlinear systems. *IEEE Trans. Auto. Contr.*, 36(11):1228–1240, November 1991.
3. D. J. Hill and P. J. Moylan. Stability results for nonlinear feedback systems. *Automatica*, 13:377–382, 1977.
4. A. Iggidr, B. Kalitine, and R. Outbib. Semidefinite lyapunov functions satbility and stabilization. *Mathtematics of Control, Signals, and Systems*, 9(2):95–106, 1996.
5. M. Krstić, I. Kanellakopoulos, and P. Kokotović. *Nonlinear and Adaptive Control Design*. Wiley, 1995.
6. I. Landau. A hyperstability criterion for model reference adaptive control systems. *IEEE Trans. Auto. Contr.*, 14:552–555, 1969.
7. I. Landau. A generalization for the hyperstability conditions for model reference adaptive systems. *IEEE Trans. Auto. Contr.*, 17:246–247, 1972.
8. I. Landau. *Adaptive Control*. Marcel Decker, 1979.
9. W. Lin. Bounded smooth state feedback and a global separation principle for non-affine nonlinear systems. *Systems and Control Letters*, 26(1):41–53, September 1995.
10. R. Lozano, B. Brogliato, and I. Landau. Passivity and global stabilization of cascaded nonlinear systems. *IEEE Trans. Auto. Contr.*, 37:1386–1388, 1992.
11. P. C. Parks. Lyapunov redesign of model reference adaptive control systems. *IEEE Trans. Auto. Contr.*, 11:362–367, 1966.
12. A. Saberi, P. Kokotovic, and H. Sussman. Global stabilization of partially linear composite systems. *IEEE Trans. Auto. Contr.*, 33:358–365, 1988.

13. R. Sepulchre, M. Janković, and P. Kokotović. *Constructive Nonlinear Control.* Springer, 1997.
14. J. Willems. Dissipative dynamical systems, parts I and II. *Arch. Rational Mechanics and Analysis*, 45:321–393, 1972.

Robust H_∞ Control of Time-Delay Systems: A Generalized Popov Theory Approach

J.M. Dion[1], *L. Dugard*[1], *D. Ivanescu*[1], *S.I. Niculescu*[2], *and V. Ionescu*[3]

[1] Laboratoire d'Automatique de Grenoble, ENSIEG, BP 46, 38402 Saint Martin d'Herès, France, e-mail: dion, dugard, dan@lag.ensieg.inpg.fr

[2] HEUDIASYC, Université de Technologie BP 529, 60205 Compiègne, France, e-mail: silviu@hds.utc.fr
On leave from Laboratoire d'Automatique de Grenoble (France);
Also with the Department of Automatic Control, University "Politehnica" Bucharest (Romania)

[3] University "Politehnica" Bucharest, Splaiul Independentei 313, 77206 Bucharest, Romania, e-mail: vlad@starnets.ro

Summary. The paper deals with the generalized Popov theory applied to state-delayed systems. Sufficient conditions for stabilizing uncertain time-delay systems as well as for γ - attenuation achievement are given in terms of algebraic properties of a Popov system via a Liapunov - Krasovskii functional. The considered approach, successfully applied to classical linear systems, is new in the context of time delay systems and gives some interesting interpretations of H^∞ memoryless control problems in terms of Popov triplets and associated objects. The proposed delay - independent criteria are very simple and easy checkable via numerical procedures.

Keywords: Time-delay Systems. Riccati theory. Robust control. Kalman - Popov - Yakubovich systems. Disturbance attenuation. Liapunov - Krasovskii functional.

1. Introduction

The stability of dynamical physical systems including delay is a problem of recuring interest in the last years, since the existence of a delay in a system can destabilize the system ([1] [5], [11], [16] [12], [21]). Of recent interest is the stabilizing problem formulated for state-delayed systems including additional performance requirements such as γ-attenuation with respect to exogenous disturbances and/or robustness with respect to parameter uncertainties.

As the investigations in this area are incipient, "simple" memoryless controllers have been studied prior to other compensation techniques. However, it is worthwhile to mention that in spite of the simplicity of the formulation of the memoryless stabilization problem, sophisticated technical machineries such as Riccati inequality and linear matrix inequality techniques have been thoroughly involved, see [14, 15, 17, 18, 19, 24].

In this paper an attempt has been made in order to apply the generalized Popov theory (see [2, 3, 4, 6, 7, 8, 9]) for simultaneously achieving closed-loop robust stability and γ-attenuation for systems described in terms of retarded functional differential equations (RFDE) ([5], [11]). The generalized Popov theory provides a unified approach to the considered problems and extends some of our previous results given in [10]. Indeed, in [10] the case without uncertainties was studied. We will consider here robustness aspects with respect to parameter uncertainties.

The stability analysis is based on a method which is less conservative than similar ones from the literature. In fact, a strictly definite positiveness of the matrices involved in the Lyapunov-Krasovskii functional is "relaxed" to a non-strictly positiveness property.

The development is essentially based on the 'generalized' Riccati theory presented by Ionescu and Weiss [6] which is an extension of the famous Popov's positivity theory [22] to the indefinite sign case, usually encountered in game-theory situations. To be more specific, our results are based on the necessary and sufficient condition for the existence of the stabilizing solution to an adequate Kalman - Popov - Yakubovich system of indefinite sign, called in '$J - form$'. This approach, combined with the Krasovskii theory for time - delay systems, lead to explicit representation formulae. Further extensions can also be done through the framework of this theory.

The paper is organized as follows: section 2 deals with the general problem formulation. Special objectives are mentioned as well. Section 3 is devoted to a brief presentation of the generalized Popov theory. Sections 4 and 5 may be seen as the core of the paper. In section 4, the generalized Popov theory is applied to the robust stabilization of time - delay systems. In section 5, the Popov Index is extended so as to cope with the robust delay case. This extension allows to analyze the γ - attenuation performance of the closed loop system. Section 6 illustrates how the developed theory works in the case of H^∞-control problem solved via memoryless controllers. Section 7 gives an illustrative example. Some concluding remarks end this paper.

2. Problem statement

Consider the following uncertain state-delayed generalized system:

$$\begin{aligned} \dot{x}(t) &= (A+\Delta A)x(t) + (A_d+\Delta A_d)x(t-\tau) + B_1u_1(t) + B_2u_2(t) \\ y_1(t) &= C_1x(t) + D_{11}u_1(t) + D_{12}u_2(t) \\ y_2(t) &= C_2x(t) + D_{21}u_1(t) \end{aligned} \tag{6.1}$$

where $x(t) \in \mathbf{R}^n$ is the state, $u_1(t) \in \mathbf{R}^{m_1}$, $u_2(t) \in \mathbf{R}^{m_2}$ are the disturbance and control inputs, $y_1(t) \in \mathbf{R}^{p_1}$, $y_2(t) \in \mathbf{R}^{p_2}$ are the controlled and measured outputs, A, A_d, ΔA, ΔA_d, B_i, C_i, D_{ij}, $i,j = 1,2$ are constant matrices of appropriate dimensions.

Here, ΔA and ΔA_d are the uncertain variables satisfying the following norm-bounded uncertainty

$$\Delta A = D_a F_a(t) E_a \qquad F_a^T(t)F_a(t) \leq I, \quad \forall t \geq t_0 \tag{6.2}$$

$$\Delta A_d = D_d F_d(t) E_d \qquad F_d^T(t)F_d(t) \leq I, \quad \forall t \geq t_0 \tag{6.3}$$

We are interested in finding a memoryless controller

$$u_2(t) = F_2 x(t) \tag{6.4}$$

that simultaneously stabilizes the sytem (6.1) and achieves the γ - attenuation property, i.e.,
$\|T_{y_1u_1}\| < \gamma$ where $T_{y_1u_1}$ is the L^2-linear bounded input-output operator defined by the closed-loop configuration obtained by coupling (6.4) to (6.1).

When u_1 is measured, (6.4) can be replaced by

$$u_2(t) = F_2x(t) + G_2u_1(t) \tag{6.5}$$

In the present paper, we will consider (6.4). Unlike the techniques previously mentioned, our interest is directed towards the tools offered by the generalized Popov theory.

In the sequel, we will drop the explicit time dependence of $x(t)$, $u_1(t)$ and $u_2(t)$ on t for brevity.

3. Some basic results on the generalized Popov theory

In this section, a brief review of the basic results of the generalized Popov theory is presented. For more details, see [6, 7, 8, 9].

Definition 3.1. *Call $\Sigma = (A, B; P)$ where $A \in \mathbf{R}^{n\times n}$, $B \in \mathbf{R}^{n\times m}$ and*

$$P = \begin{bmatrix} Q & L \\ L^T & R \end{bmatrix} = P^T \in \mathbf{R}^{(n+m)\times(n+m)}$$

a Popov triplet.

The extensive notation $\Sigma = (A, B; Q, L, R)$ can also be used.
Let $\Sigma = (A, B; Q, L, R)$ be a Popov triplet and let

$$J = \begin{bmatrix} -I_{m_1} & \\ & I_{m_2} \end{bmatrix}, \quad m_1 + m_2 = m \tag{6.6}$$

be an arbitrary sign matrix with I_m the $m-$ order identity matrix . We associate with Σ the following objects:

1. The Kalman-Popov-Yakubovich system in J form (KPYS(Σ, J))
The following nonlinear system with unknown X, V, W:

$$\begin{aligned} R &= V^T J V \\ L + XB &= W^T J V \\ Q + A^T X + XA &= W^T J W \end{aligned} \tag{6.7}$$

is usually denoted as the KPYS(Σ, J).

2. The extended Hamiltonian pencil EHP(Σ) $\lambda M - N$ where

$$M = \begin{bmatrix} I_n & 0 & 0 \\ 0 & I_n & 0 \\ 0 & 0 & 0 \end{bmatrix}, \quad N = \begin{bmatrix} A & 0 & B \\ -Q & -A^T & -L \\ L^T & B^T & R \end{bmatrix}, \tag{6.8}$$

with $M, N \in \mathbf{R}^{(2n+m)\times(2n+m)}$.

Definition 3.2. *Any triplet (X, V, W) for which (6.7) is fulfilled and in addition $X = X^T$, V is nonsingular and of lower-left block triangular form*

$$V = \begin{bmatrix} V_{11} & 0 \\ V_{21} & V_{22} \end{bmatrix} \tag{6.9}$$

partitioned in accordance with J in (6.6) and $A + BF$ is exponentially stable for

$$F = -V^{-1}W, \tag{6.10}$$

called the stabilizing feedback gain, is called a stabilizing solution to the KPYS(Σ, J).

Definition 3.3. *The EHP(Σ) is said disconjugate if it has a stable proper deflating subspace $\mathcal{V}$ of dimension n and, in addition, if*

$$V = \begin{bmatrix} V_1 \\ V_2 \\ V_3 \end{bmatrix} \begin{matrix} n \\ n \\ m \end{matrix}$$

is any basis matrix for $\mathcal{V}$ ($\mathcal{V} = span(V)$), then V_1 is nonsingular.

Recall that $\mathcal{V}$ is said a *stable proper deflating subspace* [6, 20] of an arbitrary matrix pencil $\lambda M - N$ if $NV = MVS$, MV is monic, S is Hurwitzian and $\mathcal{V} = span(V)$. A relevant result of the generalized Popov theory is:

Theorem 3.1. *[6] Let $\Sigma = (A, B; Q, L, R)$ be a Popov triplet and J any sign matrix as in (6.6). Then the following statements are equivalent:*

1. *R is nonsingular and the KPYS(Σ, J) system has a stabilizing solution (X, V, W);*
2. *The EHP(Σ) is regular and disconjugate and, in addition, if R is partitioned in accordance with J in (6.6), i.e.,*

$$R = \begin{bmatrix} R_{11} & R_{12} \\ R_{12}^T & R_{22} \end{bmatrix} \tag{6.11}$$

then

$$R_{22} > 0, \quad \mathrm{sgn} R = J. \tag{6.12}$$

If 2 is true, then (see Definition 3.3) $X = V_2 V_1^{-1}$ and $F = V_3 V_1^{-1}$.

Definition 3.4. *Let $\Sigma = (A, B; Q, L, R)$ be a Popov triplet. Then the Popov triplet $\tilde{\Sigma} = (\tilde{A}, \tilde{B}; \tilde{Q}, \tilde{L}, \tilde{R})$ is called an $\tilde{F}$-equivalent of Σ if there exists $\tilde{F} \in \mathbf{R}^{m \times n}$ such that*

$$\tilde{A} = A + B\tilde{F}, \quad \tilde{B} = B$$

and

$$\begin{bmatrix} \tilde{Q} & \tilde{L} \\ \tilde{L}^T & \tilde{R} \end{bmatrix} = \begin{bmatrix} I & \tilde{F}^T \\ 0 & I \end{bmatrix} \begin{bmatrix} Q & L \\ L^T & R \end{bmatrix} \begin{bmatrix} I & 0 \\ \tilde{F} & I \end{bmatrix}$$

or explicitly

$$\tilde{Q} = Q + L\tilde{F} + \tilde{F}^T L^T + \tilde{F}^T R \tilde{F}, \quad \tilde{L} = L + \tilde{F}^T R, \quad \tilde{R} = R.$$

Proposition 3.1. *[7] Let Σ be a Popov triplet and let $\tilde{\Sigma}$ be any $\tilde{F}$-equivalent of it. Then the following hold:*

1. *(X, V, W) is a stabilizing solution to the KPYS(Σ, J) if and only if $(X, V, W + V\tilde{F})$ is a stabilizing solution to the KPYS($\tilde{\Sigma}, J$);*
2. *The EHP(Σ) is strictly equivalent to the EHP($\tilde{\Sigma}$) and the disconjugacy, if exists, is preserved.*

4. Generalized Popov theory applied to uncertain time-delay systems

Let us consider the following uncertain state-delayed system:

$$\dot{x} = (A + \Delta A)x + (A_d + \Delta A_d)\sigma^{-\tau}x + Bu, \qquad x = \phi \text{ on } [-\tau, 0] \tag{6.13}$$

where $x \in \mathbf{R}^n$ is the state, $(\sigma^{-\tau}x)(t) := x(t-\tau)$, $u \in \mathbf{R}^m$ is the input, $A, A_d \in \mathbf{R}^{n\times n}$, $B \in \mathbf{R}^{n\times m}$, $\tau \geq 0$ is the delay and ϕ is any continuous n-valued function on $[-\tau, 0]$. Let $\Sigma = (A, B; Q, L, R)$ be a Popov triplet where the entries A and B coincide with A and B in 6.13. The uncertainty $[\Delta A \;\; \Delta A_d]$ satisfies the following norm-bounded conditions:

$$\Delta A = D_a F_a(t) E_a \qquad F_a^T(t) F_a(t) \leq I, \;\; \forall t \geq t_0 \tag{6.14}$$

$$\Delta A_d = D_d F_d(t) E_d \qquad F_d^T(t) F_d(t) \leq I, \;\; \forall t \geq t_0 \tag{6.15}$$

with known matrices $D_a \in \Re^{n\times i_a}$, $D_d \in \Re^{n\times i_d}$. The matrices E_a and E_d are given weighting matrices.

Remark 4.1. The uncertain functions ΔA and ΔA_d are gain bounded functions and they can be also described by:

$$\Delta A(x_t(0), t) = D_a \delta_a(x_t(0), t), \;\; \delta_a(y,t)^T \delta_a(y,t) \leq y^T E_e^T E_a y, \;\; \forall y \in \Re^n$$

$$\Delta A_d(x_t(-\tau), t) = D_d \delta_d(x_t(-\tau), t), \;\; \delta_d(y,t)^T \delta_d(y,t) \leq y^T E_d^T E_d y, \;\; \forall y \in \Re^n$$

Let $R_d \in \mathbf{R}^{n\times n}$ and consider the *extended* Popov triplet for the system (6.13) - (6.15):

$$\Sigma_e = (A, [A_d \;\; D_a \;\; D_d \;\; B]; Q, [0 \;\; 0 \;\; 0 \;\; L], \begin{bmatrix} R_{dp} & 0 \\ 0 & R \end{bmatrix}), \tag{6.16}$$

with $R_{dp} = \begin{bmatrix} -R_d & 0 & 0 \\ 0 & -I_{i_a} & 0 \\ 0 & 0 & -I_{i_d} \end{bmatrix}$ Let also the following (extended) sign matrix

$$J_e = \left[\begin{array}{c|c} -I_{n+i_a+i_d} & \\ \hline & J \end{array}\right] = \left[\begin{array}{c|cc} -I_{n+i_a+i_d} & & \\ \hline & -I_{m_1} & \\ & & I_{m_2} \end{array}\right], \quad (m_1 + m_2 = m) \tag{6.17}$$

be considered. Let B, L and R be partitioned in accordance with J in (6.17), i.e.,

$$B = [B_1 \;\; B_2], \;\; L = [L_1 \;\; L_2], \;\; R = \begin{bmatrix} R_{11} & R_{12} \\ R_{12}^T & R_{22} \end{bmatrix}. \tag{6.18}$$

The following matrix result will be used in the sequel:

Lemma 4.1. *Let G, L, E and $F(t)$ be real matrices of appropriate dimensions with $F^T(t)F(t) \leq I$. Then we have*
For any scalar $\varepsilon_1 > 0$

$$LF(t)E + E^T F^T(t) L^T \leq \frac{1}{\varepsilon_1} LL^T + \varepsilon_1 E^T E \tag{6.19}$$

The first result of this section corresponds to the robust (stabilisation) control and is formulated as follows:

Theorem 4.1. *Assume that the KPYS(Σ_e, J_e) (6.16) - (6.18) has a stabilizing solution (X, V_e, W_e). Let the stabilizing feedback gain F_e be partitioned in accordance with J_e in (6.17), that is,*

$$F_e = -V_e^{-1} W_e = \begin{bmatrix} F_{dp} \\ F_1 \\ F_2 \end{bmatrix} \tag{6.20}$$

Let also u be split in accordance with B in (6.18), i.e.,

$$u = \begin{bmatrix} u_1 \\ u_2 \end{bmatrix} \begin{matrix} m_1 \\ m_2 \end{matrix} . \tag{6.21}$$

Assume further that

$$X \geq 0 \tag{6.22}$$

$$R_{11} < 0 \tag{6.23}$$

$$\tilde{Q} - R_d > E_a^T E_a + E_b^T E_b \tag{6.24}$$

where

$$\tilde{Q} = Q + L_2 F_2 + F_2^T L_2^T + F_2^T R_{22} F_2 \tag{6.25}$$

Then the state feedback

$$u_2 = F_2 x \tag{6.26}$$

stabilizes (6.13), i.e.,

$$\dot{x} = \tilde{A}x + (A_d + \Delta A_d)\sigma^{-\tau} x, \quad x = \phi \text{ on } [-\tau, 0] \tag{6.27}$$

defines an exponentially stable solution for all ϕ and for all uncertainties which satisfy (6.14 - 6.15). Here $\tilde{A} = A + \Delta A + B_2 F_2$.

Proof. Since (X, V_e, W_e) is a stabilizing solution to the KPYS(Σ_e, J_e), it has the following:

$$\begin{aligned}\begin{bmatrix} R_{dp} & \\ & R \end{bmatrix} &= V_e^T J_e V_e \\ [0\ \ 0\ \ 0\ \ L] + X[A_d\ \ D_a\ \ D_d\ \ B] &= W_e^T J_e V_e \\ Q + A^T X + XA &= W_e^T J_e W_e \end{aligned} \tag{6.28}$$

Taking into account Definition 3.2 in conjugation with (6.17), the first equation in (6.28) leads to the following structure for V_e:

$$V_e = \left[\begin{array}{c|c} V_{dp} & \\ \hline & V \end{array}\right] = \left[\begin{array}{c|cc} V_{dp} & & \\ \hline & V_{11} & 0 \\ & V_{21} & V_{22} \end{array}\right], \tag{6.29}$$

where $V_{dp} = \begin{bmatrix} V_d & & \\ & V_a & \\ & & V_p \end{bmatrix}$

Let W_e be partitioned accordingly, i.e.,

$$W_e = \left[\begin{array}{c} W_{dp} \\ \hline W \end{array}\right] = \left[\begin{array}{c} W_{dp} \\ \hline W_1 \\ W_2 \end{array}\right], \tag{6.30}$$

where $W_{dp}^T = [W_d^T\ \ W_a^T\ \ W_p^T]$ Substituting (6.29), (6.30) in (6.28), (6.28) leads to the following form:

$$\begin{aligned} [-R_d\ \ I_a\ \ I_d\ \ R] &= [-V_d^T V_d\ \ V_a^T V_a\ \ V_p^T V_p\ \ V^T J V] \\ [XA_d\ \ XD_a\ \ XD_d] &= [-W_d^T V_d\ \ \ -W_a^T V_a\ \ \ -W_p^T V_p] \\ L + XB &= W^T J V \\ Q + A^T X + XA &= -W_d^T W_d - W_a^T W_a - W_p^T W_p + W^T J W \end{aligned} \tag{6.31}$$

Using (6.20) one gets

$$\begin{aligned} F_e &= \left[\begin{array}{c} F_{dp} \\ \hline F \end{array}\right] = \left[\begin{array}{c} -V_{dp}^{-1} W_{dp} \\ \hline -V^{-1} W \end{array}\right] \\ &= \left[\begin{array}{c} F_{dp} \\ \hline F_1 \\ F_2 \end{array}\right] = \left[\begin{array}{c} -V_{dp}^{-1} W_{dp} \\ \hline -V_{11}^{-1} W_1 \\ V_{22}^{-1} V_{21} V_{11}^{-1} W_1 - V_{22}^{-1} W_2 \end{array}\right] \end{aligned} \tag{6.32}$$

where the structure (6.9) has been also taken into account and where

$$A + [A_d\ \ D_a\ \ D_d\ \ B] \begin{bmatrix} F_d \\ F_a \\ F_p \\ F \end{bmatrix}$$

is exponentially stable.

Let now

$$\tilde{F}_{e,2} := \left[\begin{array}{c} 0 \\ \hline \tilde{F}_2 \end{array}\right] = \left[\begin{array}{c} 0 \\ \hline 0 \\ F_2 \end{array}\right] \tag{6.33}$$

and let $\tilde{\Sigma}_{e,2}$ be the $\tilde{F}_{e,2}$-equivalent of Σ_e in (6.16). Following Proposition 3.1 and taking into account the zero structure of $\tilde{F}_{e,2}$ in (6.33), the updated form of the last equation in (6.31) corresponding to $\tilde{\Sigma}_{e,2}$ is

$$\begin{aligned} \tilde{A}^T X + X\tilde{A} &= -\tilde{Q} - W_d^T W_d - W_a^T W_a - W_p^T W_p \\ &\quad +(W + V\tilde{F}_2)^T J(W + V\tilde{F}_2) \end{aligned} \tag{6.34}$$

Furthermore

$$\begin{aligned} W + V\tilde{F}_2 &= \left[\begin{array}{c} W_1 \\ W_2 \end{array}\right] + \left[\begin{array}{cc} V_{11} & 0 \\ V_{21} & V_{22} \end{array}\right] \left[\begin{array}{c} 0 \\ V_{22}^{-1}V_{21}V_{11}^{-1}W_1 - V_{22}^{-1}W_2 \end{array}\right] \\ &= \left[\begin{array}{c} W_1 \\ V_{21}V_{11}^{-1}W_1 \end{array}\right] \end{aligned}$$

from which

$$\begin{aligned} &(W + V\tilde{F}_2)^T J(W + V\tilde{F}_2) = \\ &= -W_1^T W_1 + W_1^T V_{11}^{-T} V_{21}^T V_{21} V_{11}^{-1} W_1 \\ &= W_1^T V_{11}^{-T}(-V_{11}^T V_{11} + V_{21}^T V_{21}) V_{11}^{-1} W_1 \end{aligned} \tag{6.35}$$

Taking into account (6.18), the second equation in (6.31) yields, by equating the $(1,1)$ entries

$$R_{11} = -V_{11}^T V_{11} + V_{21}^T V_{21} \tag{6.36}$$

With (6.36) in (6.35) and then with (6.35) in (6.34), one gets eventually

$$\begin{aligned} &\tilde{A}^T X + X\tilde{A} \\ &\quad = -\tilde{Q} - W_d^T W_d - W_a^T W_a - W_p^T W_p - W_1^T V_{11}^{-T} \tilde{R}_{11} V_{11}^{-1} W_1 \end{aligned} \tag{6.37}$$

where

$$\tilde{R}_{11} := -R_{11} > 0 \tag{6.38}$$

as follows from (6.23).

We introduce the Lyapunov functional:

$$V(x_t) = x^T(t)Xx(t) + \int_{t-\tau}^{t} x^T(s)(R_d + E_d^T E_d)x(s)ds \tag{6.39}$$

Simple computations prove that there exist two positive numbers d_1, d_2 such that

$$d_1 \parallel x(t) \parallel^2 \leq V(t, x_t) \leq d_2 \sup_{\theta\in[t-\tau,t]} \parallel x(\theta) \parallel^2 \tag{6.40}$$

(for example, $d_1 = \lambda_{min}(X)$ and $d_2 = \lambda_{max}(X) + \lambda_{max}(S)$)

Taking the Lyapunov derivative of (6.39) with respect to (6.27), one obtains:

$$\begin{aligned} \dot{V}(x_t) &= x^T(\tilde{A}^T X + X\tilde{A})x + (\sigma^{-\tau}x)^T A_d^T X x + x^T X A_d \sigma^{-\tau} x \\ &\quad + x^T(\Delta A^T X + X\Delta A)x + (\sigma^{-\tau}x)^T \Delta A_d^T X x + x^T X \Delta A_d \sigma^{-\tau} x \\ &\quad - x^T(-R_d - E_d^T E_d)x + (\sigma^{-\tau}x)^T(-R_d - E_d^T E_d)\sigma^{-\tau} x \end{aligned} \tag{6.41}$$

Since (6.31), we can write the corresponding equations of the KPYS($\tilde{\Sigma}_{e,2}, J_e$) (see the structure of $\tilde{F}_{e,2}$ in (6.33)), and using Lemma 4.1, (6.41) becomes:

$$\begin{aligned} \dot{V}(x_t) &\leq -x^T(\tilde{Q} - R_d - E_d^T E_d)x - (\sigma^{-\tau}x)^T E_d^T E_d \sigma^{-\tau} x \\ &\quad - x^T W_d^T W_d x - x^T W_a^T W_a x - x^T W_p^T W_p x \\ &\quad - x^T W_1^T V_{11}^{-T} \tilde{R}_{11} V_{11}^{-1} W_1 x \\ &\quad + x^T(X D_a F_a(t) E_a + E_a^T F_a^T(t) D_a^T X)x \\ &\quad + (\sigma^{-\tau}x)^T E_d^T F_d(t)^T D_d^T X x + x^T X D_d F_d(t) E_d \sigma^{-\tau} x \\ &\leq -x^T(\tilde{Q} - R_d - E_d^T E_d - E_a^T E_a)x \leq -\zeta\|x\|^2, \quad \zeta > 0 \end{aligned} \tag{6.42}$$

where both (6.24) and (6.37) have been used. With (6.42) and (6.40) the proof is completed via the Krasovskii stability theorem [5].

Corollary 4.1. *If all the conditions in the statement of Theorem 4.1 hold, then*

$$\dot{x} = \tilde{A}x + (A_d + \Delta A_d)\sigma^{-\tau}x + B_1 u_1, \quad x = 0 \text{ on } [-\tau, 0] \tag{6.43}$$

defines a linear bounded input-state operator from L_+^{2,m_1} into $L_+^{2,n}$.

Proof. By $L_+^{2,r}$ we mean the Hilbert space of norm square integrable r-valued functions defined on $[0, \infty)$. The proof is a direct consequence of the exponentially stable evolution defined by (6.27) (see also [5], [23]).

Taking into account Theorem 3.1, an equivalent form of Theorem 4.1 can be stated as follows:

Theorem 4.2. *Assume that the EHP(Σ_e) is disconjugate. Assume also that*

$$R_{22} > 0, \quad \text{sgn} R = J, \quad R_{dp} < 0. \tag{6.44}$$

If

$$V_2V_1^{-1} \geq 0 \tag{6.45}$$

and both (6.23) and (6.24) hold, then (6.26) stabilizes (6.13). Here

$$\begin{bmatrix} V_1 \\ V_2 \\ V_3 \end{bmatrix} \begin{matrix} n \\ n \\ n+m \end{matrix}$$

is any basis matrix for the maximal stable proper deflating subspace of $EHP(\Sigma_e)$ and

$$F_e = V_3V_1^{-1}$$

partitioned as in (6.20).

Remark 4.2. Theorem 4.2 provides easy checkable sufficient conditions for the stabilizability of the state-delayed system (6.13), see also [18].

Notice that then we have no uncertainties, we recover the result given in [10]; in this case theorem (4.1) can be reformulated as follows:

Corollary 4.2. *Assume that the $KPYS(\Sigma_e, J_e)$ has a stabilizing solution defined by (X, V_e, W_e). Let the stabilizing feedback gain F_e be partitioned in accordance with J_e, that is,*

$$F_e = -V_e^{-1}W_e = \begin{bmatrix} F_d \\ F_1 \\ F_2 \end{bmatrix} \tag{6.46}$$

Let also u be split in accordance with B in (6.18), i.e.,

$$u = \begin{bmatrix} u_1 \\ u_2 \end{bmatrix} \begin{matrix} m_1 \\ m_2 \end{matrix} .$$

Assume also that all the conditions in the statement of theorem (4.1) hold except (6.24) which is modified as

$$\tilde{Q} - R_d > 0.$$

Then the state feedback

$$u_2 = F_2 x \tag{6.47}$$

stabilizes (6.13) without uncertainties, i.e.,

$$\dot{x} = \tilde{A}x + A_d\sigma^{-\tau}x, \quad x = \phi \text{ on } [-\tau, 0] \tag{6.48}$$

defines an exponentially stable solution for all ϕ. Here $\tilde{A} = A + B_2F_2$.

Remark 4.3. Assume that the matrix A_d has a decomposition of the form

$$A_d = M_d N_d, \quad M_d \in \mathbf{R}^{n \times l}, \quad N_d \in \mathbf{R}^{l \times n},$$

with $l \leq n$, and $\mathbf{rank}(N_d) = l$. We can obtain the same results using an appropriate Lyapunov-Krasovskii functional:

$$V(x_t) = x(t)^T X x(t) + \int_{-\tau}^{0} x_t(\theta)^T N_d^T R_d N_d x_t(\theta) d\theta.$$

5. Extended Popov index

In this section, we define an "Extended Popov index" in order to obtain a criterion for γ - attenuation.

Let $\Sigma_e = (A, [A_d\ D_a\ D_b\ B];\ Q,\ [0\ L],\ diag(R_{dp},\ R))$ be the Popov triplet for the system (6.13). In order to obtain a criterium for γ - attenuation for this time-delay system, we shall associate an index defined from $\mathcal{C}([-\tau, 0]) \times L_+^{2,m}$ to $\mathbf{R}$ by

$$J_{\Sigma_e}(\phi, u) := < \begin{bmatrix} x \\ \sigma^{-\tau} x \\ \delta_a \\ \delta_d \\ u \end{bmatrix}, \begin{bmatrix} Q & 0 & 0 & 0 & L \\ 0 & -R_d & 0 & 0 & 0 \\ 0 & 0 & -I_a & 0 & 0 \\ 0 & 0 & 0 & -I_d & 0 \\ L^T & 0 & 0 & 0 & R \end{bmatrix} \begin{bmatrix} x \\ \sigma^{-\tau} x \\ \delta_a \\ \delta_d \\ u \end{bmatrix} > \tag{6.49}$$

which will be called an *Extended Popov index.*

Here (x, u) is any pair satisfying (6.13) and in addition it belongs to $L_+^{2,n} \times L_+^{2,m}$. Here $< \cdot, \cdot >$ stands for the L_+^2-inner product.

Proposition 5.1. *Assume that the KPYS(Σ_e, J_e) (6.28) has a stabilizing solution (X, V_e, W_e). Assume also that there exists a pair $(x, u) \in L_+^{2,n} \times L_+^{2,m}$ such that (6.13) is fulfilled for some ϕ. Then the following evaluation holds:*

$$\begin{aligned} J_{\Sigma_e}(\phi, u) \ &\leq\ -\|W_d x + V_d \sigma^{-\tau} x\|_2^2 + < Wx + Vu, J(Wx + Vu) > + x_0^T X x_0 \\ &\quad (x_0 = x(0)), \end{aligned} \tag{6.50}$$

The proof is straightforward using only algebraic manipulations

Let $\hat{Q}$ be any $n \times n$ symmetric matrix. Let $\Sigma = (A, B; \hat{Q}, L, R)$ be the Popov triplet constructed with $\hat{Q}$ and with entries of Σ_e. Associate with Σ the "usual" Popov index [3]

$$J_{\Sigma}(\phi, u) = < \begin{bmatrix} x \\ u \end{bmatrix}, \begin{bmatrix} \hat{Q} & L \\ L^T & R \end{bmatrix} \begin{bmatrix} x \\ u \end{bmatrix} > \tag{6.51}$$

where $(x, u) \in L_+^{2,n} \times L_+^{2,m}$ and x and u are linked via (6.13) for some ϕ

Then we have

Proposition 5.2. *Assume that*

$$\begin{bmatrix} \hat{Q} & L_2 \\ L_2^T & R_{22} \end{bmatrix} \geq 0 \tag{6.52}$$

Assume also that all the conditions in the statement of Theorem 4.1 hold except (6.24) which is modified as

$$\tilde{Q} - R_d > \tilde{\hat{Q}} + E_a^T E_a + E_d^T E_d \tag{6.53}$$

where

$$\tilde{\hat{Q}} = \hat{Q} + L_2 F_2 + F_2^T L_2^T + F_2^T R_{22} F_2 \tag{6.54}$$

If the state feedback (6.26) stabilies (4.1) and $\phi = 0$, i.e., (6.13) becomes (6.43), then there exists $\bar{\zeta} > 0$ such that

$$J_\Sigma(0, u_1) \leq -\bar{\zeta} \|u_1 - F_1 x\|_2^2, \quad \forall u_1 \in L_+^{2,m_1} \tag{6.55}$$

where

$$J_\Sigma(0, u_1) := J_\Sigma(0, u)|_{u_2 = F_2 x} \tag{6.56}$$

Proof. Notice first that

$$\begin{bmatrix} \tilde{\hat{Q}} & \tilde{L}_2 \\ \tilde{L}_2^T & R_{22} \end{bmatrix} = \begin{bmatrix} I & F_2^T \\ 0 & I \end{bmatrix} \begin{bmatrix} \hat{Q} & L_2 \\ L_2^T & R_{22} \end{bmatrix} \begin{bmatrix} I & 0 \\ F_2 & I \end{bmatrix} \tag{6.57}$$

As (6.52) holds, (6.57) shows that $\tilde{\hat{Q}} \geq 0$. Hence (6.53) implies (6.24) and consequently Corollary 4.1 works and $J_\Sigma(0, u_1)$ defined by (6.56) makes sense.

Let

$$J_{\Sigma_e}(0, u_1) := J_{\Sigma_e}(0, u)|_{u_2 = F_2 x} \tag{6.58}$$

which makes sense as well. Since

$$J_\Sigma(0, u_1) = < \begin{bmatrix} x \\ u_1 \end{bmatrix}, \begin{bmatrix} \tilde{\hat{Q}} & \tilde{L}_1 \\ \tilde{L}_1^T & R_{11} \end{bmatrix} \begin{bmatrix} x \\ u_1 \end{bmatrix} > \tag{6.59}$$

and

$$J_{\Sigma_e}(\phi, u) := < \begin{bmatrix} x \\ \sigma^{-\tau} x \\ \delta_a \\ \delta_d \\ u \end{bmatrix}, \begin{bmatrix} Q & 0 & 0 & 0 & L \\ 0 & -R_d & 0 & 0 & 0 \\ 0 & 0 & -I_a & 0 & 0 \\ 0 & 0 & 0 & -I_d & 0 \\ L^T & 0 & 0 & 0 & R \end{bmatrix} \begin{bmatrix} x \\ \sigma^{-\tau} x \\ \delta_a \\ \delta_d \\ u \end{bmatrix} > \tag{6.60}$$

it follows from (6.59) and (6.60) and (6.50) that $J_{\Sigma}(0,u_1)$ and $J_{\Sigma_e}(0,u_1)$ are linked by

$$\begin{aligned} J_{\Sigma}(0,u_1) &= J_{\Sigma_e}(0,u_1) - <\sigma^{-\tau}x, -R_d\sigma^{-\tau}x> - <x,\tilde{Q}x> \\ &\quad + <x,\tilde{\tilde{Q}}x> - <\delta_a,-\delta_a> - <\delta_p,-\delta_d> \end{aligned} \tag{6.61}$$

Set $u_2 = F_2x$ in the right-hand side of (6.50) and obtain with (6.32)

$$\begin{aligned} & J_{\Sigma_e}(0,u_1) \leq \\ \leq\; & -\|W_dx + V_d\sigma^{-\tau}x\|_2^2 \\ & + < \begin{bmatrix} W_1x + V_{11}u_1 \\ W_2x + V_{21}u_1 + V_{22}u_2 \end{bmatrix}, \begin{bmatrix} -(W_1x + V_{11}u_1) \\ W_2x + V_{21}u_1 + V_{22}u_2 \end{bmatrix} > \\ =\; & -\|W_dx + V_d\sigma^{-\tau}x\|_2^2 + < u_1 - F_1x, (-V_{11}^TV_{11} + V_{21}^TV_{21})(u_1 - F_1x) > \\ =\; & -\|W_dx + V_d\sigma^{-\tau}x\|_2^2 - < u_1 - F_1x, \tilde{R}_{11}(u_1 - F_1x) > \end{aligned} \tag{6.62}$$

where $u_2 = (V_{22}^{-1}V_{21}V_{11}^{-1}W_1 - V_{22}^{-1}W_2)x$.
Here (6.36), (6.38) have been taken into account.

Substituting (6.62) in (6.62), one obtains (see also (6.39))

$$\begin{aligned} J_{\Sigma}(0,u_1) \;\leq\; & -\|W_dx + V_d\sigma^{-\tau}x\|_2^2 - < x, (\tilde{Q} - \tilde{\tilde{Q}})x > - \|\tilde{R}_{11}^{\frac{1}{2}}(u_1 - F_1x)\|_2^2 \\ & + x_0^TXx_0 - x^T(\infty)Xx(\infty) - < x, -R_dx > + < x, -R_dx > \\ & - < \sigma^{-\tau}x, (R_d + E_d^TE_d)(\sigma^{-\tau}x) > - < x, E_d^TE_dx > \\ & + < \sigma^{-\tau}x, (R_d + E_d^TE_d)(\sigma^{-\tau}x) > + < x, E_d^TE_dx > \\ & - < \delta_a, -\delta_a > - < \delta_d, -\delta_d > \\ =\; & -\|W_dx + V_d\sigma^{-\tau}x\|_2^2 \\ & - < x, (\tilde{Q} - R_d - \tilde{\tilde{Q}} - E_a^TE_a - E_d^TE_d)x > \\ & - \int_0^{\infty} \dot{V}(x_t)dt - \|\tilde{R}_{11}^{\frac{1}{2}}(u_1 - F_1x)\|_2^2 \end{aligned} \tag{6.63}$$

As (6.53) holds, (6.55) follows from (6.63), where $\bar{\zeta} > 0$ is the least eigenvalue of $\tilde{R}_{11}$.

Proposition 5.3. *Assume that all conditions in the statement of Theorem 4.1 hold. Assume additionally that*

$$\bar{Q} - R_d > E_a^TE_a + E_d^TE_d \tag{6.64}$$

where

$$\bar{Q} := Q + LF + F^TL^T + F^TRF$$

Then

$$\begin{aligned} \dot{x} &= \tilde{A}x + (A_d + \Delta A_d)\sigma^{-\tau}x + B_1u_1 \quad x = 0 \text{ on } [-\tau, 0] \\ v_1 &= -F_1x + u_1 \end{aligned} \tag{6.65}$$

defines a linear boundedly invertible operator on L_+^{2,m_1}.

Proof. According to Corollary 4.1, $u_1 \mapsto v_1$ is an L^2_+ operator. Let

$$\begin{aligned} \dot{x} &= \bar{A}x + (A_d + \Delta A_d)\sigma^{-\tau}x + B_1 v_1 \\ u_1 &= F_1 x \qquad\qquad\qquad + v_1 \end{aligned} \tag{6.66}$$

be the inverse system of (6.65). Here $\bar{A} = \tilde{A} + B_1F_1 = A + B_1F_1 + B_2F_2 + \Delta A$. Let us show that

$$\dot{x} = \bar{A}x + A_d\sigma^{-\tau}x \quad x = \phi \text{ on } [-\tau, 0]$$

defines an exponentially stable evolution for all ϕ. Proceed similarly to the proof of Theorem 4.1 and consider $\tilde{\Sigma}_{e,1}$ – the $\tilde{F}_{e,1}$-equivalent of Σ_e where

$$\tilde{F}_{e,1} = \left[\frac{0}{F}\right] = \begin{bmatrix} 0 \\ \hline F_1 \\ F_2 \end{bmatrix}$$

Then, as $W + VF = 0$, the updated form of the last equation in (6.31) is

$$\bar{A}^T X + X\bar{A} = -\bar{Q} - W_d^T W_d - W_a^T W_a - W_p^T W_p$$

and (6.41) becomes now

$$\begin{aligned} \dot{V}(x_t) &= -x^T(\bar{Q} - R_d)x - \|W_d x - V_d\sigma^{-\tau}x\|^2 \\ &\le -x^T(\bar{Q} - R_d - E_a^T E_a + E_d^T E_d)x \le -\hat{\zeta}\|x\|^2, \quad \hat{\zeta} > 0 \end{aligned}$$

as follows from (6.64). Thus the proof ends.

Now we shall state and prove the main result of this paper, corresponding to the robust stabilisation with performance requirement.

Theorem 5.1. *Let (6.13) together with the quadratic cost defined by the right-hand side of (6.51) be given. For arbitrary m_1, m_2 such that $m_1 + m_2 = m$, let B, L and R be partitioned as in (6.18). Assume that there exists two $n \times n$ symmetric matrices Q and R_d such that the KPYS(Σ_e, J_e), where Σ_e and J_e are defined by (6.16) and (6.17), respectively, has a stabilizing solution (X, V_e, W_e) and let the stabilizing feedback F_e and u be partitioned as in (6.20 - 6.21). Assume also that the following conditions all hold:*

1). $X \ge 0$

2). $\begin{bmatrix} \hat{Q} & L_2 \\ L_2^T & R_{22} \end{bmatrix} \ge 0$

3). $R_{11} < 0$

4). $\tilde{Q} - R_d > \tilde{\hat{Q}} + E_a^T E_a + E_d^T E_d$

5). $\bar{Q} - R_d > E_a^T E_a + E_d^T E_d$

where

$$\begin{aligned}\tilde{Q} &= Q + L_2F_2 + F_2^TL_2^T + F_2^TR_{22}F_2 \\ \tilde{\hat{Q}} &= \hat{Q} + L_2F_2 + F_2^TL_2^T + F_2^TR_{22}F_2 \\ \bar{Q} &= Q + LF + F^TL^T + F^TRF\end{aligned}$$

Then

a. $u_2 = F_2x$ *stabilizes (6.13) for all uncertainties satisfying (6.14) - (6.15)*
b. There exists $c_0 > 0$ *such that*

$$J_\Sigma(0, u_1) \leq -c_0\|u_1\|_2^2 \quad \forall u_1 \in L_+^{2,m_1}$$

where $J_\Sigma(0, u_1)$ *has been defined by (6.56), (6.51).*

Proof. **a.** follows directly from Theorem 4.1 combined with **2)** and **4)** in the statement (see the proof of Proposition 5.2).

b. From Proposition 5.3 it follows that there exists $\zeta_1 > 0$ such that

$$\|v_1\|_2^2 = \|u_1 - F_1x\|_2^2 \geq \zeta_1\|u_1\|_2^2 \tag{6.67}$$

Using Proposition 5.2 the conclusion follows by substituting (6.67) in (6.55) and putting $c_0 = \zeta_1\bar{\zeta}$.

An alternative version of Theorem 5.1 expressed in terms of the EHP is:

Theorem 5.2. *Assume that the EHP(Σ_e) is disconjugate,* $R_{22} > 0$ *and* $sgnR = J$. *If* $V_2V_1^{-1} \geq 0$ *and conditions* **2)** *to* **5)** *in the statement of Theorem 5.1 all hold, then* **a.** *and* **b.** *in Theorem 5.1 are true. Here*

$$\begin{bmatrix} V_1 \\ V_2 \\ V_3 \end{bmatrix} \begin{matrix} n \\ n \\ n+m \end{matrix}$$

is a basis matrix for the maximal stable proper deflating subspace of EHP(Σ_e) and $F_e = V_3V_1^{-1}$ *is the stabilizing feedback gain.*

Proof. The proof follows trivially from Theorem 5.1 by the aid of Theorem 3.1.

6. H^∞-control for uncertain state-delayed systems

In this section the theory developed in Sections 4 and 5 will be applied for solving the H^∞-control problem formulated for uncertain state-delayed systems. Such a problem is stated as follows.

Let the system

$$\begin{aligned} \dot{x} &= (A+\Delta A)x + (A_d+\Delta A_d)\sigma^{-\tau}x + B_1u_1 + B_2u_2 \quad x=0 \text{ on } [-\tau,0] \\ y_1 &= C_1x + D_{11}u_1 + D_{12}u_2 \end{aligned} \tag{6.68}$$

by given. Here x is the state, u_1 and u_2 are the disturbance and control inputs, respectively, and y_1 is the output to be controlled. The state x is assumed to be accessible for measurement. We are looking for a state feedback law

$$u_2 = F_2x \tag{6.69}$$

which stabilizes (6.68) for all uncertainties satisfying (6.14) - (6.15) and achieves γ-attenuation property for the closed-loop system, i.e., there exists $c_0 > 0$ such that

$$-\gamma^2\|u_1\|_2^2 + \|y_1\|_2^2 \leq -c_0\|u_1\|_2^2 \quad \forall u_1 \in L_+^{2,m_1} \tag{6.70}$$

or equivalently the system

$$\begin{aligned} \dot{x} &= (A+B_2F_2+\Delta A)x + (A_d+\Delta A_d)\sigma^{-\tau}x + B_1u_1 \quad x=0 \text{ on } [-\tau,0] \\ y_1 &= (C_1+D_{12}F_2)x + D_{11}u_1 \end{aligned} \tag{6.71}$$

defines a γ-strictly contractive input-output map. Here γ is a prescribed tolerance for the attenuation level.

Introduce

$$\begin{aligned} & \quad\;\; m_1 \quad m_2 \\ & B = [B_1 \quad B_2] \\ & \hat{Q} = C_1^TC_1, \quad L = [L_1 \quad L_2] = C_1^T[D_{11} \quad D_{12}] \\ & R = \begin{bmatrix} R_{11} & R_{12} \\ R_{12}^T & R_{22} \end{bmatrix} = \begin{bmatrix} -\gamma^2I + D_{11}^TD_{11} & D_{11}^TD_{12} \\ D_{12}^TD_{11} & D_{12}^TD_{12} \end{bmatrix} \end{aligned} \tag{6.72}$$

Then we have

Theorem 6.1. *Assume that there exist two $n \times n$ symmetric matrices Q and R_d such that all the conditions of Theorem 5.1 hold with respect to the particular data (6.72). Then for F_2 given in Theorem 5.1, (6.69) is a solution to the H^∞-control problem stated above.*

Proof. Let $\Sigma = (A, B; \hat{Q}, L, R)$. Then

$$J_\Sigma = -\gamma^2\|u_1\|_2^2 + \|y_1\|_2^2 \tag{6.73}$$

as directly follows by simple computation from (6.72). Apply Theorem 5.1 to (6.73) and the conclusion follows trivially.

Notice that $\tilde{\hat{Q}}$ in Theorem 5.1 reads now as $\tilde{\hat{Q}} = C_{1F_2}^TC_{1F_2}$ where $C_{1F_2} = C_1 + D_{12}F_2$.

Remark 6.1. In [13], [25] and [24], the H^∞-control problem has been solved in terms of Riccati inequalities for the particular case $D_{11} = 0$, $D_{12} = 0$. Notice also that we have a supplementary degree of freedom through the penalty matrix R_d

7. An example

Consider the following unstable state-delayed system

$$\begin{array}{rcl} \dot{x} & = & (1+0.2sin(t))x + (1+0.2cos(t))\sigma^{-\tau}x + 0.3u_1 + u_2 \\ y_1 & = & x \qquad\qquad\qquad\qquad + u_2 \end{array} \tag{6.74}$$

Here x, u_1, u_2 and y_1 are all scalars. The problem is to find a memoryless controller

$$u_2 = F_2 x \tag{6.75}$$

that achieves simultaneously closed-loop robust stability and γ-attenuation. The prescribed tolerance is

$$\gamma = 0.5 \tag{6.76}$$

The input data are

$$\begin{array}{l} A = 1, \quad A_d = 1, \quad B_1 = 0.3, \quad B_2 = 1 \\ C_1 = 1, \qquad\quad D_{11} = 0, \quad D_{12} = 1 \\ D_a = D_d = 0.2, \qquad E_a = E_d = 1 \end{array} \tag{6.77}$$

as directly follows from (6.74). Hence

$$\hat{Q} = C_1^T C_1 = 1, \quad L = C_1^T [D_{11} \;\; D_{12}] = [L_1 \;\; L_2] = [0 \;\; 1]$$
$$R = \begin{bmatrix} R_{11} & R_{12} \\ R_{12}^T & R_{22} \end{bmatrix} = \begin{bmatrix} -\gamma^2 I + D_{11}^T D_{11} & D_{11}^T D_{12} \\ D_{12}^T D_{11} & D_{12}^T D_{12} \end{bmatrix} = \begin{bmatrix} -0.25 & 0 \\ 0 & 1 \end{bmatrix}$$

Choose $R_d = 10$. Then the extended Popov triplet (see (6.16)) is

$$\begin{array}{rcl} \Sigma_e & = & (A, [A_d \; D_a \; D_d \; B_1 \; B_2]; \\ & & Q, [0 \; 0 \; 0 \; L_1 \; L_2], \begin{bmatrix} -R_d & 0 & 0 & 0 & 0 \\ 0 & -I_a & 0 & 0 & 0 \\ 0 & 0 & -I_d & 0 & 0 \\ 0 & 0 & 0 & R_{11} & R_{12} \\ 0 & 0 & 0 & R_{12}^T & R_{22} \end{bmatrix}) \\ & = & (1, [1 \; 0.2 \; 0.2 \; 0.3 \; 1]; \\ & & Q, [0 \; 0 \; 0 \; 0 \; 1], \begin{bmatrix} -10 & 0 & 0 & 0 & 0 \\ 0 & -1 & 0 & 0 & 0 \\ 0 & 0 & -1 & 0 & 0 \\ 0 & 0 & 0 & -0.25 & 0 \\ 0 & 0 & 0 & 0 & 1 \end{bmatrix}) \end{array} \tag{6.78}$$

and (see (6.17))

$$J_e = \left[\begin{array}{c|cc} -1 & & \\ \hline & -1 & \\ & & 1 \end{array}\right] \tag{6.79}$$

But (6.31) is equivalent to the algebraic Riccati equation (ARE) associated with Σ_e, that is,

$$A^T X + XA + Q - ([0\ 0\ 0\ L_1\ L_2] + X[A_d\ D_a\ D_d\ B_1\ B_2]) \begin{bmatrix} -R_d & 0 & 0 & 0 & 0 \\ 0 & -I_a & 0 & 0 & 0 \\ 0 & 0 & -I_d & 0 & 0 \\ 0 & 0 & 0 & R_{11} & R_{12} \\ 0 & 0 & 0 & R_{12}^T & R_{22} \end{bmatrix}^{-1} \left(\begin{bmatrix} 0 \\ 0 \\ 0 \\ L_1^T \\ L_2^T \end{bmatrix} + \begin{bmatrix} A_d^T \\ D_a^T \\ D_d^T \\ B_1^T \\ B_2^T \end{bmatrix} X\right) = 0 \tag{6.80}$$

and the additional sign constraints

$$R_{22} > 0 \text{ and } \operatorname{sgn}\begin{bmatrix} R_{dp} & \\ & R \end{bmatrix} = J_e. \tag{6.81}$$

Since the conditions (6.81) are obviously fulfilled (see the data), we focus the attention in the ARE (6.80) which becomes

$$2X - [X\ 0.2X\ 0.2X\ 0.3X\ 1+X] \begin{bmatrix} -10 & 0 & 0 & 0 & 0 \\ 0 & -1 & 0 & 0 & 0 \\ 0 & 0 & -1 & 0 & 0 \\ 0 & 0 & 0 & -0.25 & 0 \\ 0 & 0 & 0 & 0 & 1 \end{bmatrix}^{-1} \begin{bmatrix} X \\ 0.2X \\ 0.2X \\ 0.3X \\ 1+X \end{bmatrix} + Q = 0$$

leading to

$$0.49X^2 - Q + 1 = 0.$$

The stabilizing solution X and the stabilizing feedback gain F_2 are

$$X = \frac{\sqrt{Q-1}}{0.7} \tag{6.82}$$

which is greater than 0 when $Q > 1$, and

$$F_2 = -1 - \frac{\sqrt{Q-1}}{0.7} \tag{6.83}$$

Let us check the conditions stated in Theorem 5.1. The first condition is fulfilled when $Q > 1$ as (6.82) shows. For the next four ones, we have

2). $\begin{bmatrix} \hat{Q} & L_2 \\ L_2^T & R_{22} \end{bmatrix} = \begin{bmatrix} 1 & 1 \\ 1 & 1 \end{bmatrix} \geq 0$

3). $R_{11} = -0.25 < 0$

4). and 5). As $L_1 = 0$, one gets (see (6.83))

$$\tilde{Q} = \bar{Q} = Q + L_2F_2 + F_2^T L_2^T + F_2^T R_{22} F_2 = Q - 1 + \frac{Q-1}{0.49}$$

and

$$\tilde{\hat{Q}} = \hat{Q} + L_2F_2 + F_2^T L_2^T + F_2^T R_{22} F_2 = \frac{Q-1}{0.49}$$

Then

$$\begin{array}{rcl} \tilde{Q} - R_d & > & \tilde{\hat{Q}} + E_a^T E_a + E_d^T E_d \\ \bar{Q} - R_d & > & 0 + E_a^T E_a + E_d^T E_d \end{array}$$

whenever $Q > 13$.

Therefore all the conditions stated in Theorem 6.51 hold for $Q > 13$ and consequently (see (6.83))

$$u_2 = -(1 + \frac{\sqrt{Q-1}}{0.7})x$$

is the desired feedback law.

Let us check first the stability of the closed-loop system

$$\begin{array}{rcl} \dot{x} & = & (A + B_2F_2 + \Delta A)x \ + (A_d + \Delta A_d)\sigma^{-\tau}x + B_1u_1 \\ y_1 & = & (C_1 + D_{12}F_2)x \end{array}$$

In our case, the closed-loop system is (for sin(t)=1)

$$\begin{array}{rcl} \dot{x} & = & (0.2 - \frac{\sqrt{Q-1}}{0.7})x + 1.2\sigma^{-\tau}x + 0.3u_1 \\ y_1 & = & -\frac{\sqrt{Q-1}}{0.7}x \end{array} \tag{6.84}$$

Now it is easily checkable that

$$s + \frac{\sqrt{Q-1}}{0.7} - 1.2e^{-\tau s} = 0 \tag{6.85}$$

has all the roots in the open left half of the complex plane for any $Q > 13$ and all $\tau > 0$. Indeed (6.85) yields with $s = \alpha + j\beta$

$$\alpha + \frac{\sqrt{Q-1}}{0.9} - e^{-\alpha\tau}\cos\beta\tau = 0$$

which is fulfilled only for $\alpha < 0$.

We check now the γ - attenuation property:
The resultant transfer function is

$$H_{y_1u_1}(s) = 0.3\frac{-\frac{\sqrt{Q-1}}{0.7}}{s+\frac{\sqrt{Q-1}}{0.7}-1.2e^{-\tau s}}.$$

We notice that the above transfer function is a monotone one. Therefore we can check that

$$\frac{1}{|H_{y_1u_1}(0)|} = \frac{\sqrt{Q-1}-0.7}{0.3\sqrt{Q-1}} > \frac{1}{\gamma} = 2$$

for any $Q > 13$ and hence the γ-attenuation condition is fulfilled.

8. Conclusions

An extension of the generalized Popov theory to the case of uncertain state-delayed systems has been done. Our interest has been focused on the memoryless controller applied to H^∞-control problem stated for systems described through RFDE. The proposed results recover some of the existing results in the literature for particular cases. Our delay-independent criteria are very simple and easily checkable via numerical procedures. As future research directions we suggest: a) the case of time-varing delay systems b). investigation of observer-based compensation techniques.

References

1. *Stability and Control of Time-Delay Systems* , L.Dugard, E.I. Verriest Eds, LNCIS 228 Springer-Verlag Berlin , 1997
2. *Commande Optimale: Conception Optimisée des Systèmes,* (in French), J. M. Dion and D. Popescu eds., Diderot, Paris, 1996.
3. A. Halanay and V. Ionescu, "Generalized discrete-time Popov-Yakubovich theory," *Systems & Control Letters*, **20**, 1–6, 1993.
4. A. Halanay and V. Ionescu, *Time-Varying Discrete Linear Systems,* Birkhauser, 1994.
5. J. K. Hale and S. M. Verduyn Lunel, *Introduction to Functional Differential Equation,* Applied Math. Sciences, **99**, Springer Verlag, 1991.
6. V. Ionescu and M. Weiss, "Continuous and Discrete-time Riccati theory: A Popov function approach," *Linear Algebra and Its Appl.*, **193**, 173–205, 1993.
7. V. Ionescu and C. Oară, "Generalized discrete-time Riccati theory," *SIAM J. on Control and Optimization*, **34**(2), 601–619, 1996.
8. V. Ionescu, "Reverse discrete-time Riccati equation and extended Nehari problem," *Linear Algebra and Its Appl.*, **236**, 59–94, 1996.
9. V. Ionescu, C. Oară and M. Weiss, *Generalized Riccati Theory,* John Wiley, to appear.
10. V. Ionescu, S.-I. Niculescu, J.-M. Dion, L. Dugard, H. Li, "Generalized Popov Theory Applied To State-Delayed Systems," *Proc. IFAC Conf. System Structure and Control,* Nantes , july 1998.

11. V. B. Kolmanovskii and V. R. Nosov, *Stability of Functional Differential Equations,* Mathematics in Science and Engineering, **180**, Academic Press, NY, 1986.
12. Lehman, B.: Stability of chemical reactions in a CSTR with delayed recycle stream. *Proc. 1994 Amer. Contr. Conf.*, Baltimore, Maryland, U.S.A., (1994) 3521-3522.
13. J. H. Lee, S. W. Kim and W. H. Kwon, "Memory-less H_∞ controllers for state delayed systems," *IEEE Trans. on Automatic Contr.*, **39**, 159–162, 1994.
14. H. Li, S. I. Niculescu, L. Dugard and J. M. Dion, "Robust H_∞ control for uncertain linear time-delay systems: A linear matrix inequality approach. Part I," *Proc. 35th IEEE Conf. Decision and Contr.*, Kobe, Japan, 1996.
15. H. Li, S. I. Niculescu, L. Dugard and J. M. Dion, "Robust H_∞ control for uncertain linear time-delay systems: A linear matrix inequality approach with guaranteed α-stability. Part II," *Proc. 35th IEEE Conf. Decision and Contr.*, Kobe, Japan, 1996.
16. *Time-delay systems. Qualitative aspects on stability and stabilization* (in French), Diderot Eds., Paris, 'Nouveaux Essais' series , 1997
17. S. I. Niculescu, "H_∞ memoryless control with an α-stability constraint for time-delay systems: An LMI approach," *Proc. 34th IEEE Conf. Decision and Contr.*, New-Orleans, U.S., 1507–1512, 1995.
18. S. I. Niculescu and V. Ionescu, "On delay-independent stability criteria: A matrix pencil approach," *IMA J. Math. Contr. Information,* 1997.
19. S. I. Niculescu, C. E. de Souza, J. M. Dion and L. Dugard, "Robust H_∞ memoryless control for uncertain linear systems with time-varying delay," *Proc. 3rd European Contr. Conf.*, Rome, Italy, 1814–1818, 1995.
20. C. Oară, "Proper deflating subspaces: Properties, algorithms and application," *Numerical Algorithms*, **7**, 335–377, 1994.
21. D. Perlmutter, *Stability of chemical reactors*, Prentice Hall, New Jersey, 1972.
22. V.M. Popov, "Hyperstability of Control Systems," *Springer Verlag*, 1973
23. V. Răsvan, *Absolute Stability of Time-Delay Control Systems,* (in Russian), Nauka, 1983.
24. L. Xie and C. E. de Souza, "Robust stabilization and disturbance attenuation for uncertain delay systems," *Proc. 2nd European Contr. Conf.*, Groningen, The Netherlands, 667–672, 1993.
25. S. Wang, Y. Gu, J. Qian, "Robust H^∞ control for a class of linear uncertain time-delay systems" *Proc. 4th IFAC Conf. System Structure and Control*, Bucharest, Romania 1997.

Passivity based control of the inverted pendulum

R. Lozano and I. Fantoni

Heudiasyc, UTC, UMR CNRS 6599, BP 20529, 60205 Compiègne, France, e-mail: rlozano,isabelle.fantoni@hds.utc.fr

Summary. The inverted pendulum has been used as a benchmark for motivating the study of nonlinear control techniques. We propose a simple controller for balancing the inverted pendulum and rise it to its upper equilibrium position. The control strategy is based on the passivity properties of the cart and pendulum system.

Keywords : Nonlinear systems, Passivity, Inverted pendulum.

1. Introduction

The inverted pendulum is a very popular experiment used for educational purposes in modern control theory. It is basically a pole which has a pivot on a cart that can be moved horizontally. The pole moves freely around the cart and the control objective is to bring the pole to the upper unstable equilibrium position by moving the cart on the horizontal plane. Since the angular acceleration of the pole can not be controlled directly, the inverted pendulum is an underactuated mechanical system. Therefore, the techniques developed for fully-actuated mechanical robot manipulators can not be used to control the inverted pendulum.

The cart and pole system is also very popular because the standard nonlinear control techniques are ineffective to control it. Indeed the relative degree [2] of the system is not constant (when the output is chosen to be the swinging energy of the pendulum), the system is not input-output linearizable. [3] have shown that the inverted pendulum is not feedback linearizable. An additional difficulty comes from the fact that when the pendulum swings past the horizontal (i.e. $\pm\pi$) the controllability distribution does not have a constant rank.

[9] presented a control strategy decomposed in a sequence of steps to bring the pendulum from its lower stable equilibrium position to its unstable equilibrium position having a restricted travel. [1] proposed a nonlinear state feedback control law to regulate the cart position as well as the swinging energy of the pendulum. The resulting closed-loop system possesses a locally stable periodic orbit, though the region of attraction has not been determined. [5] proposed a linear controller that stabilizes the linearized model of the inverted pendulum having restricted travel. The region of attraction when the controller is applied to the nonlinear model of the inverted pendulum is still to be determined.

[6] presented a control law based on the technique consisting on adding integrators. Their technique can be used to stabilize the inverted pendulum in its upper equilibrium position when the pendulum is initially above the horizontal plane.

[7] and [8] proposed a strategy to control the inverted pendulum by swinging it up to its unstable equilibrium position. The stability analysis is carried out by using a Lyapunov technique. The stabilization algorithm proposed here has been inspired by the work in [7] and [8]. The difference relies on the fact that in the present approach the control algorithm is obtained by considering the passivity properties of the inverted pendulum. The control algorithm as well as the convergence analysis turn out to be very simple as compared to the existing control strategies. The performance of the proposed control law is shown in a simulation example.

2. System Dynamics

Consider the cart and pendulum system as shown in Fig (7.1). We will consider the standard assumptions, *i.e.* massless rod, point masses, no friction, etc.. M is the mass of the cart, m the mass of the pendulum, concentrated in the bob, θ the angle that the pendulum makes with the vertical and l the length of the rod. The following equations of motion can be obtained either by applying Newton's second law or by the Euler-Lagrange formulation (see Appendix):

$$\begin{bmatrix} M+m & ml\cos\theta \\ ml\cos\theta & ml^2 \end{bmatrix}\begin{bmatrix} \ddot{x} \\ \ddot{\theta} \end{bmatrix} + \\ + \begin{bmatrix} 0 & -ml\sin\theta\dot{\theta} \\ 0 & 0 \end{bmatrix}\begin{bmatrix} \dot{x} \\ \dot{\theta} \end{bmatrix} + \begin{bmatrix} 0 \\ -mgl\sin\theta \end{bmatrix} = \begin{bmatrix} f \\ 0 \end{bmatrix} \tag{7.1}$$

System (7.1) can also be written as:

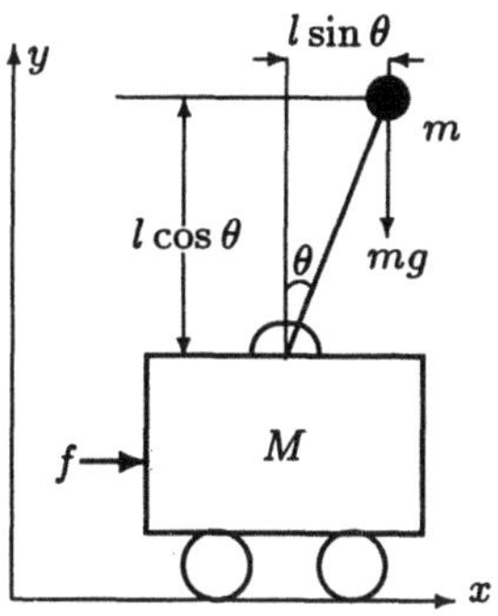

Fig. 7.1. The cart pendulum system

$$M(q)\ddot{q} + C(q,\dot{q})\dot{q} + G(q) = \tau \tag{7.2}$$

where:

$$q = \begin{bmatrix} x \\ \theta \end{bmatrix} \quad M(q) = \begin{bmatrix} M+m & ml\cos\theta \\ ml\cos\theta & ml^2 \end{bmatrix} \quad C(q,\dot{q}) = \begin{bmatrix} 0 & -ml\sin\theta\dot{\theta} \\ 0 & 0 \end{bmatrix} \tag{7.3}$$

$$G(q) = \begin{bmatrix} 0 \\ -mglsin\theta \end{bmatrix} \quad \text{and} \quad \tau = \begin{bmatrix} f \\ 0 \end{bmatrix} \tag{7.4}$$

Note that $M(q)$ is symmetric and

$$det(M(q)) = (M+m)ml^2 - m^2l^2\cos^2\theta = Mml^2 + m^2l^2\sin^2\theta > 0 \tag{7.5}$$

Therefore $M(q)$ is positive definite for all q. From (7.3) it follows that

$$\dot{M}(q) - 2C(q,\dot{q}) = \begin{bmatrix} 0 & ml\sin\theta\dot{\theta} \\ -ml\sin\theta\dot{\theta} & 0 \end{bmatrix} \tag{7.6}$$

which is an antisymmetric matrix. An important property of antisymmetric matrices which will be used in establishing the passivity property of the inverted pendulum is:

$$z^T(\dot{M}(q) - 2C(q,\dot{q}))z = 0 \qquad \forall z \tag{7.7}$$

The pendulum potential energy can be defined as $P = mgl(\cos\theta - 1)$. Note that P is related to $G(q)$ as follows:

$$G(q) = \frac{\partial P}{\partial q} = \begin{bmatrix} 0 \\ -mgl\sin\theta \end{bmatrix} \tag{7.8}$$

2.1 Passivity of the inverted pendulum

The total energy of the cart and pole system is given by

$$\begin{aligned} E &= \tfrac{1}{2}\dot{q}^T M(q)\dot{q} + P(q) \\ &= \tfrac{1}{2}\dot{q}^T M(q)\dot{q} + mgl(\cos\theta - 1) \end{aligned} \tag{7.9}$$

Therefore, from (7.1), (7.2), (7.3), (7.4), (7.6), (7.7) and (7.8) we obtain:

$$\begin{aligned} \dot{E} &= \dot{q}^T M(q)\ddot{q} + \tfrac{1}{2}\dot{q}^T \dot{M}(q)\dot{q} + \dot{q}^T G(q) \\ &= \dot{q}^T(-C(q,\dot{q})\dot{q} - G(q) + \tau) + \tfrac{1}{2}\dot{q}^T \dot{M}(q)\dot{q} + \dot{q}^T G(q) \\ &= \dot{q}^T\tau = \dot{x}f \end{aligned} \tag{7.10}$$

Integrating both sides of the above equation we obtain

$$\int_0^t \dot{x}f\,dt = E(t) - E(0) \geq -2mgl - E(0) \tag{7.11}$$

Therefore the system having f as input and $\dot{x}$ as output is passive. Note that for $f = 0$ the system (7.1) has two equilibrium points. $(x,\dot{x},\theta,\dot{\theta}) = (0,0,0,0)$ is an unstable equilibrium point and $(x,\dot{x},\theta,\dot{\theta}) = (0,0,\pi,0)$ is a stable equilibrium point. The total energy $E(q,\dot{q})$ is equal to 0 for the unstable equilibrium point and to $-2mgl$ for the stable equilibrium point. The control objective is to stabilize the system around its unstable equilibrium point, *i.e.* to bring the pendulum to its upper position and the cart displacement to zero simultaneously.

3. Stabilizing control law

Let us first note that in view of (7.9) and (7.3), if $\dot{x} = 0$ and $E(q,\dot{q}) = 0$ then

$$\frac{1}{2}ml^2\dot{\theta}^2 = mgl(1 - \cos\theta) \tag{7.12}$$

The above equation defines a very particular trajectory. Note that $\dot{\theta} = 0$ only when $\theta = 0$. This means that the pendulum angular position moves clockwise or counter-clockwise until it reaches the equilibrium point $(\theta, \dot{\theta}) = (0, 0)$. Thus our objective can be reached if the system can be brought to the orbit (7.12) for $\dot{x} = 0$ and $x = 0$. The passivity property of the system suggests us to use the total energy E in (7.9) in the controller design. Since we wish to bring to zero $x, \dot{x}$ and E we propose the following Lyapunov function candidate:

$$V(q, \dot{q}) = \frac{1}{2}E(q, \dot{q})^2 + \frac{k_v}{2}\dot{x}^2 + \frac{k_x}{2}x^2 \tag{7.13}$$

where k_v and k_x are strictly positive constants to be defined later. Note that $V(q, \dot{q})$ is a positive semi-definite function. Differentiating V and using (7.10) we obtain

$$\begin{aligned} \dot{V} &= E\dot{E} + k_v\dot{x}\ddot{x} + k_x x\dot{x} \\ &= E\dot{x}f + k_v\dot{x}\ddot{x} + k_x x\dot{x} \\ &= \dot{x}(Ef + k_v\ddot{x} + k_x x) \end{aligned} \tag{7.14}$$

Let us now compute $\ddot{x}$ from (7.2). The inverse of $M(q)$ can be obtained from (7.3) and (7.5) and is given by:

$$M^{-1}(q) = \frac{1}{ml^2(M + m\sin^2\theta)} \begin{bmatrix} ml^2 & -ml\cos\theta \\ -ml\cos\theta & M+m \end{bmatrix} \tag{7.15}$$

Therefore we have

$$\begin{bmatrix} \ddot{x} \\ \ddot{\theta} \end{bmatrix} = [det(M(q))]^{-1} \left(\begin{bmatrix} 0 & m^2l^3\dot{\theta}\sin\theta \\ 0 & -m^2l^2\dot{\theta}\sin\theta\cos\theta \end{bmatrix} \begin{bmatrix} \dot{x} \\ \dot{\theta} \end{bmatrix} + \begin{bmatrix} -m^2l^2 g\sin\theta\cos\theta \\ (M+m)mgl\sin\theta \end{bmatrix} + \begin{bmatrix} ml^2 f \\ -mlf\cos\theta \end{bmatrix} \right)$$

$\ddot{x}$ can thus be written as

$$\ddot{x} = \frac{1}{M + m\sin^2\theta}\left[m\sin\theta(l\dot{\theta}^2 - g\cos\theta) + f\right] \tag{7.16}$$

Introducing the above in (7.14) one has

$$\dot{V} = \dot{x}\left[f\left(E+\frac{k_v}{M+m\sin^2\theta}\right) + \frac{k_v m\sin\theta(l\dot{\theta}^2 - g\cos\theta)}{M+m\sin^2\theta} + k_x x\right] \tag{7.17}$$

For simplicity and without loss of generality we will consider $M = m = l = 1$, thus

$$\dot{V} = \dot{x}\left[f\left(E+\frac{k_v}{1+\sin^2\theta}\right) + \frac{k_v\sin\theta(\dot{\theta}^2 - g\cos\theta)}{1+\sin^2\theta} + k_x x\right] \tag{7.18}$$

We propose a control law such that

$$f\left(E+\frac{k_v}{1+\sin^2\theta}\right) + \frac{k_v\sin\theta(\dot{\theta}^2 - g\cos\theta)}{1+\sin^2\theta} + k_x x = -\dot{x} \tag{7.19}$$

which will lead to

$$\dot{V} = -\dot{x}^2 \tag{7.20}$$

The control law in (7.19) will have no singularities provided that

$$\left(E+\frac{k_v}{1+\sin^2\theta}\right) \neq 0 \tag{7.21}$$

The above condition will be satisfied if for some $\epsilon > 0$

$$|E| \leq \frac{k_v - \epsilon}{2} \leq \frac{k_v}{1+\sin^2\theta} \tag{7.22}$$

Note that when using the control law (7.19), the pendulum can get stuck at the (lower) stable equilibrium point $(x, \dot{x}, \theta, \dot{\theta}) = (0, 0, \pi, 0)$. In order to avoid this singular point which occurs when $E = -2mgl$, we require $|E| < 2mgl$ *i.e.* $|E| < 2g$ (for $m = 1$, $l = 1$). Taking also (7.22) into account, we require

$$|E| < c = min(2g, \frac{k_v - \epsilon}{2}) \tag{7.23}$$

Since V is a non-increasing function (see (7.20)), (7.23) will hold if the initial conditions are such that

$$V(0) \leq \frac{c^2}{2} \tag{7.24}$$

The above defines the region of attraction as will be shown in the next section. Note that by choosing k_x small, the initial displacement $x(0)$ can be large. (7.24) imposes however limits on the initial values of $\dot{x}$ and $\dot{\theta}$.

Finally, with this condition, the control law can be written:

$$f = \frac{k_v \sin\theta \left(g\cos\theta - \dot{\theta}^2\right) - \left(1 + \sin^2\theta\right)\left(k_x x + \dot{x}\right)}{k_v + \left(1 + \sin^2\theta\right) E} \tag{7.25}$$

4. Stability Analysis

The stability analysis will be based on LaSalle´s invariance theorem (see for instance [4], page 117). In order to apply LaSalle´s theorem we require to define a compact (closed and bounded) set Ω with the property that every solution of system (7.1) which starts in Ω remains in Ω for all future time . Since $V(q, \dot{q})$ in (7.13) is a non-increasing function, (see (7.20)), then x, $\dot{x}$, and $\dot{\theta}$ are bounded. Since $\cos\theta$ and $\sin\theta$ are bounded functions, we can define the state z of the closed loop system as being composed of x, $\dot{x}$, $\cos\theta$, $\sin\theta$ and $\dot{\theta}$. Therefore, the solution of the closed loop system $z = F(z)$ remains inside a compact set Ω that is defined by the initial state values. Let Γ be the set of all points in Ω such that $\dot{V}(z) = 0$. Let M be the largest invariant set in Γ. LaSalle´s theorem insures that every solution starting in Ω approaches M as $t \to \infty$. Let us now compute the largest invariant set M in Γ.

In the set Γ (see (7.20)), $\dot{V} = 0$ and $\dot{x} = 0$ which implies that x and V are constant. From (7.13) it follows that E is also constant. Comparing (7.14) and (7.20) it follows that the control law has been chosen such that

$$-\dot{x} = Ef + k_v\ddot{x} + k_x x \tag{7.26}$$

From the above equation we conclude that Ef is constant in Γ. Since E is also constant, it follows that the control input f is also constant. If the force input f was constant and different from zero, the displacement x would diverge. Since this would lead to a contradiction we conclude that $f = 0$ in Γ. From (7.26) it follows that $x = 0$ in Γ.

With $f = 0, \dot{x} = 0, \ddot{x} = 0$, the system (7.1) becomes:

$$ml\ddot{\theta}\cos\theta - ml\dot{\theta}^2\sin\theta = 0 \tag{7.27}$$

$$ml^2\ddot{\theta} - mgl\sin\theta = 0 \tag{7.28}$$

Introducing (7.28) into (7.27), we obtain:

$$\frac{g}{l}\sin\theta\cos\theta - \dot{\theta}^2\sin\theta = 0 \tag{7.29}$$

Thus we have either

$$\dot{\theta}^2 = \frac{g}{l}\cos\theta \tag{7.30}$$

or

$$\sin\theta = 0 \tag{7.31}$$

Let us study each case separately. If $\sin\theta = 0$, then θ is either equal to 0 or π. $\theta = \pi$ has been excluded by imposing conditions (7.23) and (7.24). Therefore (7.31) implies $\theta = 0$.

Let us now study (7.30). Differentiating (7.30) we obtain

$$2\dot{\theta}\ddot{\theta} = -\frac{g}{l}\dot{\theta}\sin\theta \tag{7.32}$$

We will next (7.32) study in two different cases:

- Case 1. If $\dot{\theta} \neq 0$, (7.32) becomes $2\ddot{\theta} = -\frac{g}{l}\sin\theta$. Combining this equation with (7.28) we conclude that $\sin\theta = 0$ which implies that $\theta = 0$ as proved above.
- Case 2. If $\dot{\theta} = 0$ then $\ddot{\theta} = 0$ which together with (7.28) implies that $\sin\theta = 0$ which implies that $\theta = 0$ as proved above.

Therefore the largest invariant set M is given by the point $(x, \dot{x}, \theta, \dot{\theta}) = (0, 0, 0, 0)$. Provided that the state initial conditions satisfy (7.23) and (7.24) all the solutions converge to the origin.

5. Simulation results

In order to observe the performance of the proposed control law based on passivity we have performed simulations on MATLAB, using SIMULINK.

We have considered the system taking $M = 1$, $m = 1$ and $l = 1$, and $g = 9.8m.s^{-2}$. We must not forget that our control law requires initial conditions such that (7.23) and (7.24) are satisfied. We have some freedom in the choice of the coefficients k_v and k_x. For, instance, if $x(0)$ is large, we can choose k_x small such that the conditions (7.23) and (7.24) basically impose restrictions only on $E(0)$ and $\dot{x}(0)$. Our algorithm allows us to bring the inverted pendulum close to the origin but the inverted pendulum will remain swinging while getting closer and closer to the origin. Once the system is close enough to the origin, we could switch to a linear control law as has being proposed in previous papers on the subject. Figures 2 and 3 show the results for $k_v = 10, k_x = 10^{-2}$ and for an initial position:

$$x = -10 \qquad v = 1$$

$$\theta = \frac{\pi}{4} \qquad \omega = 0$$

Simulations showed that our control law brings the state of the system to the origin even when $x(0)$ is far from the origin. Note that the energy E, goes to zero too and that the Lyapunov function V is always decreasing and converges to zero.

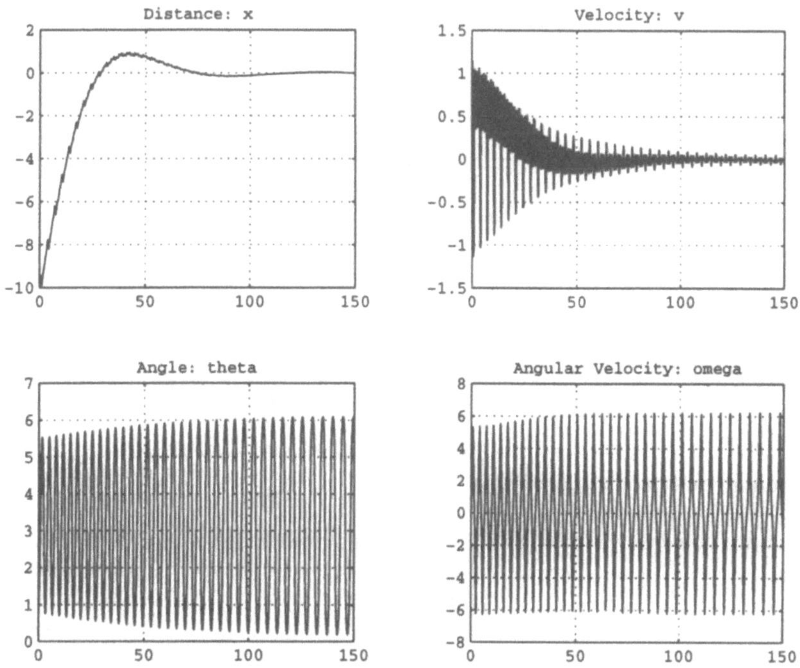

Fig. 7.2. States of the System.

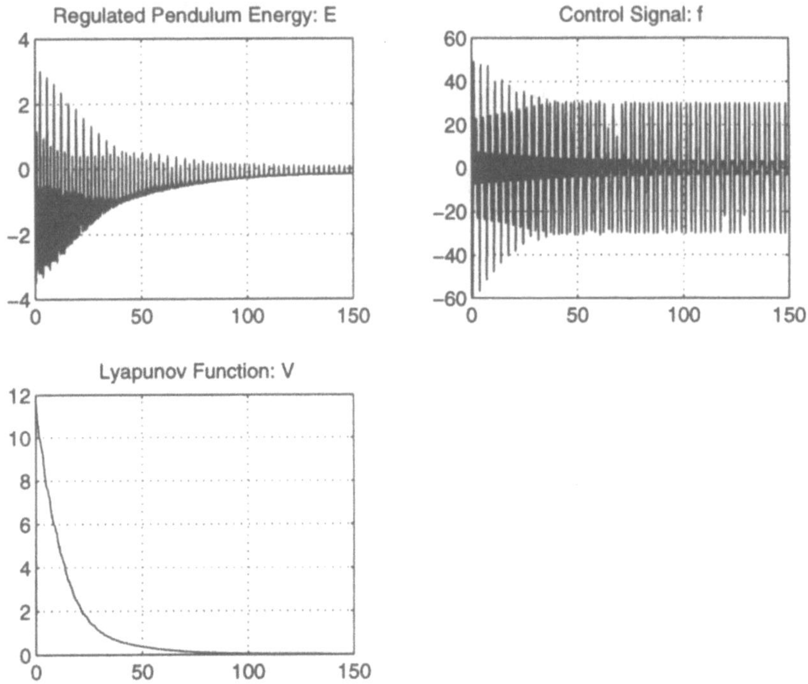

Fig. 7.3. Performance of the System.

Appendix

In this Appendix we obtain the equations of motion of the inverted pendulum via the Newton's second law and also by the Euler-Lagrange formulation.

Using Newton's second law

Consider the system in Fig (7.1). The coordinates of the pendulum (x_G, y_G) are:

$$\begin{aligned} x_G &= x + l\sin\theta \\ y_G &= l\cos\theta \end{aligned}$$

Applying Newton's second law in the x direction:

$$
\begin{aligned}
M\frac{d^2x}{dt^2} + m\frac{d^2x_G}{dt^2} &= f \\
M\frac{d^2x}{dt^2} + m\frac{d^2}{dt^2}(x + l\sin\theta) &= f \\
M\ddot{x} + m\left(\ddot{x} + l\frac{d}{dt}\cos\theta\dot{\theta}\right) &= f \\
M\ddot{x} + m\ddot{x} + ml\left(\cos\theta\ddot{\theta} - \sin\theta\dot{\theta}^2\right) &= f \\
(M+m)\ddot{x} - ml(\sin\theta)\dot{\theta}^2 + ml(\cos\theta)\ddot{\theta} &= f \qquad (7.33)
\end{aligned}
$$

Applying Newton's second law to the rotational motion:

$$
\begin{aligned}
m\frac{d^2x_G}{dt^2}l\cos\theta - m\frac{d^2y_G}{dt^2}l\sin\theta &= mgl\sin\theta \\
\left[m\frac{d^2}{dt^2}(x + l\sin\theta)\right]l\cos\theta - & \\
-\left[m\frac{d^2}{dt^2}(l\cos\theta)\right]l\sin\theta &= mgl\sin\theta \\
m\left[\ddot{x} - l(\sin\theta)\dot{\theta}^2 + l(\cos\theta)\ddot{\theta}\right]l\cos\theta + & \\
+m\left[l(\cos\theta)\dot{\theta}^2 + l(\sin\theta)\ddot{\theta}\right]l\sin\theta &= mgl\sin\theta \\
ml\ddot{x}\cos\theta + ml^2\ddot{\theta} &= mgl\sin\theta \qquad (7.34)
\end{aligned}
$$

Finally, from (7.33) and (7.34) we obtain the system in (7.1).

Via Lagrange's Equations

Using the system described in Fig 7.1, the mathematical model can also be derived by using Lagrange's equations. We first present the kinetic and potential energies which are used to compute the Lagrangian function.

The kinetic energy of the cart is $K_1 = \frac{M\dot{x}^2}{2}$. The kinetic energy of the pendulum is $K_2 = \frac{m\dot{x_G}^2}{2} + \frac{m\dot{y_G}^2}{2}$ where $x_G = x + l\sin\theta$ and $y_G = l\cos\theta$. The total kinetic energy is

$$
K = K_1 + K_2 = \frac{1}{2}(M+m)\dot{x}^2 + ml\dot{x}\dot{\theta}\cos\theta + \frac{1}{2}ml^2\dot{\theta}^2
$$

The total potential energy is $P = mgl(\cos\theta - 1)$. The Lagrangian is given by

$$
\begin{aligned}
L &= K - P \\
L &= \frac{1}{2}(M+m)\dot{x}^2 + ml\dot{x}\dot{\theta}\cos\theta + \frac{1}{2}ml^2\dot{\theta}^2 - mgl(\cos\theta - 1)
\end{aligned}
$$

The corresponding equations of motion are derived using Lagrange's equations:

$$\frac{d}{dt}\left(\frac{\partial L}{\partial \dot{q}}(q,\dot{q})\right) - \frac{\partial L}{\partial q}(q,\dot{q}) = \tau \tag{7.35}$$

where $q = (q_1, ...q_n)^T$ represents the generalized variables, one for each degree of freedom of the system, $\tau = (\tau_1, ..., \tau_n)^T$ denotes forces that are externally applied to the system.

In our case the generalized variables are x and θ, *i.e.* $q = (x, \theta)^T$. We therefore have:

$$\begin{aligned}
\left(\frac{\partial L}{\partial \dot{x}}\right) &= (M+m)\dot{x} + ml\dot{\theta}\cos\theta \\
\left(\frac{\partial L}{\partial x}\right) &= 0 \\
\left(\frac{\partial L}{\partial \dot{\theta}}\right) &= ml\dot{x}\cos\theta + ml^2\dot{\theta} \\
\left(\frac{\partial L}{\partial \theta}\right) &= mgl\sin\theta - ml\dot{x}\dot{\theta}\sin\theta
\end{aligned}$$

From Lagrange's equations (7.35), we finally obtain the equations of motion (7.33), (7.34) and the system (7.1).

References

1. Chung C.C and J. Hauser. 1995. Nonlinear control of a swinging pendulum. *Automatica*, Vol 31, No 6, 851-862.
2. Isidori A. 1989, *Nonlinear Contrcl Systems: An Introduction*, 2nd ed. Springer-Verlag, Berlin.
3. Jakubczyk, B., W. Respondek, 1980. On the linearization of control systems. *Bull. Acad. Polon. Sci. Math.*, Vol 28, 517-522.
4. Hassan K. Khalil. 1992. Non-Linear Systems. *MacMillan Publishing.*
5. Z. Lin, A. Saberi, M. Gutmann and Y. A. Shamash. 1996. Linear controller for an inverted pendulum having restricted travel: A high-and-low approach. *Automatica*, Vol 32, No 6, 933-937.
6. Mazenc F. and L. Praly. 1996. Adding integrators, saturated controls, and stabilization for feedforward systems.*IEEE Transactions on Automatic Control*, Vol 41, No 11, pp 1559-1578. November 1996.
7. Praly L.. 1995. Stabilisation du système pendule-chariot: Approche par assignation d'energie. Personal communication.
8. Spong M. and L. Praly. 1996. Control of underactuated mechanical systems using switching and saturation. *Proceedings of the Block Island Workshop on Control Using Logic Based Switching, Springer-Verlag.*

9. Wei Q., W.P. Dayawansa and W.S. Levine. 1995. Nonlinear controller for an inverted pendulum having restricted travel. *Automatica*, Vol 31, No 6, 841-850.

Feedback Linearization, Integrator Backstepping and Passivity–Based Controller Designs: A Comparison Example

M. Torres and R. Ortega

Laboratoire des Signaux et Systèmes, C.N.R.S. - Supélec - Université Paris-Sud, Plateau de Moulon, 91192 Gif-sur-Yvette, France

This work is dedicated to Ioan Landau on his 60th birthday for his continuing support in the battle for world passivation.

Summary. The objective of this paper is to compare in a simple example three controller design techniques for stabilization of nonlinear systems which have emerged in the last few years: 1) Feedback linearization, which as the name says aims at achieving a linear system in closed–loop via nonlinearities cancellation, is probably the best well known and easier to understand; 2) Integrator backstepping, which is another widely publicized technique that has proven very succesful for systems with special triangular structures; 3) Passivity–based control, which has traditionally received a wide acceptance among practicioning engineers, achieves system stabilization solving an, apparently more natural, passivation problem. Application of these, seemingly unrelated, methodologies will typically lead to the definition of different control schemes. For some specific examples these differences blurr and some interesting connections and similarities between the controller design techniques emerge. The careful study of such cases will improve our understanding of their common ground fostering cross–fertilization. In this paper we investigate these questions, both analytically and via simulations, for the simple problem of stabilization of a magnetically levitated ball. Our motivation in choosing this particular example stems, not just from the fact that due to its simplicity the connections between the controllers are best revealed, but also that such an equipment is available

in many engineering schools, hence experimental work can be easily carried out to complement our studies.
Keywords: Nonlinear systems, nonlinear control, passivity, stability theory.

1. Introduction

Among the various controller design techniques for stabilization of nonlinear systems that have emerged in the last few years we can distinguish three which are probably the most widely applicable and systematic: feedback linearization control (FLC), integrator backstepping control (IBC) and passivity–based control (PBC). The three techniques have evolved from different considerations and exploit different properties of the system. While in FLC we aim at a linear system, in PBC we are satisfied with assigning a certain passivity property to a suitably defined map, –consequently enforcing a certain storage function to the closed–loop dynamics–. In IBC we aim at a somehow intermediate objective (more ambitious than passivation but less demanding than linearization) of assigning a *strict* Lyapunov function to the closed–loop.

It is clear then that the three techniques will typically lead to the definition of different control schemes. For some specific examples these differences blurr and some interesting connections and similarities between the controller design techniques emerge. It is our belief that the detailed study of such cases will improve our understanding of their common ground fostering cross–fertilization. In this paper we compare these three controller design techniques in an example which is simple enough to reveal the connections between the controllers, but yet requires the application of nonlinear control to achieve a satisfactory performance. We choose to study a magnetically levitated ball which is available in many engineering schools.

Magnetic bearing systems with several electromagnets have been studied from the perspective of flatness and feedback linearization in [9] and with a backstepping approach in [15]. An output feedback adaptive IBC for the levitated ball studied here, albeit with a simpler model, has been reported in [8]. In [13] several PBCs are developed for this system. Here, we adapt for our simplified example the schemes of [9] for FLC, of [15] for IBC, and of [13] for PBC. We refer the reader to [5], [12] for background material on FLC, to [8] for an exhaustive coverage of IBC and to [13] for PBC, in particular of Euler–Lagrange systems.

The remaining of the paper is organized as follows. In section 2 we present the model, the control problem and the controller structure that we adopt for the three techniques. Namely, a nested–loop controller configuration with a fast inner–loop around the electrical dynamics and a slower mechanical outer–loop. In sections 3, 4 and 5 we derive the FLC, PBC and IBC, respectively. Their comparison is presented in section 6. Simulations are given in section

7, and we wrap up the paper with concluding remarks and open problems in section 8.

Before closing this introduction a word on style is in order. Since our main objective with this paper is pedagogical, –that is clarify the connections between the three techniques–, we have adopted an informal format of presentation without theorems and proofs. For this reason we have also favoured simpler, as opposed to more powerful or novel, solutions to the problem.

Consider the system of Fig. 8.1 consisting of an iron ball in a vertical magnetic field created by a single electromagnet. We denote by y the position of the ball measured with respect to the nominal position with the y-axis oriented upwards. The equation of motion for the mechanical subsystem is obtained from Newton's second law

$$m\ddot{y} = F(y,i) - mg$$

where m is the mass of the ball, mg is the gravitation force, $F(y,i)$ is the force produced by the electromagnet, with i its electrical current. Adopting the flux linkage model [10], $F(y,i)$ can be computed as

$$F(y,i) = \frac{\partial}{\partial y}\int_0^i \lambda(y,i)di$$

where $\lambda(y,i)$ is the flux linkage in the electromagnet. The dynamics of the electrical subsystem follow from Faraday's and Kirchhoff's voltage laws

$$\dot{\lambda} + Ri = u$$

where $R > 0$ is the electrical resistance and u is the input voltage.

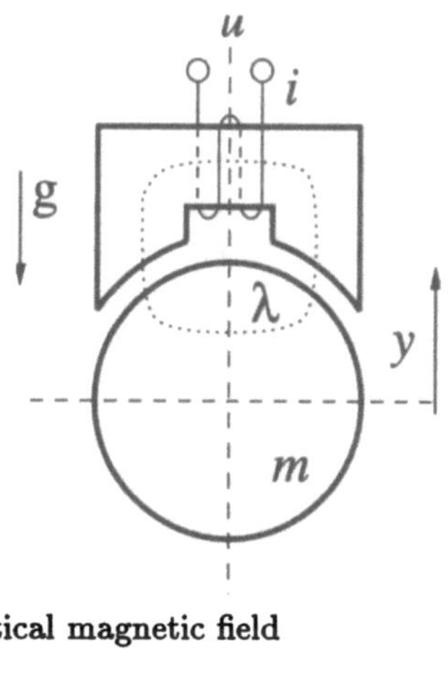

Fig. 8.1. A ball in a vertical magnetic field

If we assume linearity of the magnetic circuit the flux linkage is given by $\lambda(y,i) = L(y)i$, with $L(y)$ the inductance. A suitable approximation for the latter, in the domain $-\infty < y \le c_2 - \epsilon$ where $c_2 > 0$ is the nominal air gap, is given by

$$L(y) = \frac{c_1}{c_2 - y}$$

where ϵ is some small positive number, and c_1 is some positive constant that depends on the number of coil turns, air permeability and the cross–sectional area of the electromagnet. To simplify the presentation in the sequel we will assume $c_2 = 1$ and denote $c_1 = c$. Collecting all the equations above and rearranging terms we obtain the full model

$$\Sigma : \begin{cases} \dot{\lambda} &= -\frac{R}{c}(1-y)\lambda + u \\ F &= \frac{1}{2}\lambda^2 \\ m\ddot{y} &= F - mg \end{cases} \tag{8.1}$$

The *control objective* is to track a bounded reference signal $y_*(t)$ (with known and bounded first, second and third order derivatives) with all signals bounded. We will assume in the first instance that the *full state* is available for measurement. Then we will show that for PBC we only require the measurement of y and $\dot{y}$.

Notice that the system Σ can be decomposed as a feedback interconnection of an electrical (Σ_e) and a mechanical (Σ_m) subsystem as depicted in Fig. 8.2. Since the latter is just a double integrator it is reasonable to concentrate our attention on the problem of controlling force F, in the understanding that if force is suitably regulated then position y will be easy to control with classical linear techniques, e.g., PI control. An additional motivation for this approach stems from the fact that typically there is a natural time–scale decomposition between the electrical and mechanical dynamics. For these reasons the so–called nested–loop control is the prevailing structure in practical applications of electromechanical systems which will be adopted in the FLC, PBC and IBC designed below.

The nested–loop control configuration is shown in Fig. 8.3, where C_{il}, C_{ol} are the inner and outer–loop controllers, respectively. The inner–loop controller is designed to achieve asymptotic tracking of the desired force F_d. The outer–loop controller takes care of the position tracking, and essentially generates a signal F_d such that *if* $F \to F_d$ then position tracking, i.e., $y \to y_*$, is ensured. In applications C_{ol} is typically a simple PI around position errors. (In some cases, for instance if Σ_m is not LTI as in [14], an interlaced design of C_{il} and C_{ol} is needed.)

Remark 1.1. Another fundamental property of electromechanical systems, first reported in [3], is that we can always define a feedback decomposition

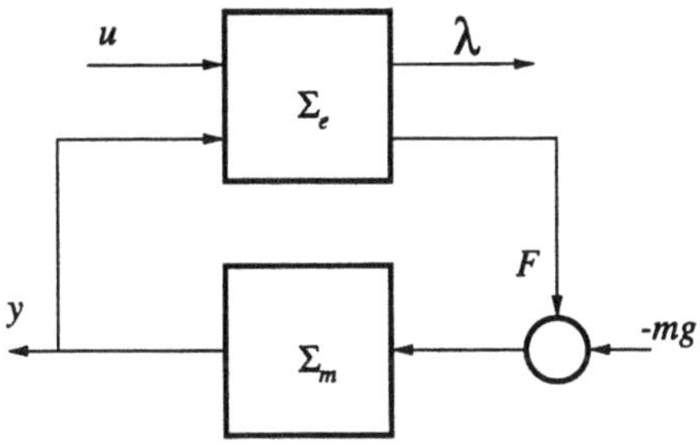

Fig. 8.2. Feedback decomposition of the system

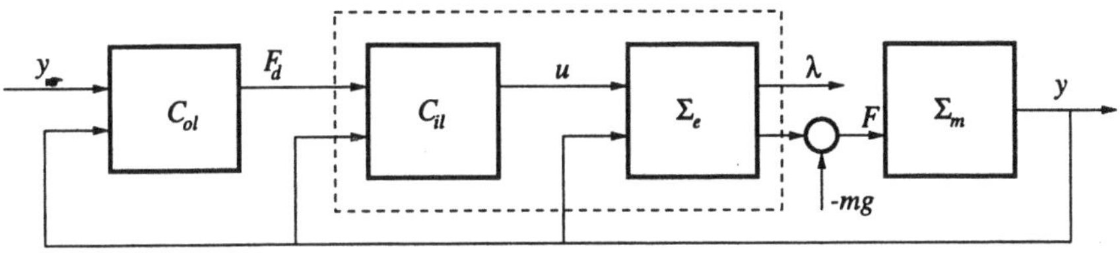

Fig. 8.3. Nested–loop control configuration.

such that each subsystem is *passive*. This feature, which is crucial for the succesful design of PBC of *underactuated* electrical subsystems, is not required in this simple fully actuated example. See the recent book [13] for a detailed exposition of PBC.

2. Feedback–linearization control

We start our study with FLC since it is clearly the best well–known and, in this particular example, the simplest to derive and understand. To this end, we find convenient to introduce a change of coordinates and we write the system dynamics in terms of force, position and velocity as

$$\begin{aligned} \dot{F} &= -\tfrac{2R}{c}(1-y)F + \sqrt{\tfrac{2F}{c}}u \\ m\ddot{y} &= F - mg \end{aligned} \tag{8.2}$$

Differentiating once more $\ddot{y}$ we get

$$\begin{aligned} my^{(3)} &= \dot{F} \\ &= -R(1-y)\frac{2F}{c} + \sqrt{\frac{2F}{c}}u \end{aligned}$$

It is clear from the equations above that, taking as output y and provided $F > 0$, the system has relative degree 3 and no zero dynamics. Thus, an FLC can be immediately defined as $u = u_{\text{FL}}$ with

$$u_{\mathrm{FL}} = \sqrt{\frac{c}{2F}} m v_{\mathrm{FL}}(y, \dot{y}, F) + R(1-y)\sqrt{\frac{2F}{c}} \tag{8.3}$$

which yields

$$y^{(3)} = v_{\mathrm{FL}}(y, \dot{y}, F)$$

A suitable choice for the outer loop signal is then

$$v_{\mathrm{FL}}(y, \dot{y}, F) = y_*^{(3)} - k_2[\underbrace{(\frac{1}{m}F - g)}_{\ddot{y}} - \ddot{y}_*] - k_1\dot{\tilde{y}} - k_0\tilde{y} \tag{8.4}$$

where we have defined the tracking error $\tilde{y} \triangleq y - y_*$.

The FLC (8.3), (8.4) is a static state feedback law which ensures in closed–loop the third order linear dynamics

$$D(p)\tilde{y} = 0$$

where $p \triangleq \frac{d}{dt}$, the characteristic polynomial is given as

$$D(p) \triangleq p^3 + k_2p^2 + +k_1p + k_0 \tag{8.5}$$

and the coefficients $k_i > 0$, $i = 1, 2, 3$ are chosen to ensure $D(p)$ is a Hurwitz polynomial.

Remark 2.1. The controller (8.3) has a singularity in $F = 0$. As we will see later PBC and IBC are also only locally defined. This problem is, of course, related with the poor controllability properties of levitation systems with just one electromagnet. Since the focus of the present paper is on *structural* similarities of various controllers instead of their precise stability domains we will concentrate on *local* properties and assume throughout that the system operates away from singularity regions.

Remark 2.2. Since in closed–loop $\dot{F} = mv_{\mathrm{FL}}(y, \dot{y}, F)$, it is useful – for the sake of comparison with the other controllers – to think of $mv_{\mathrm{FL}}(y, \dot{y}, F)$ as a desired value for $\dot{F}$.

3. Passivity–based control

In PBC the stabilization objective is recasted in terms of a passivation problem. That is, we will ensure stability of the closed–loop system by rendering passive (actually strictly passive to enforce *asymptotic* stability) a suitably defined map. Depending on the particular passivity property that we want to impose to the closed–loop there are several ways to carry out the design.

This choice is, of course, intimately tied with the open–loop properties of the system, in particular its relative degree and stable invertibility [2]. Notice however that we have freedom in the choice of outputs and corresponding storage functions.

As discussed above we will adopt the nested loop (*i.q.* cascaded) controller structure of Fig. 8.3 to design our PBC.[1] The inner–loop controller is designed following the principles of energy shaping and damping injection explained below. To facilitate the comparison between the controllers we take the outer–loop controller $\mathcal{C}_{ol}$ as for the FLC above.

Controller design

The procedure to design the cascaded PBC consists of the following steps: 1) Choice of u to ensure flux tracking, i.e., $\lambda \to \lambda_d$; 2) Choice of λ_d so that $\lambda \to \lambda_d \Rightarrow F \to F_d$; 3) Proof that $F \to F_d \Rightarrow y \to y_*$ with all signals bounded. These steps are detailed below.

1. Flux tracking

First, we prove that the electrical subsystem satisfies some *passivity* properties. This fundamental step will identify the output which is "easy to control" and its corresponding storage function. An incremental version of this function will be the desired storage function used to enforce passivity to the closed–loop.

In systems described by Euler-Lagrange equations the passivity property is always satisfied with storage function the total energy. This feature makes PBC a natural approach to control this class of systems. We propose then the (scaled) magnetic energy in the inductance as storage function candidate

$$H = \frac{1}{2}\lambda^2$$

whose derivative along the trajectories of (8.1) satisfies

$$\begin{aligned} \dot{H} &= -\frac{R}{c}(1-y)\lambda^2 + \lambda u \\ &\leq -\alpha\lambda^2 + \lambda u \end{aligned}$$

with $\alpha \triangleq \frac{R\epsilon}{c} > 0$. Integrating the last inequality from 0 to t and using the fact that $H \geq 0$ yields

$$\int_0^t u(s)\lambda(s)ds \geq \alpha \int_0^t \lambda^2(s)ds + \beta$$

where we have set $\beta \triangleq -H(0)$. This proves (local) *output strict passivity* of the map $u \mapsto \lambda$.

[1] In [13] other PBCs for this system are reported.

The passivity property above suggests to try to assign the to the closed–loop the *desired storage function*

$$H_d = \frac{1}{2}\tilde{\lambda}^2 \tag{8.6}$$

where $\tilde{\lambda} \triangleq \lambda - \lambda_d$, with λ_d the desired flux that induces on the mechanical subsytem the force required to regulate y, this signal is defined in the next step. With some abuse of notation this step is commonly refered as *energy shaping.*

It is easy to see that if we set $u = u_{\text{PB}}$ with

$$u_{\text{PB}} = \dot{\lambda}_d + \frac{R}{c}(1-y)\lambda_d + v \tag{8.7}$$

then we get the error dynamics

$$\dot{\tilde{\lambda}} = -\frac{R}{c}(1-y)\tilde{\lambda} + v$$

Proceeding as above we can establish the bound

$$\dot{H}_d \leq -\alpha\tilde{\lambda}^2 + \tilde{\lambda}v \tag{8.8}$$

which proves that the map $v \mapsto \tilde{\lambda}$ is output strictly passive as desired. Of course, if we set $v \equiv 0$ we have that $\tilde{\lambda} \to 0$, exponentially fast. Given that the rate of convergence α may be small to enhance performance it is convenient to add a so–called *damping injection* term into the electrical subsytem, setting for instance

$$v = -R_{\text{DI}}\tilde{\lambda}$$

with $R_{\text{DI}} > 0$. It is clear that this additional term will speed up the electrical transient. However, as will be shown in the simulations of section 7, increasing R_{DI} does not necessarily leads to transient performance improvement. This stems from the fact that the overall closed–loop system under PBC is still nonlinear and we are confronted with unpredictable *peaking* effects.

2. From flux tracking to force tracking

The force of magnetical origin F can be written in terms of the flux error $\tilde{\lambda}$ and the desired flux λ_d as

$$F = \frac{1}{2c}[\lambda_d^2 + \tilde{\lambda}(\tilde{\lambda} + 2\lambda_d)] \tag{8.9}$$

¿From the analysis above we have that $\tilde{\lambda} \to 0$, hence it is reasonable to choose λ_d as the solution of

$$F_d = \frac{1}{2c}\lambda_d^2 \tag{8.10}$$

where $F_d > 0$ is some desired force, whose derivative is assumed to be known. We can now write the control (8.7) in terms of F_d and $\dot{F}_d$ as

$$u_{\text{PB}} = \sqrt{\frac{c}{2F_d}}\dot{F}_d + R(1-y)\sqrt{\frac{2F_d}{c}} + v \tag{8.11}$$

If F_d is bounded we can conclude from (8.9) and (8.10) that the force tracking controller (8.11) ensures $F \to F_d$. The problem is that in position tracking problems F_d will not be *a priori* bounded, since it is generated by C_{ol}. This leads to the third step of our design.

3. From force tracking to position tracking

In this final step we have to design C_{ol} and prove convergence of the position error to zero. We start by defining F_d which, following the approach taken for the FLC of the previous section, we set as

$$F_d = m[\ddot{y}_* - k_2\dot{\tilde{y}} - k_1\tilde{y} - k_0\int_0^t \tilde{y}(s)ds] \tag{8.12}$$

with $k_0, k_1, k_2 > 0$ chosen as in section 3. Notice that $\dot{F}_d$ is used in (8.11). From the definition of F_d above, this implies that $\ddot{y}$ (or equivalently F) is required for implementation of the PBC (8.11), (8.12). Remark also that we have introduced in (8.12) an integral term. This takes care of the steady-state error due to the constant disturbance mg, but it could be removed and simply replaced by g. As we will see in the simulations, and is well known in applications, the inclusion of the integral term robustifies the loop. Furthermore it simplifies the comparison of the various schemes because in this way the closed–loop mechanical dynamics is of the same order.

Let us now analyse the *stability* of the overall system (8.1), (8.11), (8.12) with $u = u_{\text{PB}}$ and $v \equiv 0$ (or $v = -R_{\text{DI}}\tilde{\lambda}$) . We know already from the analysis of the flux tracking step that $\tilde{\lambda} \to 0$ exponentially fast. To study the mechanical dynamics let us replace (8.9), (8.10), and (8.12) in (8.1) to get the error equation

$$G(p)\tilde{y} = \frac{1}{2mc}\tilde{\lambda}(\tilde{\lambda} + 2\lambda_d) - g$$

where $G(p) \triangleq \frac{1}{p}D(p)$ and $D(p)$ as in (8.5). This system admits a state–space realization of the form[2]

$$\dot{x} = Ax + B\frac{1}{2mc}\tilde{\lambda}(\tilde{\lambda} + 2\lambda_d) \tag{8.13}$$

with $x \triangleq [\tilde{y},\ \dot{\tilde{y}},\ z_2]^T$, $\dot{z}_2 = \tilde{y}$, $B \in \mathcal{R}^3$, and the characteristic polynomial of $A \in \mathcal{R}^{3\times 3}$ equal to $D(p)$. Now, $\tilde{\lambda}$ is an exponentially decaying term and from (8.10), (8.12) and boundedness of $\ddot{y}_*$ we see that λ_d satisfies the bounds

[2] Notice the presence of the zero at the origin that takes care of the constant term g.

$$\begin{aligned} |\lambda_d| &\leq \alpha_1 + \alpha_2 |F_d| \\ &\leq \alpha_3 + \alpha_4 |x| \end{aligned}$$

for some suitable constants $\alpha_i > 0,\ i = 1, \cdots, 4$. Consequently the forcing term satisfies the bound

$$|\frac{1}{2mc}\tilde{\lambda}(\tilde{\lambda} + 2\lambda_d)| \leq (\alpha_5 + \alpha_6 |x|)\tilde{\lambda} \tag{8.14}$$

for some $\alpha_5,\ \alpha_6 > 0$. The proof of asymptotic stability of (8.13) follows considering the quadratic function

$$V_1 = \frac{1}{2} x^T P x \tag{8.15}$$

with P the symmetric positive–definite solution of

$$A^T P + PA = -Q,\ Q = Q^T > 0$$

and invoking standard arguments for exponentially stable systems perturbed by linearly bounded terms multiplied by an exponential; see, e.g., [6].

Output–feedback passivity–based control

As discussed in [13], one of the main advantages of PBC is the possibility of avoiding state measurements. In this subsection we develop a PBC that requires only the measurement of y and $\dot{y}$ preserving asymptotic stability. To this end, we recall that $\ddot{y}$ is needed only in the evaluation of $\dot{F}_d$, hence we propose to replace F_d in (8.12) by

$$F_d = m[\ddot{y}_* - k_2 z_1 - k_1 \tilde{y} - k_0 \int_0^t \tilde{y}(s) ds] \tag{8.16}$$

where z_1 is the *approximate derivative* of $\tilde{y}$, that is

$$\dot{z}_1 = -a z_1 + b\dot{\tilde{y}} \tag{8.17}$$

with $a,\ b > 0$. This modification does not affect the flux tracking property $\tilde{\lambda} \to 0$, nor the control signal (8.11). The key point is that now u_{PB} can be implemented *feeding–back only y and $\dot{y}$.*

The stability analysis of this new scheme mimicks the developments above. That is, we obtain a state space realization of the mechanical dynamics of the form (8.13) but with the augmented state $x \triangleq [\tilde{y},\ \dot{\tilde{y}},\ z_1,\ z_2]^T$. The new A matrix is

$$A = \begin{bmatrix} 0 & 1 & 0 & 0 \\ -k_1 & 0 & -k_2 & -k_0 \\ 0 & b & -a & 0 \\ 1 & 0 & 0 & 0 \end{bmatrix}$$

and the forcing term still satisfies the bound (8.14). To study the stability of the unforced equation let us introduce the partition

$$A \triangleq \begin{bmatrix} A_1 & \vdots & B_1 \\ \cdots & \cdot & \cdots \\ -C_1^T & \vdots & 0 \end{bmatrix}$$

with $A_1 \in \mathcal{R}^{3\times 3}$, $B_1, C_1 \in \mathcal{R}^{3\times 1}$ defined in an obvious manner. We can then prove that

$$\det(sI - A) = \det(sI - A_1)[C_1^T(sI - A_1)^{-1}B_1 + s]$$

and $\det(sI - A_1)$ is a Hurwitz polynomial for all $k_1, k_2, a, b > 0$. Hence it only remains to choose the integral gain k_0 to ensure stability of A. Stability of the forced equation follows *verbatim* from the analysis of the full state feedback case.

4. Integrator backstepping control

As pointed out in section 8.3 of [8], where a simplified version of our model is considered, the system (8.1) is *not in any of the special forms* required for application of integrator backstepping. However, as shown in [15] (see also [8]) this does not preclude a backstepping–like design. Since almost any design can be interpreted as backstepping–like we point out from the outset that we will take–off from the approach used in [15] for a magnetic bearing system as applied to our simple levitated ball. Motivated by our final objective of comparison of techniques we will also present a variation of the basic scheme of [15] that incorporates features of the PBC. A similar combination of PBC and IBC has been reported in [4] where ideas of [11] are added to an IBC to remove singularities in induction motor control.

In IBC we start from the mechanical equation of (8.1) and, *assuming F* is the control, define an F_d that stabilizes this subsytem. For the sake of comparison we choose F_d as in FLC and PBC, that is (8.12). Adding and substracting $\dot{F}_d$ to $my^{(3)} = \dot{F}$ yields the first error equation

$$\dot{x} = Ax + \frac{1}{m}B(F - F_d)$$

where x, A and B are as in (8.13). We start constructing our Lyapunov function with the quadratic function (8.15), whose derivative yields

$$\begin{aligned} \dot{V}_1 &= -\frac{1}{2}x^T Qx + \frac{1}{m}x^T PB(F - F_d) \\ &= -\frac{1}{2}x^T Qx + \frac{1}{2mc}x^T PB\tilde{\lambda}(\lambda + \lambda_d) \end{aligned}$$

where we have used the definitions of F and λ_d in (8.10) to get the last equation. Let us now look at the dynamic equation of $\tilde{\lambda}$, that is

$$\dot{\tilde{\lambda}} = -\frac{R}{c}(1-y)\lambda + u - \dot{\lambda}_d$$

At this stage we decide a control u that stabilizes $\tilde{\lambda}$ plus a term to compensate for the cross term in $\dot{V}_1$ above. We have here the choice of adopting an FLC–like approach cancel the term $-\frac{R}{c}(1-y)\lambda$ and add a damping term $-R_{\text{DI}}\tilde{\lambda}$, or follow the PBC route, use the existing damping, and set $u = u_{\text{IB}}$ with

$$u_{\text{IB}} = u_{\text{PB}} \tag{8.18}$$

and u_{PB} as in (8.7). We can complete our Lyapunov function with H_d (8.6) as

$$V = V_1 + \frac{1}{\beta}H_d$$

where we have added a tuning parameter $\beta > 0$. Differentiating V and using (8.8) we get

$$\dot{V} \leq -\frac{1}{2}x^T Qx - \frac{\alpha}{\beta}\tilde{\lambda}^2 + \tilde{\lambda}[\frac{1}{\beta}v + \frac{1}{2mc}x^T PB(\lambda + \lambda_d)]$$

The IBC design is completed setting

$$v = -\frac{\beta}{2mc}x^T PB(\lambda + \lambda_d) \tag{8.19}$$

which removes the cross terms and yields the desired strict Lyapunov function

$$\dot{V} \leq -\frac{1}{2}x^T Qx - \frac{\alpha}{\beta}\tilde{\lambda}^2$$

5. Comparison of the schemes

We will now compare FLC (8.3), (8.4), PBC (8.11), (8.12) (without damping injection, i.e., with $v = 0$), and IBC (8.11), (8.18), (8.19). To this end, we notice that, (8.12), which we repeat here for ease of reference

$$F_d = m[\ddot{y}_* - k_2\dot{\tilde{y}} - k_1\tilde{y} - k_0\int_0^t \tilde{y}(s)ds]$$

is related to (8.4) as

$$\dot{F}_d = mv_{\text{FL}}(y, \dot{y}, F)$$

Hence, the three control inputs can be written as

$$\begin{aligned} u_{\mathrm{FL}} &= \sqrt{\frac{c}{2F}}mv_{\mathrm{FL}}(y,\dot{y},F) + R(1-y)\sqrt{\frac{2F}{c}} \\ u_{\mathrm{PB}} &= \sqrt{\frac{c}{2F_d}}mv_{\mathrm{FL}}(y,\dot{y},F) + R(1-y)\sqrt{\frac{2F_d}{c}} \\ u_{\mathrm{IB}} &= u_{\mathrm{PB}} - \frac{\beta}{\sqrt{2mc}}[\tilde{y},\ \dot{\tilde{y}},\ \int_0^t \tilde{y}(s)ds]PB(\sqrt{F}+\sqrt{F_d}) \end{aligned}$$

The following remarks are in order

- FLC and PBC differ only on the utilisation of the forces in some terms: desired forces F_d for PBC instead of the actual forces F for FLC.
- One additional difference comes from the generation of F_d, which involves the inclusion of an additional state. Hence, PBC and IBC are *dynamic* state feeback controllers while FLC is static.
- The IBC equals the PBC plus the signal that achieves the cancellation of the cross terms in $\dot{V}$. The two schemes approach as $\beta \to 0$.
- In contrast with FLC, the closed–loop equations for PBC and IBC are nonlinear. In PBC we get a *cascade* structure where the first subsystem is exponentially stable and some growth conditions on the input to the second subsystem allows us to complete the stability proof. In IBC we get a closed–loop system of the form

$$\begin{aligned} \begin{bmatrix} \dot{\tilde{\lambda}} \\ \dot{x} \end{bmatrix} &= \begin{bmatrix} -\frac{R}{c}(1-y) & -\frac{\beta}{\sqrt{2c}}(\sqrt{F}+\sqrt{F_d})B^T P \\ \frac{1}{\sqrt{2c}}(\sqrt{F}+\sqrt{F_d})B & A \end{bmatrix} \begin{bmatrix} \tilde{\lambda} \\ x \end{bmatrix} \\ &\triangleq A_{\mathrm{DI}} + A_{\mathrm{SK}}) \begin{bmatrix} \tilde{\lambda} \\ x \end{bmatrix} \end{aligned}$$

  where the block–diagonal matrix A_{DI} is "stable" and A_{SK} satisfies the skew–symmetry property

$$\begin{bmatrix} \frac{1}{\beta} & 0 \\ 0 & P \end{bmatrix} A_{\mathrm{SK}} = -A_{\mathrm{SK}}^T \begin{bmatrix} \frac{1}{\beta} & 0 \\ 0 & P \end{bmatrix}$$

  This pattern is present in all IBC designs.
- Given that their closed–loop dynamics are nonlinear, predicting the effect of the tuning parameters for PBC and IBC is less obvious than for FLC.
- While there is an *output feedback* version of PBC, the presence of the extra term in IBC hampers this possibility. However, if we want to add damping to the PBC we need the full state.
- As explained in the previous section an FLC–like version of IBC may be obtained as

$$u_{\text{IB}} = \sqrt{\frac{c}{2F_d}} m v_{\text{FL}}(y, \dot{y}, F) + R(1-y)\sqrt{\frac{2F}{c}}$$
$$-\frac{\beta}{\sqrt{2c}}\left[\tilde{y},\ \dot{\tilde{y}},\ \int_0^t \tilde{y}(s)ds\right] PB(\sqrt{F}+\sqrt{F_d}) - R_{\text{DI}}(\sqrt{F}-\sqrt{F_d})$$

Notice that it combines terms from PBC and FLC plus the damping injection and the additional "decoupling" term.

6. Simulation results

In this section we present some simulation results of the levitated ball model (8.1) in closed–loop with FLC, PBC and IBC. To facilitate the comparison, and unless otherwise stated, we have in all cases fixed $D(p) = (p+10)^3$. We tried a series of small step references in position. As expected, when we applied larger steps we cross through controller singularities. This happened for all schemes with basically the same size of "admissible" references. Away from singularities the control signals do not differ significantly, therefore we show in all cases only the position response

To motivate the utilisation of nonlinear control we designed first a pole placement linear controller for the linearized approximation of (8.1) around the zero equlibrium. In Fig. 8.4 we show the step responses for various references. As expected the performance is degraded as we move away from the domain of validity of the modeling approximation. The response of FLC (8.3), (8.4), which is shown for the sake of reference in Fig. 8.5, is of course the same for all reference levels. It is interesting to note that we observed very little performance degradation when the actual parameters R and c were replaced by estimates within practically meaningful ranges of $\pm 50\%$. Comparable robustness properties were observed for the other schemes.

PBC was tried in different variations. Fig. 8.6 shows the response of (8.11), (8.12) without damping injection, i.e., with $v = 0$ and without integral action $k_0 = 0$. In this case the mechanical dynamics is order two and we fixed the characteristic polynomial at $(p+10)^2$. Even though the errors eventually converge to zero, the transient response is extremely poor even when the poles of the mechanical dynamics were pushed farther left. Fig. 8.7 shows the effect of adding an integral action, while for Fig. 8.8 we included also the damping injection. Notice that for large values of R_{DI} the response overshoots, due to the peaking phenomenon. The response of the output–feedback version of PBC (8.16), (8.17) is shown in Fig. 8.9. The bandwidth of the approximate differentiator was chosen five times faster than the mechanical dynamics. Finally in Fig. 8.10 we present the responses of the IBC (8.11), (8.18), (8.19) for a value of $\beta = 0.05$.

In summary we can say that in must of the cases, and as long as we kept away from the singularity region, the observed responses were consistent with the theoretical predictions. Also, despite the significant (analytical)

differences of the three controllers that we pointed out above we could not observe big discrepancies in the simulation responses. We believe this stems from the fact that the system is rather benign (it is passive with respect to λ, and y is a flat output!) particularly for the choice of parameters given here.

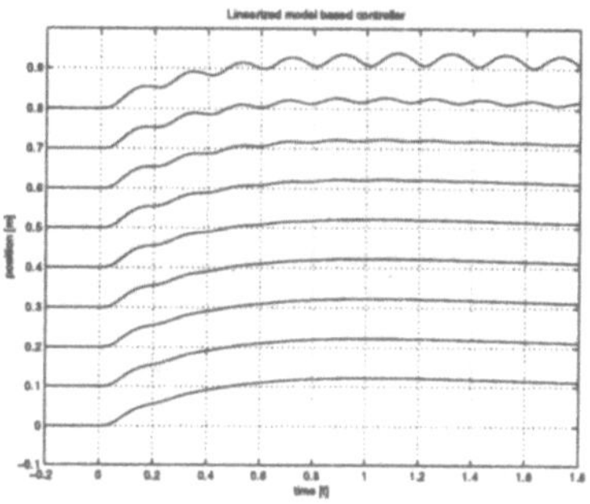

Fig. 8.4. Linearized model based controller

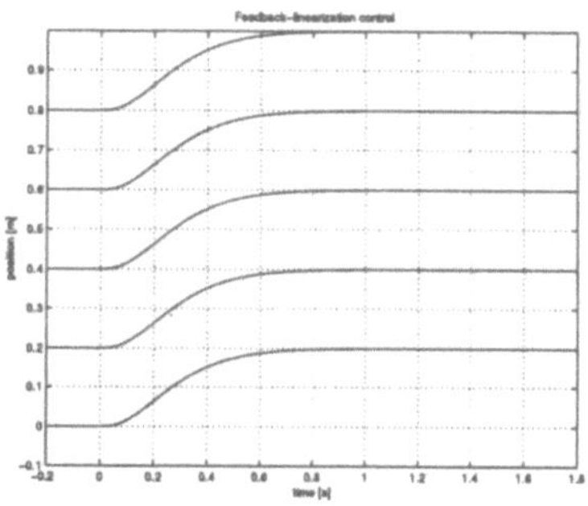

Fig. 8.5. Feedback-linearization control

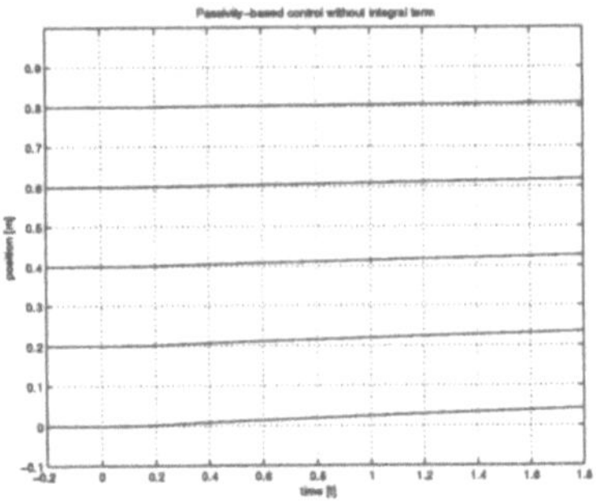

Fig. 8.6. Passivity-based controller without integral term

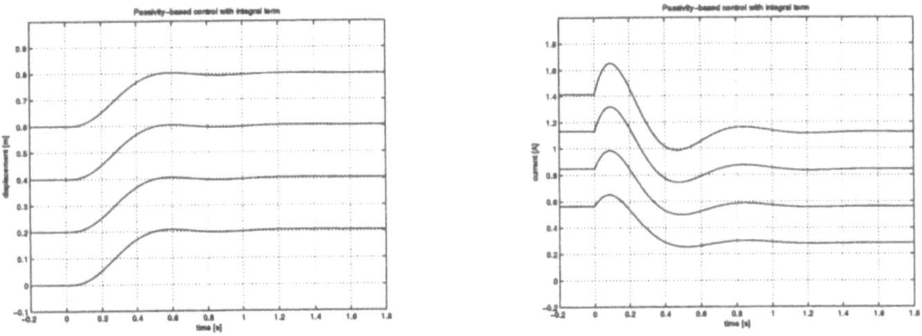

Fig. 8.7. Passivity-based controller with integral term

7. Conclusions and further research

The following concluding remarks concerning general aspects of FLC, PBC and IBC are in order

- While in FLC we aim at a linear system, in PBC we are satisfied with assigning a certain passivity property to a suitably defined map, – consequently enforcing a certain storage function to the closed–loop dynamics –. In IBC we aim at a somehow intermediate objective (more ambitious than passivation but less demanding than linearization) of assigning a *strict* Lyapunov function to the closed–loop.
- Both, in IBC and PBC, the main question is "Which storage–Lyapunov function you (can) want to assign?" In PBC, at least of Euler–Lagrange systems, the answer is provided by the systems energy function (either of the full system or just a part of it). The power of IBC is that, for a certain class of cascaded systems, the Lyapunov function can be *recursively* constructed. The example above shows that IBC can be profitably combined with PBC to answer this question.
- It is often argued and reasonable to expect, though hard to prove rigourously, that avoiding cancellation of nonlinearities enhances the ro-

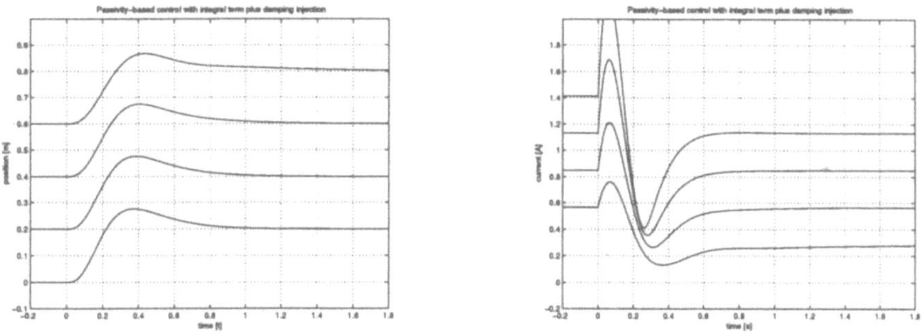

Fig. 8.8. Passivity-based controller with integral term plus damping injection

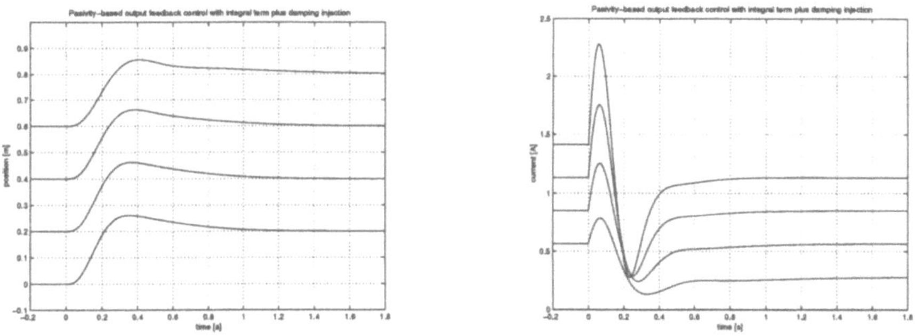

Fig. 8.9. Output feedback passivity-based controller

bustness of the scheme in the face of parameter uncertainties.[3] Our simulation evidence in the present example, however, showed that all schemes are highly insensitive to these uncertainties.

The three schemes will in general exhibit different transients and posses different robustness properties, a challenging research problem is to establish some common framework to compare their robustness and performance properties. One particularly interesting question is to assess the degrees of freedom provided to the designer to enhance the systems response.

The skew–symmetry property of IBC is related, though not so clear how, with the so–called workless–forces of Euler–Lagrange systems. In PBC we essentially disregard these forces and concentrate on the damping injection to increase the convergence rate. In this class of systems these coupling terms are related with the transformation of energy from one form to the other, e.g., from electrical to magnetical in electrical circuits or from potential to kinetic in mechanical systems. In a recent interesting paper [1] the possibility of "shaping" also these forces for performance improvement is explored. This is

[3] See [7] for an induction motor example where the superiority of PBC is unquestionably established, both theoretically and experimentally.

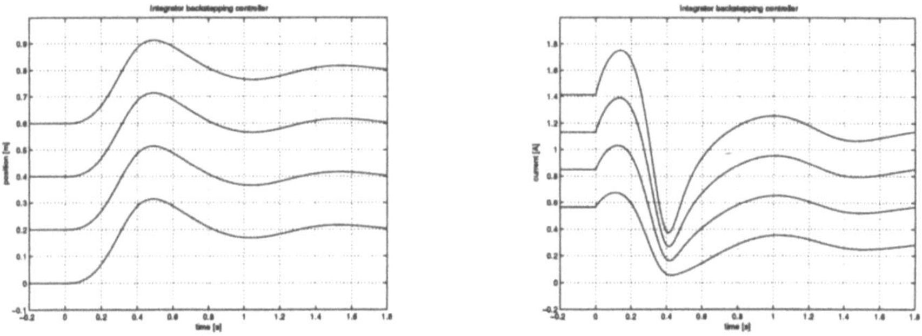

Fig. 8.10. Integrator backstepping-based controller

an intriguing, and certainly quite reasonable, proposition whose effect cannot be captured with our classical convergence rate analysis.

Acknowledgement. The authors are greatful to Laurent Praly for many helpful discussions that helped clarified the issues discussed in this paper.

References

1. O. Bas, V. Davidkovich, A. M. Stankovich and G. Tadmor, Passivity–based sensorless control of a smooth rotor permanent magnet synchronous motor, *36th IEEE CDC*, San Diego, USA, 1997.
2. C. Byrnes, A. Isidori, and J.C. Willems, "Passivity, Feedback Equivalence and the Global Stabilization of Minimum Phase Nonlinear Systems", *IEEE Trans. Aut. Cont.*, Vol. 36, No. 11, pp. 1228-1240, 1991.
3. G. Espinosa and R. Ortega, State Observers are Unnecessary for Induction Motor Control, *Syst. & Cont. Letters*, Vol 23, pp. 315-323, 1994.
4. J. Hu, D. Dawson, and Y. Ou, A global adaptive link position tracking controller for robot manipulators driven by induction motors, *34th IEEE CDC*, New Orleans, USA, 1995.
5. A. Isidori, *Nonlinear Control Systems*, Springer–Verlag, 2nd Edition, 1995.
6. H. Khalil, *Nonlinear Systems*, Second Ed. Prentice–Hall, New Jersey, 1996.
7. K. Kim, R. Ortega, A. Charara, and J.P. Vilain, Theoretical and Experimental Comparison of Two Nonlinear Controllers for Current-fed Induction Motors, *IEEE Trans. Cont. Syst. Techn.*, **3**, 1–11, (1997).
8. M. Krstic, I. Kanellakopoulos, and P. Kokotovic, *Nonlinear and Adaptive Control Design*, John Wiley, 1995.
9. J. Levine, J. Lottin, and J. Ponsart, A nonlinear approach to the control of magnetic bearings, *IEEE Trans. on Cont. Syst. Techn.*, Vol. 4, No. 5, Sept. 1996.
10. J. Meisel, *Principles of electromechanical energy conversion*, McGraw Hill, 1966.
11. P.J. Nicklasson, R. Ortega, and G. Espinosa, On output feedback tracking control of the general rotating machine, *IEEE Trans. Automat. Contr.*, **42**, 5, 1997.

12. H. Nijmeijer and A. van der Schaft, *Nonlinear Dynamical Control Systems*, Springer Verlag, 1990.
13. R. Ortega, A. Loria, H. Sira, and P. Nicklasson, *Passivity-based control of Euler–Lagrange systems*, Springer–Verlag, Berlin, Communications and Control Engineering, (to appear in 1998).
14. E. Panteley and R. Ortega, Cascaded Control of Feedback Interconnected Systems: Application to Robots with AC Drives, *Automatica*, Nov 1997.
15. M. Queiroz and D. Dawson, Nonlinear control of magnetic bearing: A backstepping approach, *IEEE Trans. on Cont. Syst. Techn.*, Vol. 4, No. 5, Sept. 1996.

A Paradox in Pole Placement Design

K. J. Åström

Department of Automatic Control, Lund Institute of Technology, Box 118, S-221 00 Lund, Sweden, e-mail: kja@control.lth.se

Summary. It is well known that the pole placement problem can be solved for a system that is controllable and observable. this seems to contradict the fact that there are limitations on the performance of a closed loop system for a process with poles and zeros in the right half plane. This paradox is investigated and resolved in the paper. As a result we also obtain rules for choosing closed loop poles in a pole-placement design. The results are relevant not only to conventional design problems but particularly to adaptive control where the design is done automatically.

1. Introduction

Much research in automatic control has been devoted to design methods, some examples are found in [11], [6], [10], [1], [4], [5], [3] and [8]. A very attractive feature of the systematic design methods is that starting with specifications the methods will deliver a controller that satisfies the specifications if one exists. One drawback with most methods is that the specifications only cover some aspects of the design while others are neglected. The result is that very poor control can be obtained from seemingly reasonable specifications. Some extreme examples are presented in [7]. One reason for poor design is that the fundamental limitations imposed by the presence of right half plane poles and zeros are neglected. In [2] it is shown that these limitations can be expressed in terms of inequalities on the crossover frequency.

In this paper we will consider a particular design method namely pole-placement. By investigating the paradox we find that in order to design a robust controller it is necessary to choose the closed loop poles carefully. For systems with poles and zeros in the right half plane it is necessary that the dominant poles are chosen to respect the fundamental performance inequalities. It is important to select the closed loop poles carefully also for systems with no poles and zeros in the right half plane. Some guidelines are given in the paper.

2. The Paradox

Consider a single-input single-output time invariant linear system that is controllable and observable. Let $P(s)$ be the transfer function of the system. Factor the transfer function as

$$P(s) = P_{mp}(s)P_{nmp}(s) \tag{9.1}$$

where P_{mp} is the minimum phase part and P_{nmp} is the non-minimum phase part. The factorization is normalized so that $|P_{nmp}(i\omega)| = 1$ and $\arg P_{nmp}(i\omega) \leq 0$. Let the system be controlled with a controller which has no singularities in the right half plane, and let the gain crossover frequency be ω_{gc}. In [2] it is shown that in order to have a closed loop system the crossover frequency must be such that the phase lag of the non-minimum phase part P_{nmp} at ω_{gc} is not too large, i.e.

$$\arg P_{nmp}(i\omega_{gc}) > -\varphi_0$$

The value of φ_0 depends on the stability margins required, a ball park figure is $\varphi_0 = \pi/2$. This implies that for a system with a right half plane zero at $s = z$ we have $\omega_{gc} < z$, and for a system with a right half plane pole at $s = p$ we have $\omega_{gc} > p$.

Since the system is controllable the closed loop poles can be placed arbitrarily and we have the paradox: How can the crossover frequency be limited when the closed loop poles can be made arbitrarily fast?

3. System with a Zero in the Right Half Plane

To get some insight into the paradox we will investigate a specific example. Consider a process with the transfer function

$$P(s) = \frac{as + 1}{s(s+1)} \tag{9.2}$$

The zero is unstable if $a < 0$. It is assumed that $a = -10$ which means that the system has a zero in the right half plane at $s = 0.1$. According to the inequality (2.) the crossover frequency for robust control is limited to 0.1 rad/s.

To carry out a pole-placement design assume that it is desirable to have a closed loop system with the characteristic polynomial

$$(s+\alpha\omega_0)(s^2+\omega_0 s+\omega_0^2) \tag{9.3}$$

with $\alpha = 1$ this corresponds to a Butterworth pattern. The desired closed loop characteristic equation can be obtained with the controller

$$C(s) = \frac{s_0 s + s_1}{s+r} \tag{9.4}$$

where

$$s(s+1)(s+r) + (as+1)(s_0 s + s_1) = (s+\alpha\omega_0)(s^2+\omega_0 s+\omega_0^2)$$

Equating equal powers of s gives a linear equation which has the solution

$$\begin{aligned} r &= \frac{-\alpha a^2\omega_0^3 + (1+\alpha)a\omega_0^2 - (1+\alpha)\omega_0 + 1}{a-1} \\ s_0 &= \frac{\alpha a\omega_0^3 - (1+\alpha)\omega_0^2 + (1+\alpha)\omega_0 - 1}{a-1} \\ s_1 &= \alpha\omega_0^3 \end{aligned} \tag{9.5}$$

To obtain a fast closed loop system we choose $\omega_0 = 10$ and $\alpha = 1$. This gives closed loop poles that are much faster than the achievable crossover frequency $\omega_{gc} = 0.1$. Equation (9.5) then gives the controller parameters $r = 9274.5$, $s_0 = 925.5$ and $s_1 = 1000$. The loop transfer function is

$$L(s) = \frac{(as+1)(s_0 s+s_1)}{s(s+1)(s+r)} = -9255\frac{(s-0.1)(s+1.0805)}{s(s+1)(s+9274)}$$

The process pole at $s = -1$ is almost canceled by the controller zero at $s = -s_1/s_0 = -1.0805$. The Bode diagram of the loop transfer function is shown in Figure 9.1 and the Nyquist curve is shown in Figure 9.2.

The loop transfer function has a low frequency asymptote that intersects the line $\log|L(i\omega)| = 0$ at $\omega = -1/a$, i.e. at the slow unstable zero. The magnitude then becomes close to one and it remains so until the the break point, $\omega = r \approx \alpha a\omega_0^3$, at the controller pole. The phase is also close to $-180°$ over that frequency range which means that the stability margin is very poor. The crossover frequency is 6.58 and the phase margin is $\varphi_m = 0.15°$. The maximum sensitivities are $M_s = 678$ and $M_p = 677$ which also shows that the system is extremely sensitive. The slope of the magnitude curve at crossover is also very small which is another indication of the poor robustness of the system. The example illustrates clearly the danger of using a design method in a routine manner. It also shows that it is not sufficient to only check controllability. For this particular problem there are severe limitations

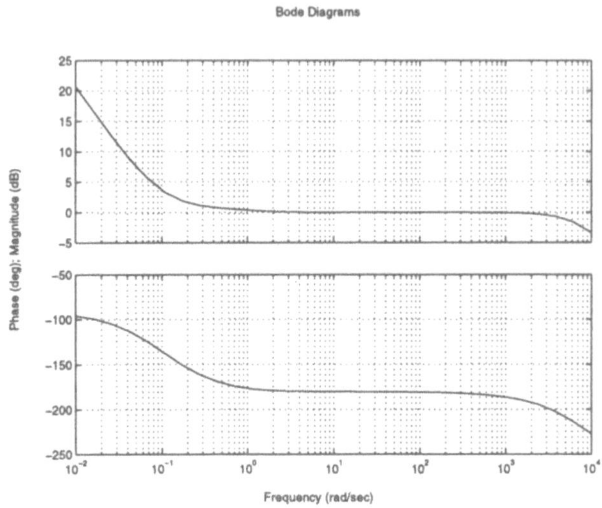

Fig. 9.1. Bode diagram for the loop transfer function of a system with a slow unstable zero at $s = 0.1$, where the specifications are $\omega_0 = 10$ and $\alpha = 1$.

caused by the right half plane zero. Trying to make designs which violate these limitations by making a closed loop system that is too fast we obtain a closed loop system that has very poor stability margins even if the closed loop poles are quite well damped. Also notice that even if the gain crossover frequency is 6.58 rad/s the sensitivity becomes larger than one for $\omega = 0.107$, which is close to the right half plane zero. Feedback is not effective for disturbances having higher frequencies than 0.107, because disturbances will be amplified by the feedback. The example shows that it is important to be aware of the limitations when designing control systems. It is of course possible to obtain sensible control designs for example by choosing smaller values of ω_0.

4. A Slow Stable Zero

Next we will investigate what happens if a system has a slow but stable zero. For this purpose we consider the system with the transfer function (9.2) but we will now assume that a is positive which means that the pole is in the left half plane. To be specific we assume $a = 10$, which means that the system has a zero at $s = -0.1$.

4.1 Bad Choice of Closed Loop Poles

We will first choose closed loop poles with the Butterworth configuration used in Section 3.. The desired closed loop characteristic polynomial is then given by (9.3) with $\omega_0 = 10$ and $\alpha = 1$. It follows from Equation (9.5) that

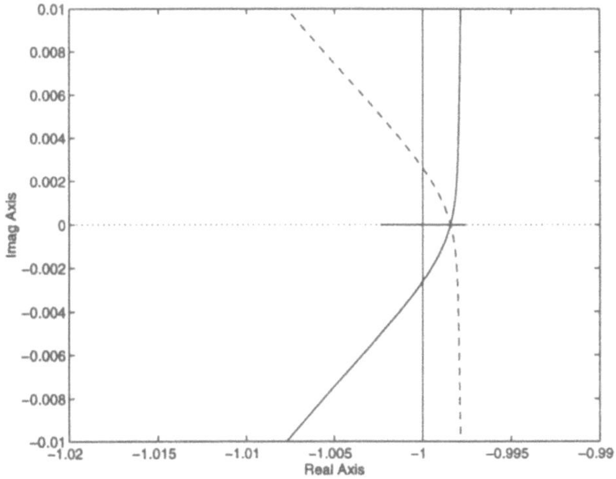

Fig. 9.2. Nyquist curve for the loop transfer function of a system with a slow unstable zero at $s = 0.1$. The specifications for the controller are, $\omega_0 = 10$ and $\alpha = 1$.

the controller parameters are $r = -10891$, $s_0 = 1091$ and $s_1 = 1000$. The controller thus has a very fast unstable pole $p = 10891$ in the right half plane. Since the magnitude of this pole is very large it follows from Bode's integral that the closed loop system is very sensitive. The loop transfer function is.

$$L(s) = \frac{(as+1)(s_0 s + s_1)}{s(s+1)(s+r)} = 1091\frac{(s+0.1)(s+0.9166)}{s(s+1)(s-10891)}$$

The process pole at $s = -1$ is almost canceled by the controller zero at $s = -s_1/s_0 = -0.9166$. The Bode diagram of the loop transfer function is shown in Figure 9.3. The amplitude curve is similar to the one in Figure 9.1. The system has a very small phase margin and it is extremely sensitive. The design is thus very poor. One indication of this is the presence of the unstable mode in the controller. Next we will show that the reason for this is a poor choice of the desired closed loop poles. The fact that a bad choice of closed loop poles can result in unstable controller has been observed before, see e.g. [12] and [9].

4.2 An Improved Design

We will now investigate if the difficulty encountered can be avoided by changing the desired closed loop poles. We choose $\omega_0 = 10$ to have fast dominant poles but we will change the value of α. It follows from Equation (9.5) that the controller coefficients are given by.

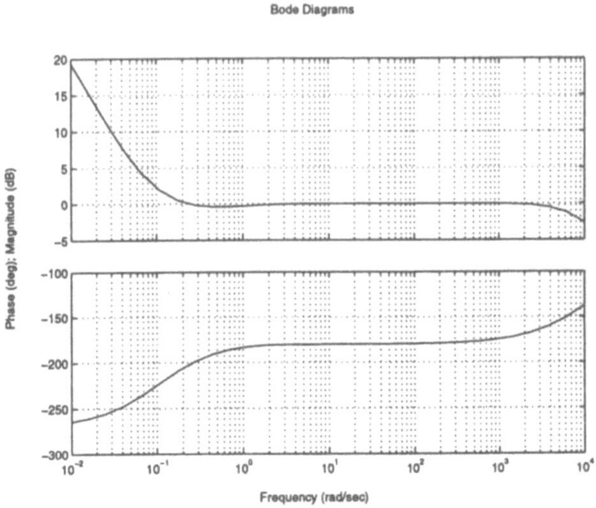

Fig. 9.3. Bode diagram for the loop transfer function of a system with a slow stable zero at $s = -0.1$. The specifications are $\omega_0 = 10$ and $\alpha = 1$.

$$\begin{aligned} r &= \frac{991 - 99010\alpha}{9} \\ s_0 &= \frac{-91 + 9910\alpha}{9} \\ s_1 &= 100\alpha \end{aligned}$$

These equations show that the value of α is critical. To have a stable controller α must be less than 0.01001, and to avoid that the controller has a zero in the right half plane α has to be larger than 0.00918. The range of α that give good controllers is thus quite narrow range. Let us choose $\alpha = 0.01$, ($\alpha = 1/a\omega_0$ in general). This means that the slow zero is canceled by a slow controller pole. The controller parameters are

$$\begin{aligned} r &= \frac{1}{a} \\ s_0 &= \omega_0 - 1 \\ s_1 &= \omega_0^2 \end{aligned}$$

and the loop transfer function becomes

$$L(s) = \frac{(\omega_0 - 1)s + \omega_0^2}{s(s+1)}$$

The loop transfer function has a low frequency asymptote that intersects the line $|L(i\omega)| = 1$ at $s = \omega_0$. There is a pole at $s = 1$ and a zero at $s = -\omega_0^2/(\omega_0 - 1)$. With $\omega_0 = 10$ and $\alpha = 0.01$ the controller parameters become $r = 0.1$, $s_0 = 0.9$ and $s_1 = 10$. The Bode diagram of the system in in Figure 9.4 indicates that the design is quite robust. The phase margin is $\varphi_m = 52°$. The example shows that the selection of closed loop poles has

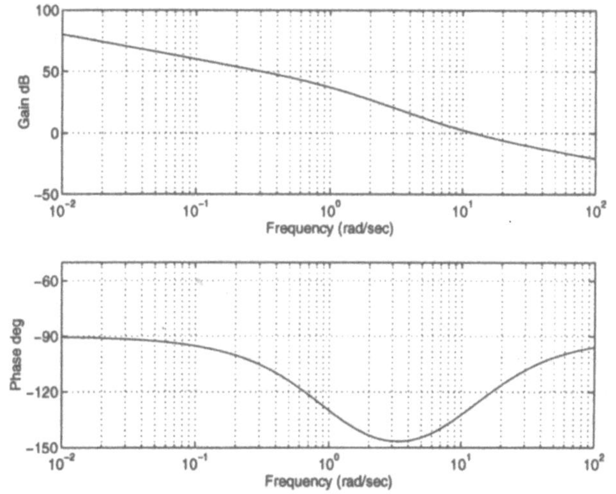

Fig. 9.4. Bode diagram for the loop transfer function of a system with a slow stable zero, at $s = -0.1$. The specifications are, $\omega_0 = 10$ and $\alpha = 0.01$.

a drastic influence on the robustness of the closed loop system obtained. In the particular we found that the closed loop system was very sensitive to parameter variations if the closed loop poles were selected as a Butterworth pattern. A very good design was obtained by choosing one closed loop pole at the slow process zero or close to it. Notice that this is not possible if the slow zero lies in the right half plane.

5. Properties of the Sensitivity Functions

A good insight into the phenomena observed is obtained by investigating the sensitivity function and the complementary sensitivity function. Let the process transfer function be $P = N_p/D_p$, and let the controller transfer function be $C = N_c/D_c$. The sensitivity functions then becomes.

$$S = \frac{D_pD_c}{N_pN_c + D_pD_c} = \frac{D_pD_c}{D_{cl}}$$
$$T = \frac{N_pN_c}{N_pN_c + D_pD_c} = \frac{N_pN_c}{D_{cl}}$$

where D_{cl} denotes the characteristic polynomial of the closed loop system.

First consider the case when the process has a slow zero and the closed loop poles are considerably faster. Consider the Bode diagram of the complementary sensitivity function T. For low frequencies the function has a horizontal asymptote. The first break point occurs at the slow process zero. At that zero the sensitivity function has an asymptote with slope 1 and the complementary sensitivity function increases. If there are more slow zeros the function increases even faster. The magnitude curve starts to decrease for frequencies corresponding to the closed loop poles. It thus follows that the complementary sensitivity can be very large if there is a wide gap between the slowest process zero and the slowest closed loop pole.

Consider the case when the process or the controller has a pole that is faster than the closed loop poles. The sensitivity function has the property $S(\infty) = 1$. At high frequencies the Bode diagram has a horizontal asymptote. If there is a fast process or controller pole the sensitivity will increase until the frequency is corresponds to the closed loop poles when the sensitivity function decreases again. The sensitivity function will thus be very large if the fast process pole is much larger than the fastest closed loop pole.

Consider for example the system in Section 4. The sensitivity functions are

$$S(s) = \frac{s(s+1)(s+9274)}{(s+10)(s^2+10s+100)}$$
$$T(s) = \frac{9255(s-0.1)(s+1.0805)}{(s+10)(s^2+10s+100)}$$

The Bode diagrams of these transfer functions are shown in Figure 9.5. Consider for example the Bode diagram of complementary sensitivity which is unity at low frequencies. The sensitivity starts to increase with slope 1 at the first break point which corresponds to the slow zero at $\omega = 0.1$. The next break point is at $\omega =$ 1sensitivity function then increases with slope 1 until the next break point approximately at $\omega = 1$ where the slope increases to 2. The function has its maximum approximately at $\omega = 10$. An crude estimate based on the asymptotes thus give a peak of about 1000. The actual maximum of the complementary sensitivity is $M_p = 677$.

Next we will consider the sensitivity function. We will explore the magnitude curve starting at high frequencies. The first break point occurs at $\omega = 9274$ then sensitivity function then increases until the next break point which occurs at $\omega = 10$. An crude estimate based on the asymptotes thus give a peak of about 927. The actual maximum of the complementary sensitivity is $M_s = 678$.

6. Conclusions

The fact that a system is controllable and observable does not mean that robust closed loop systems with arbitrary closed loop poles can be obtained.

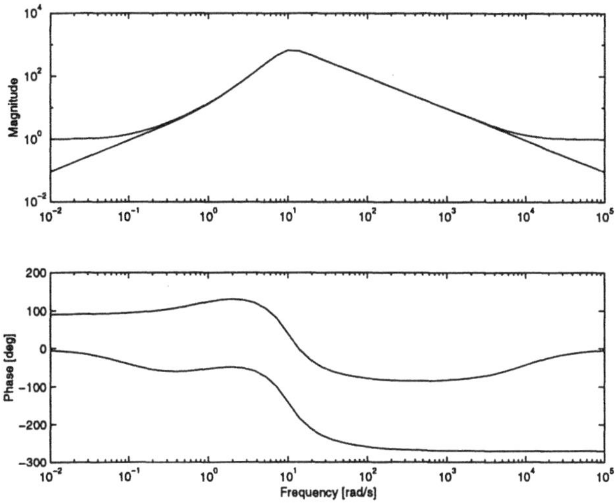

Fig. 9.5. Bode diagrams of the sensitivity function and the complementary sensitivity function.

For a system with poles and zeros in the right half plane there are severe limitations that must be considered to obtain a robust closed loop system. Attempts to design systems which violates these limitations gives closed loop systems that are very sensitive to parameter variations.

We also find that the closed loop poles should be chosen carefully in a pole placement design in order to obtain a robust closed loop system. Based on the insight derived from the examples the following rules can be suggested:

For minimum phase systems which have zeros that are slower than the dominant closed loop poles we should position additional closed loop poles which are slightly faster than the process zeros. If there are open loop poles that are significantly faster than the dominant poles we should position additional closed loop poles that are slightly slower than the fast poles. This means that there are controller poles close to the zeros. In such a case a two-degree of freedom configuration should be chosen so that the slow controller poles are not excited by set point changes.

For non-minimum phase systems the design inequalities should be used to select the dominant closed loop poles. Slow unstable zeros and fast unstable poles impose severe limitations on what can be achieved.

The result is of interest in control system design and particularly for adaptive control where the design is executed automatically.

It is a particular pleasure for me to dedicate this paper to my old friend Ioan Landau on the occasion of his sixtieth birthday and thank him for many of the stimulating discussions we have had on identification, control design and adaptation.

Acknowledgments

The work reported in this paper was supported by the Swedish Research Council for Engineering Science under contract 95-759. This support is gratefully acknowledged.

References

1. B. D. O. Anderson and J. B. Moore. *Optimal Control Linear Quadratic Methods.* Prentice Hall, Englewood Cliffs, N.J., 1990.
2. Karl Johan Åström. Fundamental limitations of control system performance. In A. Paulraj, V.Roychowdhury, and C. D. Schaper, editors, *Communications, Computation, Control and Signal Processing—A Tribute to Thomas Kailath,* pages 355–363. Kluwer, Boston, 1997.
3. Karl Johan Åström and Björn Wittenmark. *Adaptive Control.* Addison-Wesley, Reading, Massachusetts, second edition, 1995.
4. Stephen P. Boyd and Craig H. Barratt. *Linear Controller Design - Limits of Performance.* Prentice Hall Inc., Englewood Cliffs, New Jersey, 1991.
5. Michael Green and D. J. N. Limebeer. *Linear Robust Control.* Prentice Hall, Englewood Cliffs, N.J., 1995.
6. I. M. Horowitz. *Synthesis of Feedback Systems.* Academic Press, New York, 1963.
7. L.H Keel and S. P. Bhattacharyya. Robust, fragile, or optimal. *IEEE Transactions on Automatic Control,* 42(8):1098–1105, August 1997.
8. I. D Landau. Competition, interaction, control. In G. Bastin and M. Gevers, editors, *Plenary Lectures and Minicourses, European Control Conference,* Brussels, Belgium, July 1997.
9. Mats Lilja and Karl Johan Åström. Approximate pole placement approach. *Journal of Guidance, Control, and Dynamics,* 15(5):1082–1086, 1992.
10. J. M. Maciejowski. *Multivariable Feedback Design.* Addison-Wesley, Reading, Massachusetts, 1989.
11. George C. Newton, Jr, Leonard A. Gould, and James F. Kaiser. *Analytical Design of Linear Feedback Controls.* John Wiley & Sons, 1957.
12. A. Wallenborg and Karl Johan Åström. Limit cycle oscillations in high performance robot drives. In *Preprints Control '88,* UK, 1988. University of Oxford.

Stabilization by output feedback revisited

A. Isidori

Department of Systems Science and Mathematics, Washington University, St.Louis, MO 63130, USA
and
Dipartimento di Informatica e Sistemistica, Università di Roma "La Sapienza", 00184 Rome, Italy, e-mail: isidori@giannutri.caspur.it

Summary. [1] Simplicity, elegance and robustness of design methods based on root-locus properties are hardly matched by any other method for feedback design. However, these methods tend to loose their effectiveness in the challenging problem of (robustly) stabilizing, via output feedback, an unstable and non-minimum phase system. In this paper, we illustrate a simple concept leading to a recursive design method for stabilization by output feedback, which inherits most of the benefits, and overcomes a number of drawbacks, of the classical design methods based on root-locus properties. Of course, this does not lead to any specific breakthrough in the case of linear systems, where the solution of a diophantine equation would suffice, but opens a totally new perspective in the problem of robustly stabilizing, via output feedback, a possibly unstable and non-minimum phase nonlinear system.

Keywords: Robust stabilization, output feedback, unstable zero-dynamics

1. Introduction

One of the basic fundamental issues in control theory is the ability to design a feedback law to the purpose of robustly stabilizing a system, in the presence of structured uncertainties, such as parameter variations, and/or unstructured uncertainties, such as unmodeled dynamics.

[1] Work supported in part by NSF under grant ECS-9707891, by AFOSR under grant F49620-95-1-0232, and by MURST.

In the case of finite-dimensional, time-invariant, linear models, a variety of techniques are available, which range form elementary methods based on the properties of the root locus (which lends itself to a very intuitive characterization of the "stability margin" inherent to certain design techniques, such as "small-gain" and/or "high-gain") to more sophisticated methods such as, in the case of parameter uncertainties, those relying upon the analysis of the closed-loop characteristic polynomial (whose uncertain coefficients range over prescribed intervals) or the design of a feedback controller imposing a fixed (parameter independent) Lyapunov function. In the case of unstructured perturbations, a rather common standpoint is to look at the controlled system as to the feedback interconnection of two subsystems, only one of which is accurately modeled, and then to seek a control law able to bring the L_2-gain of the loop below a fixed threshold (see, e.g. [3]). In this way a problem of robust stabilization is cast as a problem of lowering the L_2-gain between a disturbance input and a regulated output of an appropriate controlled subsystem.

The actual setup in which these robust design methods have to be implemented is that in which only a set of variables (and not the full state) is available for feedback. In this case the entire design problem becomes of course more demanding, in view of the need of including some kind of "state observer" in the feedback architecture. As it happens for any design scheme dealing with the selection of specific "measured outputs", the possibility of implementing the above-mentioned methods for feedback design is deeply influenced by the asymptotic properties of the so-called "transmission zeros" of the system in question. As a matter of fact, it is well-known that in a closed-loop involving a system having zeros in the right-half plane (a *non-minimum phase* system) the gain margin is finite and the lowest achievable L_2-gain between a disturbance input and the measured output is bounded away from zero.

In the last few years, similar methods have been gradually developed also for nonlinear systems. Fundamental in this respect were the methods for robust design based on the theory of differential games and the corresponding L_2-gain analysis developed by Basar and Bernhard [2], Van der Schaft [28], Ball *et al.* [1], the notion and the properties of input-to-state stable system developed by Sontag *et al.* [23] [24] [16], the methods of adaptive output feedback controllers studied by Marino and Tomei [19] [20], the methods for nonlinear adaptive control developed by Krstic *et al.* [15], the methods for robust feedback stabilization proposed by Freeman and Kokotovic [5], [6], and by Sepulchre *et al.* [22], the methods for output feedback stabilization developed by A.Teel and L.Praly [25] [26], the methods for stabilization of interconnected uncertain systems proposed by Jiang *et al.* [10] [11].

A common goal of these research efforts was the development of constructive design methods for the synthesis of (state and/or output) feedback controllers yielding asymptotic stability (with prescribed region of attraction). A major step forward in the design of output feedback stabilizers was the recent result by Teel and Praly [25], who have proved that global stabilizability via state feedback and a property of uniform observability (previously introduced by Bornard and Hammouri) imply the possibility of semiglobal stabilization via output feedback. To cope with the restricted information structure, Teel-Praly's stabilization scheme includes an approximate state observer (whose role is actually that of producing approximate estimates of a number "higher order" derivatives of the output) earlier developed by Khalil [14] to cope with a similar (though more restricted) stabilization problem and also very successfully used by Khalil himself [13] in the problem of asymptotic tracking/rejection of given families of commands/disturbances.

In one form or in another, most of the methods indicated above directly apply only to nonlinear systems having a stable zero dynamics (we recall that the *zero dynamics* of a system are the internal dynamics which arise when input and initial conditions are chosen in such a way as to constrain the output to remain identically zero), i.e. to a class of nonlinear systems which is the analogue of the class of linear minimum phase systems. The main reason why this occurs is that most of the methods in question use feedback in order to overcome certain unwanted terms in the dynamics and this in turn tends to render the system poorly observable, with the consequence of enforcing an internal behavior which is very close to that of the zero dynamics. As a result, stabilization may occur only if the latter is asymptotically stable, i.e. if the system is minimum phase. For example, in the stabilization of a system of the form

$$\begin{aligned} \dot{z} &= f_0(z, y, p) \\ \dot{y} &= f_1(z, y, p) + u \end{aligned}$$

robust stability (with respect to the uncertain parameter p) is usually achieved by choosing the control u in such a way as to offset the (possibly unpleasant) presence on the term $f_1(x, y, p)$ in the derivative of a candidate Lyapunov function of the form

$$U(z, y) = V(z) + (y - y^*(z))^2 .$$

If only y is available for feedback, this can always be achieved (in semiglobal terms) whenever $y^*(z) = 0$ and $V(z)$ is a Lyapunov function for the zero dynamics

$$\dot{z} = f_0(z, 0, p) ,$$

by means of a control of the form $u = -ky$, where k is a sufficiently large number. If, on the contrary, the zero dynamics are unstable, some (output-based) estimate of a nontrivial $y^*(z)$ is needed, and this might not be compatible with a control action which, by dominating the term $f_1(x, y, p)$, would render z poorly observable from y.

In this paper, we describe the rationale of a simple design method by which it is possible to robustly stabilize, using output feedback, a rather wide class of nonlinear systems whose *zero dynamics are unstable*, and in which the difficulty pointed out in the previous example is somewhat overcome. Since the underlying basic idea owes its motivation to a natural observation which arises in the problem of stabilizing a (possibly unstable and non-minimum phase) linear system using (root-locus)-type methods, we focus the presentation on the case of linear systems, leaving to a different paper the details of the analysis in the corresponding nonlinear version (see [9]).

2. Recursive output feedback stabilization of linear systems

Consider a single-input single-output linear system with no input-output feed-through. Let n denote the dimension of its state space and let r denote its relative degree. It is well-known that, after a suitable change of coordinates in the state space, such a system can always be put in the following form

$$\begin{aligned} \dot{z} &= F_0 z + G_0 x_1 \\ \dot{x}_1 &= x_2 \\ &\cdots \\ \dot{x}_{r-1} &= x_r \\ \dot{x}_r &= H_0 z + a_1 x_1 + \cdots + a_{r-1} x_{r-1} + a_r x_r + bu \\ y &= x_1 \, , \end{aligned} \tag{10.1}$$

where $z \in \mathbb{R}^{n-r}$.

With this system, we associate an *auxiliary subsystem*, with state

$$x_{\mathrm{a}} = \mathrm{col}(z, x_1, \ldots, x_{r-1})$$

input u_{a} and output y_{a}, defined as follows

$$\begin{aligned} \dot{z} &= F_0 z + G_0 x_1 \\ \dot{x}_1 &= x_2 \\ &\cdots \\ \dot{x}_{r-1} &= u_{\mathrm{a}} \\ y_{\mathrm{a}} &= H_0 z + a_1 x_1 + \cdots + a_{r-1} x_{r-1} \, . \end{aligned} \tag{10.2}$$

Observe that the state space of this system has dimension $n-1$. In what follows, we will show that if the auxiliary system thus defined is stabilizable by output feedback, so is the original system (10.1).

Proposition 2.1. *If the auxiliary system (10.2) is stabilizable by output feedback, then also the original system (10.1) is stabilizable by output feedback.*

The proof of this result is constructive, i.e. it will be shown that from the knowledge of a stabilizer for the auxiliary subsystem (10.2) it is possible to construct a stabilizer for the original system (10.1). Since the former is $n-1$ dimensional, the problem of finding a stabilizer may result, in principle, in a much simpler problem. For instance, suppose the original system (10.1) is a non-minimum phase system (the case in which we are interested the most). Then by definition

$$\dot{z} = F_0 z$$

is unstable and, indeed, (10.2) is an unstable system. However, (10.2) itself might well be a *minimum-phase* system. If this were the case, then a dynamic output feedback stabilizer for (10.2) could easily be determined by means of elementary (root-locus based) arguments consisting of addition of stable zeros, high-gain and far-off stable poles.

Proof of Proposition 2.1. Suppose, for the moment, that x_r, namely the $(r-1)$-th derivative of the output y of (10.1) is available for feedback (this is only a temporary hypothesis that will be removed later) and consider a dynamic stabilizer of the form

$$\begin{aligned} \dot{\eta} &= A\eta + Bx_r \\ u &= \frac{1}{b}[-a_r x_r + C\eta + Dx_r] \,. \end{aligned} \tag{10.3}$$

This yields a closed loop system

$$\begin{aligned} \dot{z} &= F_0 z + G_0 x_1 \\ \dot{x}_1 &= x_2 \\ &\cdots \\ \dot{x}_{r-1} &= x_r \\ \dot{x}_r &= H_0 z + a_1 x_1 + \cdots + a_{r-1} x_{r-1} + C\eta + Dx_r \\ \dot{\eta} &= A\eta + Bx_r \,. \end{aligned}$$

Relabeling the state variable x_r as ζ, it is trivially observed that the controller (10.3) stabilizes the original plant (10.1) if the controller

$$\begin{aligned} \dot{\zeta} &= C\eta + D\zeta + y_a \\ \dot{\eta} &= A\eta + B\zeta \\ u_a &= \zeta \end{aligned} \tag{10.4}$$

stabilizes the *auxiliary subsystem* (10.2) (see Figure 10.1).

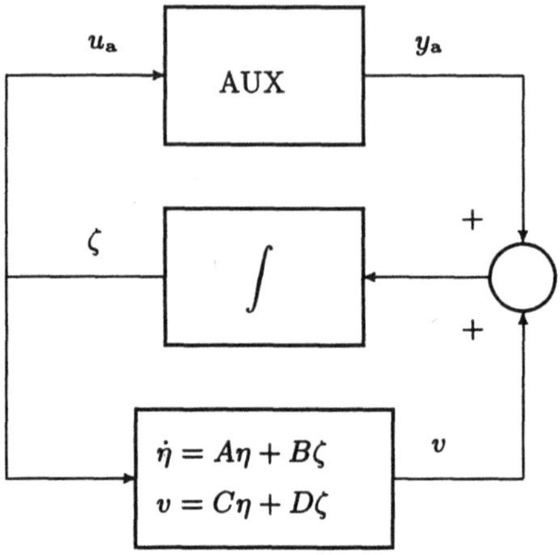

Fig. 10.1.

Note that system (10.4) has a special *structural constraint*: it has relative degree one and its high-frequency gain is equal to 1. Thus, the issue is to determine, for the auxiliary system (10.2), a dynamic output feedback stabilizer having relative degree one and high-frequency gain equal to 1. The following construction explains how this can be achieved.

Let

$$\begin{aligned} \dot\eta &= L\eta + My_a \\ u_a &= N\eta \end{aligned} \tag{10.5}$$

be *any* dynamic output feedback stabilizer for (10.2) and consider the controller

$$\begin{aligned} \dot\eta &= L\eta + M\xi \\ \dot\xi &= -k\xi + ky_a \\ u_a &= N\eta + \tfrac{1}{k}\xi\,, \end{aligned} \tag{10.6}$$

where k is a positive number.

Note that the controller thus defined (namely controller (10.6)) does have, as required, relative degree one and high-frequency gain equal to 1. In order to check that controller (10.6) stabilizes the auxiliary plant (10.2), let the latter be expressed in the form

$$\begin{aligned} \dot x_a &= A_a x_a + B_a u_a \\ y_a &= C_a x_a \end{aligned} \tag{10.7}$$

and consider the corresponding closed-loop system

$$\begin{aligned} \dot x_a &= A_a x_a + B_a[N\eta + \tfrac{1}{k}\xi] \\ \dot\eta &= L\eta + M\xi \\ \dot\xi &= -k\xi + kC_a x_a\,, \end{aligned}$$

which is also depicted in Figure 10.2.

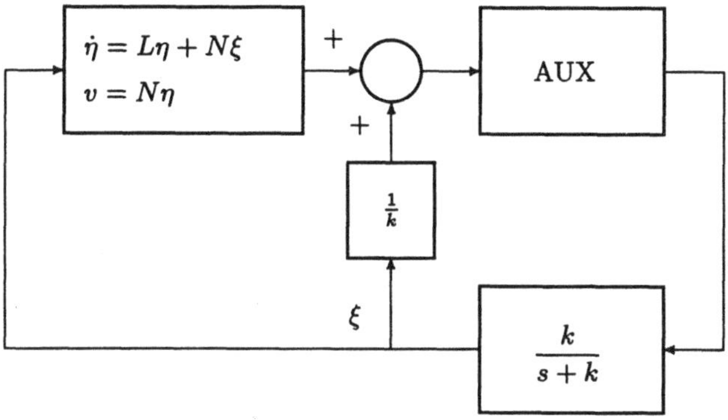

Fig. 10.2.

From the structure of the block-diagram of Figure 2, it is intuitively understood that the stability goal can be achieved, if k is large enough. In fact, for large values of k, the shortcut from ξ to u_a becomes negligible and the system reduces of the feedback interconnection of Figure 10.3a, which in turn differs, from the basic feedback interconnection of Figure 10.3b, stable by hypothesis (recall that (10.5) stabilizes the auxiliary plant (10.2)), only by the insertion of a first order low pass filter whose transfer function

$$\frac{1}{\varepsilon s + 1}$$

has unit gain and small time constant $\varepsilon = \frac{1}{k}$. The latter it is known not to compromise the stability of the interconnection if ε is small enough.

Formally, the proof that system in Figure 2 is stable proceeds as follows. Define

$$w = \xi - C_a x_a$$

to obtain

$$\begin{aligned} \dot{x}_a &= A_a x_a + B_a N\eta + \tfrac{1}{k}(B_a w + B_a C_a x_a) \\ \dot{\eta} &= L\eta + M C_a x_a + M w \\ \dot{w} &= -kw - C_a[A_a x_a + B_a N\eta + \tfrac{1}{k}(B_a w + B_a C_a x_a)] \,. \end{aligned}$$

The latter can be viewed (set, for convenience, $\varepsilon = \frac{1}{k}$) as the feedback interconnection of

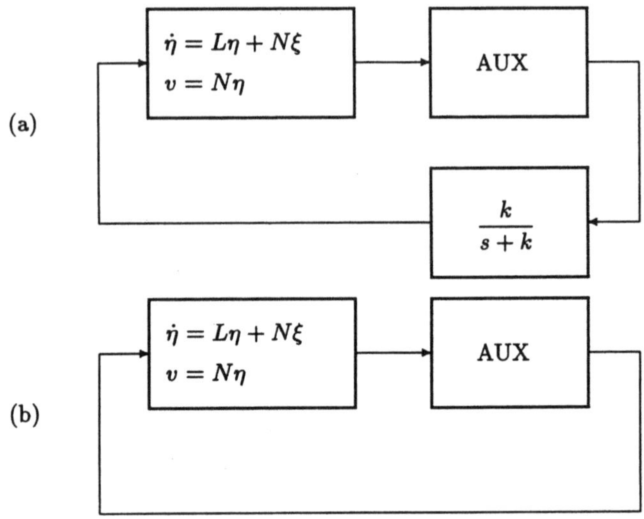

Fig. 10.3.

$$\begin{aligned} \begin{pmatrix} \dot{x}_a \\ \dot{\eta} \end{pmatrix} &= \begin{pmatrix} (A_a + \varepsilon B_a C_a) & B_a N \\ M C_a & L \end{pmatrix} \begin{pmatrix} x_a \\ \eta \end{pmatrix} + \begin{pmatrix} \varepsilon B_a \\ M \end{pmatrix} w \\ v &= \begin{pmatrix} C_a(A_a + \varepsilon B_a C_a) & C_a B_a N \end{pmatrix} \begin{pmatrix} x_a \\ \eta \end{pmatrix} \end{aligned} \tag{10.8}$$

and

$$\dot{w} = -(\frac{1}{\varepsilon} + \varepsilon C_a B_a) w - v \,. \tag{10.9}$$

Noting that system (10.8) is stable by hypothesis at $\varepsilon = 0$ (see Figure 3b), let γ denote its H_∞-gain. By continuity, for any $\delta > 0$, there is an ε^* such that, for all $0 < \varepsilon < \varepsilon^*$, system (10.8) is stable and its H_∞-gain does not exceed $\gamma + \delta$. On the other hand, for small ε, system (10.9) is stable and has an H_∞-gain equal to

$$\gamma(\varepsilon) = \frac{\varepsilon}{1 + \varepsilon^2 C_a B_a}$$

where $\gamma(\varepsilon) \to 0$ as $\varepsilon \to 0$. As a consequence, by the small gain theorem, if ε is small enough the feedback interconnection of (10.8) and (10.9) is stable.

The previous construction shows how to construct, starting from any given dynamic output feedback stabilizer for (10.2), another dynamic output feedback stabilizer having relative degree one and high-frequency gain equal to 1. This is indeed of the form (10.4), as it can be seen via the change of variables

$$\zeta = N\eta + \frac{1}{k}\xi$$

in (10.6), which yields

$$\begin{aligned} \dot{\eta} &= (L - kMN)\eta + kM\zeta \\ \dot{\zeta} &= (NL - kNMN + kN)\eta + k(NM - 1)\zeta + y_{\mathrm{a}} \\ u_{\mathrm{a}} &= \zeta\,. \end{aligned} \tag{10.10}$$

At this point, we need to remove the hypothesis that x_r is available for feedback. To this end, set (as in [14])

$$\dot{\xi} = \begin{pmatrix} \dot{\xi}_1 \\ \dot{\xi}_2 \\ \cdots \\ \dot{\xi}_{r-1} \\ \dot{\xi}_r \end{pmatrix} = \begin{pmatrix} \xi_2 + gc_{r-1}(y - \xi_1) \\ \xi_3 + g^2 c_{r-2}(y - \xi_1) \\ \cdots \\ \xi_r + g^{r-1}c_1(y - \xi_1) \\ g^r c_0(y - \xi_1) \end{pmatrix} := P\xi + Qy \tag{10.11}$$

where $c_0, c_1, \ldots, c_{r-1}$ are the coefficients of an Hurwitz polynomial $p(\lambda) = c_0 + c_1\lambda + \cdots + c_{r-1}\lambda^{r-1} + \lambda^r$, and let ξ_r replace x_r in (10.3). In other words, consider the controller

$$\begin{aligned} \dot{\eta} &= A\eta + B\xi_r \\ \dot{\xi} &= P\xi + Qy \\ u &= \frac{1}{b}[-a_r\xi_r + C\eta + D\xi_r]\,. \end{aligned} \tag{10.12}$$

Define $e_i = g^{r-i}(x_i - \xi_i)$, for $i = 1, 2, \ldots, r$ and observe that the interconnection of (10.1) and (10.3) becomes

$$\begin{aligned} \dot{x}_{\mathrm{a}} &= A_{\mathrm{a}}x_{\mathrm{a}} + B_{\mathrm{a}}x_r \\ \dot{x}_r &= C_{\mathrm{a}}x_{\mathrm{a}} + C\eta + Dx_r - De_r + a_r e_r \\ \dot{\eta} &= A\eta + Bx_r - Be_r \\ \dot{e} &= g\tilde{A}e + \tilde{B}(C_{\mathrm{a}}x_{\mathrm{a}} + C\eta + Dx_r - De_r + a_r e_r)\,. \end{aligned} \tag{10.13}$$

where $\tilde{A}$ is a Hurwitz matrix (independent of g) and $\tilde{B} = \mathrm{col}(0, 0, \ldots, 0, 1)$. This can be viewed as the feedback interconnection of two subsystems. One of these, with input e_r, internal state (x, x_r, η) and output $C_{\mathrm{a}}x + C\eta + Dx_r$ is stable by construction and has an H_∞-gain bounded by a fixed number independent of g, if $g > 1$. The other one, with input $C_{\mathrm{a}}x + C\eta + Dx_r$, internal state e and output e is stable for large g and, as an elementary calculation shows, has an H_∞-gain that decreases to 0 as $g \to \infty$. Thus, again by the small gain theorem, the interconnection is stable for large g. ◁

In summary, the proof of Proposition 1 shows that, starting from the matrices L, M, N which characterize a dynamic output feedback stabilizer

$$\begin{aligned} \dot{\eta} &= L\eta + My_{\rm a} \\ u_{\rm a} &= N\eta \end{aligned}$$

for the auxiliary subsystem (10.2), it is possible to construct a dynamic output feedback stabilizer for the original plant (10.1). The latter has the following expression

$$\begin{aligned} \dot{\eta} &= (L - kMN)\eta + kM\xi_r \\ \dot{\xi} &= P\xi + Qy \\ u &= \frac{1}{b}[-a_r\xi_r + (NL - kNMN + kN)\eta + k(NM - 1)\xi_r] \,, \end{aligned}$$

in which

$$P = \begin{pmatrix} -gc_{r-1} & 1 & 0 & \cdots & 0 \\ -g^2c_{r-2} & 0 & 1 & \cdots & 0 \\ \cdot & \cdot & \cdot & \cdots & \cdot \\ -g^{r-1}c_1 & 0 & 0 & \cdots & 1 \\ -g^rc_0 & 0 & 0 & \cdots & 0 \end{pmatrix}, \qquad Q = \begin{pmatrix} -gc_{r-1} \\ -g^2c_{r-2} \\ \cdot \\ -g^{r-1}c_1 \\ -g^rc_0 \end{pmatrix},$$

where $c_0, c_1, \ldots, c_{r-1}$ are the coefficients of an Hurwitz polynomial, and k, g are large positive numbers.

At this point, the question naturally arises of how general is the approach outlined above to the problem of stabilization by output feedback, or – what is the same – how much one can count on the hypothesis, basic to the whole construction, that the auxiliary subsystem (10.2) is stabilizable by output feedback. In what follows, we answer this question in positive terms, by showing that this hypothesis *must necessarily hold* (modulo a memoryless output feedback) if the original system is stabilizable by output feedback. In other words, there is no loss of generality in approaching the problem of stabilization via output feedback in the manner suggested above, i.e. by recursively "reducing" the problem to that of stabilizing an auxiliary lower-dimensional subsystem.

Proposition 2.2. *Suppose the system*

$$\begin{aligned} \dot{x} &= Ax + Bu \\ y &= Cx \end{aligned} \tag{10.14}$$

is stabilizable by output feedback. Then there exists a number k such that the auxiliary subsystem associated to

$$\begin{aligned} \dot{x} &= (A + kBC)x + Bu \\ y &= Cx \end{aligned} \tag{10.15}$$

is stabilizable by output feedback.

Remark. System (10.15) is obtained from system (10.14) via a memoryless output feedback of the form

$$u \mapsto u + ky$$

and this indeed leaves the property of stabilizability by output feedback unchanged. The auxiliary system associated with (10.14) may fail to be stabilizable by output feedback but, for some choice of k, the auxiliary system associated with (10.15) does have this property. Actually, as the proof of Proposition 2 shows, this is true so long as k does not render

$$\det \begin{pmatrix} F_0 & G_0 \\ H_0 & a_1 + k \end{pmatrix} = 0 \, .$$

Since the latter is a first order equation in k, there is only one value of k for which the auxiliary system associated with (10.15) may fail to be stabilizable by output feedback. ◁

Proof of Proposition 2.2. Suppose system (10.14) is stabilizable by output feedback. Then, trivially, also system (10.15), which differs from system (10.14) by a memoryless output feedback, is stabilizable by output feedback, for any k. Without loss of generality, suppose system (10.14) is put in the form (10.1) and consider the matrix

$$\begin{pmatrix} A + kBC - \lambda I & B \\ C & 0 \end{pmatrix} = \begin{pmatrix} F_0 - \lambda I & G_0 & 0 & \cdots & 0 & 0 \\ 0 & -\lambda & 1 & \cdots & 0 & 0 \\ 0 & 0 & -\lambda & \cdots & 0 & 0 \\ \cdot & \cdot & \cdot & \cdots & \cdot & \cdot \\ 0 & 0 & 0 & \cdots & 1 & 0 \\ H_0 & a_1 + k & a_2 & \cdots & a_r - \lambda & 1 \\ 0 & 1 & 0 & \cdots & 0 & 0 \end{pmatrix} .$$

Since (10.15) is stabilizable by state feedback and detectable, standard use of Hautus's stabilizability and detectability tests shows that

$$\operatorname{rank} \begin{pmatrix} F_0 - \lambda I & G_0 \end{pmatrix} = \operatorname{rank} \begin{pmatrix} F_0 - \lambda I \\ H_0 \end{pmatrix} = n_0$$

for every λ in the closed right-half plane, where n_0 is the dimension of the vector z in (10.1). In other words, (F_0, G_0) is a stabilizable pair and (F_0, H_0) is a detectable pair.

Now, choose k so that

$$\det \begin{pmatrix} F_0 & G_0 \\ H_0 & a_1 + k \end{pmatrix} \neq 0 \, . \qquad (10.16)$$

This is always possible. In fact, the determinant in (10.16), a polynomial of first degree in k, is not identically vanishing because, since (F_0, G_0) is a stabilizable pair and (F_0, H_0) is a detectable pair,

$$\det \begin{pmatrix} F_0 & G_0 \\ H_0 & 0 \end{pmatrix} \neq 0 .$$

We want to show that the auxiliary subsystem associated with (10.15) is stabilizable by state feedback and detectable. To this end, observing that the latter is given by

$$\begin{aligned} \dot{z} &= F_0 z + G_0 x_1 \\ \dot{x}_1 &= x_2 \\ &\cdots \\ \dot{x}_{r-1} &= u_a \\ y_a &= H_0 z + (a_1 + k)x_1 + \cdots + a_{r-1}x_{r-1} , \end{aligned} \tag{10.17}$$

we construct the matrix

$$\begin{pmatrix} F_0 - \lambda I & G_0 & 0 & \cdots & 0 & 0 \\ 0 & -\lambda & 1 & \cdots & 0 & 0 \\ 0 & 0 & -\lambda & \cdots & 0 & 0 \\ \cdot & \cdot & \cdot & \cdots & \cdot & \cdot \\ 0 & 0 & 0 & \cdots & 1 & 0 \\ 0 & 0 & & \cdots & -\lambda & 1 \\ H_0 & a_1 + k & a_2 & \cdots & a_{r-1} & 0 \end{pmatrix} \tag{10.18}$$

and we check that its first $n-1$ rows (respectively, its first $n-1$ columns) are independent for every λ in the closed right-half plane. The independence of the first $n-1$ rows is guaranteed by the structure of (10.18) and by the fact that (F_0, G_0) is a stabilizable pair. The independence of the first $n-1$ columns is guaranteed by the structure of (10.18), by the fact that (F_0, H_0) is a detectable pair and, if $\lambda = 0$, by the fact that k is assumed to satisfy (10.16). ◁

3. Robust output feedback stabilization of nonlinear systems

Consider now a nonlinear system modeled by equations of the form

$$
\begin{aligned}
\dot{z} &= f_0(z, x_1, \ldots, x_{r-1}, x_r, p) \\
\dot{x}_1 &= x_2 \\
&\cdots \\
\dot{x}_{r-1} &= x_r \\
\dot{x}_r &= h_0(z, x_1, \ldots, x_{r-1}, p) + a(x_1, \ldots, x_r) + b(x_1)u \\
y &= x_1 \, ,
\end{aligned}
\tag{10.19}
$$

where $z \in \mathbb{R}^{n-r}$ and p is a (possibly vector-valued) unknown parameter, ranging over a compact set $\mathcal{P}$.

With this system, we associate an *auxiliary plant*

$$
\begin{aligned}
\dot{x}_{\mathrm{a}} &= f_{\mathrm{a}}(x_{\mathrm{a}}, u_{\mathrm{a}}, p) \\
y_{\mathrm{a}} &= h_{\mathrm{a}}(x_{\mathrm{a}}, p)
\end{aligned}
\tag{10.20}
$$

in which

$$
x_{\mathrm{a}} = \begin{pmatrix} z \\ x_1 \\ \cdots \\ x_{r-2} \\ x_{r-1} \end{pmatrix}, \qquad f_{\mathrm{a}}(x_{\mathrm{a}}, u_{\mathrm{a}}, p) = \begin{pmatrix} f_0(z, x_1, \ldots, x_{r-1}, u_{\mathrm{a}}, p) \\ x_2 \\ \cdots \\ x_{r-1} \\ u_{\mathrm{a}} \end{pmatrix},
$$

and

$$
h_{\mathrm{a}}(x_{\mathrm{a}}, p) = h_0(z, x_1, \ldots, x_{r-1}, p) \, .
$$

Motivated by the arguments presented in the previous section, we assume that the auxiliary plant thus defined is semiglobally practically robustly stabilizable by output feedback. More precisely, we assume the following (in the sequel we denote, regardless of the actual value of the integer k, by $\mathcal{B}_R$ the closed cube $\mathcal{B}_R = \{x \in \mathbb{R}^k : |x_i| \leq R, 1 \leq i \leq k\}$).

Assumption A1. *For each (arbitrarily large) number $R > 0$, and each (arbitrarily small) number $\varepsilon > 0$, there exists a dynamic output feedback controller*

$$
\begin{aligned}
\dot{\eta} &= L(\eta) + M y_{\mathrm{a}} \\
u_{\mathrm{a}} &= N(\eta) + J y_{\mathrm{a}}
\end{aligned}
\tag{10.21}
$$

where $N(\eta)$ is such that $\lim_{\eta \to 0} N(\eta) = 0$ uniformly in ε for small ε and J is independent of ε, and a C^1 positive definite proper function $V(x_{\mathrm{a}}, \eta)$ with the following property. Let

$$
\Omega_c = \{(x_{\mathrm{a}}, \eta) : V(x_{\mathrm{a}}, \eta) \leq c\} \, ,
$$

and

$$\dot{V}_{(10.20)-(10.21)}(x_a,\eta) = \\ = \frac{\partial V}{\partial x_a}[f_a(x_a,N(\eta)+Jh_a(x_a,p),p)] + \frac{\partial V}{\partial \eta}[L(\eta)+Mh_a(x_a,p)] \quad (10.22)$$

For some $c>0$, $d>0$ such that $\Omega_d \subset \mathcal{B}_\varepsilon \subset \mathcal{B}_R \subset \Omega_c$, and some class $\mathcal{K}_\infty$ function α, the inequality

$$\dot{V}_{(10.20)-(10.21)}(x_a,\eta) \le -\alpha(V(x_a,\eta)) \quad (10.23)$$

holds, for all $(x_a,\eta)\in \Omega_{c+1}\setminus \Omega_{d/2}$. ◁

Concerning system (10.19), it will also be assumed that, for each $R>0$, there exists $S>0$ such that,

$$|f_0(z,x_1,\ldots,x_{r-1},x_r',p) - f_0(z,x_1,\ldots,x_{r-1},x_r'',p)| \le S|x_r'-x_r''| \\ |a(x_1,\ldots,x_{r-1},x_r') - a(x_1,\ldots,x_{r-1},x_r'')| \le S|x_r'-x_r''|$$

for all $(x_1,\ldots,x_{r-1})\in\mathcal{B}_R$, for all $(x_r',x_r'')\in I\!R^2$ and for all $p\in\mathcal{P}$ (i.e. a global Lipschitz condition *in the argument* x_r).

Consider now a feedback law defined as follows

$$\begin{aligned} \dot\eta &= L(\eta)+M\lambda := \varphi(\eta,\lambda)\\ \dot\lambda &= -k\lambda + k\sigma_\rho(kx_r - kN(\eta) - kJ\lambda) := \psi(\eta,\lambda,x_r)\\ u &= \frac{1}{b(x_1)}\Big[\frac{\partial N}{\partial\eta}[L(\eta)+M\lambda] + J[-k\lambda+k\sigma_\rho(kx_r-kN(\eta)-kJ\lambda)]\\ &\quad + kN(\eta)+kJ\lambda - kx_r - a(x_1,\ldots,x_r)\Big] := \gamma(\eta,\lambda,x_1,\ldots,x_r), \end{aligned} \quad (10.24)$$

where k is a positive number, and $\sigma_\rho(\cdot)$ is a saturation function

$$\sigma_\rho(a) = \begin{cases} a & \text{if } |a|<\rho \\ \dfrac{a}{|a|} & \text{if } |a|\ge\rho\,. \end{cases}$$

This feedback law is able to semiglobally practically stabilize plant (10.19).

Theorem 3.1. *For any $R>0$ and any $\bar\varepsilon>0$, there is a quadruplet $L(\eta), M, N(\eta), J$ and there are values of ρ, k such that, in the feedback interconnection of (10.19) and (10.24), namely in the system*

$$\begin{aligned} \dot x_a &= f_a(x_a,x_r,p)\\ \dot x_r &= h_a(x_a,p) + a(x_1,\ldots,x_r) + b(x_1)\gamma(\eta,\lambda,x_1,\ldots,x_r)\\ \dot\eta &= \varphi(\eta,\lambda)\\ \dot\lambda &= \psi(\eta,\lambda,x_r)\,, \end{aligned} \quad (10.25)$$

any initial condition satisfying $\{x_a(0),x_r(0),\eta(0),\lambda(0)\}\in\mathcal{B}_R$ produces a trajectory that converges, in finite time, to the set $\mathcal{B}_{\bar\varepsilon}$.

The controller (10.24) needs access to the state variables $x_1, \ldots, x_r$. If these are not directly availbale for feedback, it is possible to replace them – using a technique due to Khalil-Esfandiari [14] – with appropriate estimates $\xi_1, \ldots, \xi_r$ provided by a bench of "approximate differentiators", of the form (10.11).

In doing this, another saturation is needed, in order to avoid the occurrence of finite escape times for large values of k. This yields the following controller

$$\begin{aligned}
\dot{\eta} &= L(\eta) + M\lambda \\
\dot{\lambda} &= -k\lambda + k\sigma_\rho(k\xi_r - kN(\eta) - kJ\lambda) \\
u &= \frac{1}{b(x_1)}\Big[\frac{\partial N}{\partial \eta}[L(\eta) + M\lambda] + J[-k\lambda + k\sigma_\rho(k\xi_r - kN(\eta) - kJ\lambda)] \\
&\qquad + \sigma_\mu(kN(\eta) + kJ\lambda - k\xi_r - a(\xi_1, \ldots, \xi_r))\Big] \\
\dot{\xi} &= P\xi + Qy\,,
\end{aligned} \tag{10.26}$$

where P and Q are as in (10.11).

The proof that a controller with this structure is able to yield (if g and μ are large enough) semiglobal practical robust stability is similar to that of Theorem 1.

References

1. J. Ball, J.W. Helton, and M.L. Walker, H_∞ control for nonlinear systems via output feedback, *IEEE Trans. Autom. Control*, **38**: pp. 546–559, 1993
2. T. Basar and P. Bernhard, *H_∞-optimal control and related Minimax design problems*, Birkhauser, 1990.
3. P. Dorato and R.K.Yedavalli, *Recent advances in Robust Control*, IEEE Press, 1990.
4. F. Esfandiari and H.K. Khalil, Output feedback stabilization of fully linearizable systems, *Int. J. Contr.*, **56**: pp. 1007–1037, 1992.
5. R.A. Freeman and P.V. Kokotović, Design of 'softer' robust nonlinear control laws, *Automatica*, **29**: pp. 1425–1437, 1993.
6. R.A. Freeman and P.V. Kokotović, Tracking controllers for systems linear in the unmeasured states, *Automatica*, **32**: pp. 735–746, 1996.
7. J. Imura, T. Sugie, and T. Yoshikawa, Global robust stabilization of nonlinear cascaded systems, *IEEE Trans. Automat. Contr.*, **AC-39**: pp. 1084–1089, 1994.
8. A. Isidori, *Nonlinear Control Systems*, Springer-Verlag, New York, third edition, 1995.
9. A. Isidori, Semiglobal practical stabilization of uncertain non-minimum-phase nonlinear systems via output feedback, *IFAC Symp. NOLCOS 98*, 1998.
10. Z.P. Jiang and I. Mareels, A small gain control method for nonlinear cascaded systems with dynamic uncertainties, *IEEE Trans. Automat. Contr.*, **AC-42**: pp. 292–308, 1997.

11. Z.P.Jiang, I.M.Mareels and Y.Wang, A Lyapunov formulation of the nonlinear small-gain theorem for interconnected ISS systems, *Automatica*, **32**: pp. 1211–1215, 1996.
12. Z.P. Jiang, A. Teel and L. Praly, Small-gain theorem for ISS systems and applications, *Mathematics of Control, Signals and Systems*, **7**: pp. 95–120, 1994.
13. H.Khalil, Robust servomechanism output feedback controllers for feedback linearizable systems, *Automatica*, **30**: pp. 1587–1599, 1994.
14. H.K. Khalil and F. Esfandiari, Semiglobal stabilization of a class of nonlinear systems using output feedback, *IEEE Trans. Automat. Contr.*, **AC-38**: pp. 1412–1415, 1993.
15. M.Krstic, I.Kanellakopoulos, P.Kokotovic, *Nonlinear Adaptive Control Design*, J.Wiley (New York), 1995.
16. Y. Lin, *Lyapunov Function Techniques for Stabilization*, Ph.D. Dissertation, Rutgers University, 1992.
17. Z. Lin and A. Saberi, Robust semi-global stabilization of minimum-phase input-output linearizable systems via partial state and output feedback, *IEEE Trans. Automat. Contr.*, **AC-40**: pp. 1029–1041, 1995.
18. N.A. Mahmoud and H.K. Khalil, Asymptotic regulation of minimum phase nonlinear systems using output feedback, *IEEE Transactions on Automatic Control*, **AC-41**, pp. 1402–1412, 1997.
19. R. Marino, P.Tomei, Global adaptive output feedback control of nonlinear systems, part I: linear parametrization, *IEEE Trans. Autom. Control* **AC-38**: pp. 17–32, 1993.
20. R. Marino, P.Tomei, Global adaptive output feedback control of nonlinear systems, part II: nonlinear parametrization, *IEEE Trans. Autom. Control* **AC-38**: pp. 33–48, 1993.
21. F. Mazenc and L. Praly, Adding integrations, saturated controls and stabilization for feedforward systems, *IEEE Trans. Automat. Contr.*, **AC-41**: pp. 1559–1577, 1996.
22. R.Sepulchre. M.Jankovic, P.V.Kokotovic, *Constructive Nonlinear Control*, Springer Verlag, 1996.
23. E.D.Sontag, On the input-to-state stability property, *European J. Contr.*, **1**: pp. 24–36, 1995.
24. E.Sontag, Y.Wang, On characterizations of the input-to-state stability property, *Syst. Contr. Lett.* **24**: pp. 351–359, 1995.
25. A. Teel and L. Praly, Global stabilizability and observability imply semi-global stabilizability by output feedback, *Systems Contr. Lett.*, **22**: pp. 313–325, 1994.
26. A. Teel and L. Praly, Tools for semiglobal stabilization by partial state and output feedback, *SIAM J. Control & Optimization*, **33**: pp. 1443–1488, 1995.
27. A. Tornambe, Output feedback stabilization of a class of non-minimum phase nonlinear systems, *Systems Contr. Lett.*, 19:193–204, 1992.
28. A.J. Van der Schaft, L_2-gain analysis of nonlinear systems and nonlinear H_∞ control, *IEEE Trans. Autom. Control*, **AC-37**: pp.770–784, 1992.

Regulation of non-minimum phase outputs: a flatness based approach

M. Fliess[1], *H. Sira-Ramírez*[2], *and R. Marquez*[1,2]

[1] Laboratoire des Signaux et Systèmes, C.N.R.S. - Supélec - Université Paris-Sud, Plateau de Moulon, 91192 Gif-sur-Yvette, France, e-mail: fliess,marquez@lss.supelec.fr
[2] Departamento Sistemas de Control, Escuela de Ingeniería de Sistemas, Universidad de Los Andes, Mérida, Venezuela, e-mail: isira@ing.ula.ve

À Ioan Landau,
en hommage amical pour
son soixantième anniversaire.

Summary. [1] We present a flatness based control synthesis for non-minimum phase outputs of linear and nonlinear systems. Three concrete examples are illustrating our approach.

Introduction

We want to regulate a non-minimum phase controllable and observable SISO system, with input u and output y, given by the transfer function $\frac{P(s)}{Q(s)}$, where

[1] Two authors (MF & RM) were partially supported by the P.R.C.-G.D.R. *Automatique* and by the European Commission's Training and Mobility of Researchers (TMR) Contract ERBFMRXT-CT970137. Two authors (HSR & RM) were partially supported by the National Council for Scientific and Technological Research of Venezuela (CONICIT) and by the Programme de Coopération Postgradué of the French Government.

- $P(s), Q(s) \in \mathbf{R}[s]$ are coprime,
- some of the transmission zeros, i.e., some of the zeros of $P(s)$, have non-negative real parts.

By Bézout's theorem there exists $A, B \in \mathbf{R}[s]$ such that $AP + BQ = 1$. Introduce a new system variable z by

$$z = A(\frac{d}{dt})y + B(\frac{d}{dt})u \tag{11.1}$$

The transfer function of the system with input u and output z is $\frac{1}{Q(s)}$. The quantities u, y and z satisfy

$$\begin{array}{rcl} u & = & Q(\frac{d}{dt})z \\ y & = & P(\frac{d}{dt})z \end{array} \tag{11.2}$$

Equations (11.2) tell us that u and y are expressed as linear combinations of z and its derivatives up to some finite order. According to (11.1), z is, conversely, expressed as a linear combination of u and y and their derivatives up to some finite order. Those properties are reminiscent to *flatness* (cf. [19, 22]) and z is called a *flat output*. We will exploit those properties for our non-minimum phase tracking (see [1] and [7] for other flatness based control strategies of linear systems). Over a finite time interval $T_1 \leq t \leq T_2$, a suitable open loop strategy for $u(t)$, $y(t)$ and $z(t)$ may be given by taking polynomial time functions which satisfy (11.1)-(11.2). This polynomial specification is far from being the only possible one. On an infinite time interval $t \geq T$, one could choose finite Fourier series for tracking some periodic reference trajectory. The feedback stabilization around the reference trajectory may be achieved by classic techniques. Robustness could be tackled as in [1]. Many of the well-known performance limitations related to non-minimum phase systems (see, e.g., [40]) are thus bypassed.

The zero-dynamics, which was introduced by Byrnes and Isidori [5] (see, also, [26, 36] and the references therein), allows checking the non-minimum phase character of input-output nonlinear systems. The control of such non-minimum phase systems has already been the subject of a rich literature (see, e.g., [10, 24, 29, 32, 37, 50, 52][2]).

Here we attack the problem from the point of view of flatness, which has already been utilized in [32, 37]. A (*differentially*) *flat* system [19, 22] is equivalent to a controllable linear system via a special type of dynamic feedback, called *endogenous*. It is best understood without making any distinction between the system variables:

1. Every system variable may be expressed as a function of the components of a finite set $z = (z_1, \ldots, z_m)$ and of a finite number of their time-derivatives.

[2] The thesis of van Nieuwstadt [37] contains an interesting bibliographical analysis.

2. Every component of z may be expressed as a function of the system variables and of a finite number of their time-derivatives.
3. The components of z are differentially independent, i.e., they are not related by any differential relation.

The fictitious output z is called a *flat output*[3]. In all known examples it may be chosen with a clear engineering and physical meaning. More formally, the notion of flat output may be introduced for linear systems using the language of modules (see [3, 13, 14, 17, 19]) and for nonlinear systems via differential fields (see [12, 14, 19]). This last approach could have been replaced by the differential geometry of infinite jets and prolongations (see [21, 22, 38, 39] and the references therein).

Three concrete examples are examined and simulations are provided. The first one is an experimental flexible structure given by a transfer function [8, 9]. The second one, which has already been studied by one of the authors [46, 47], is a dc-to-dc power converter, which is static state-feedback linearizable. The last one is a PVTOL aircraft [24] which is not static state-feedback linearizable but flat, (see [32, 33] and [22]). In all those three cases we have given a stabilizing feedback around the reference trajectory. Note that for the power converter this is a passivity based output feedback [48].

ACKNOWLEDGEMENTS. The authors would like to thank Prof. H. Bourlès, Prof. D. Claude and Dr. E. Delaleau for some helpful comments.

1. Linear systems

1.1 Modules

Let k be a field and $k[\frac{d}{dt}]$ be the commutative principal ideal ring of polynomials of the form $\sum_{finite} a_\alpha \frac{d^\alpha}{dt^\alpha}$, $a_\alpha \in k$. Let M be a $k[\frac{d}{dt}]$-module. An element $m \in M$ is said to be *torsion* if, and only if, there exists a polynomial $\pi \in k[\frac{d}{dt}]$, $\pi \neq 0$, such that $\pi m = 0$. The set tM of all torsion elements of M is a submodule of M; it is said to be *trivial* if, and only if, $tM = \{0\}$. A $k[\frac{d}{dt}]$-module is said to be *torsion* if, and only if, all its elements are torsion; it is said to be *torsion-free* if, and only if, tM is trivial.

A finitely generated $k[\frac{d}{dt}]$-module M is said to be *free* if, and only if, there exists a *basis*, i.e., a finite set $b = (b_1, \ldots, b_m)$ such that

– any element of M depends $k[\frac{d}{dt}]$-linearly on b,
– the components of b are $k[\frac{d}{dt}]$-linearly independent.

The *rank* of this free module is m.

Here are some standard properties of finitely generated modules over principal ideal rings (see, e.g., [30]).

[3] If m independent channels have been distinguished, this last statement is equivalent to saying that z possesses m components.

1. A finitely generated $k[\frac{d}{dt}]$-module M may be written
$$M \simeq tM \oplus \mathcal{F} \tag{11.3}$$
where tM is the torsion submodule and $\mathcal{F} \simeq M/tM$ is a free module. The *rank* of M is, by definition, the rank of $\mathcal{F}$.
2. For a finitely generated $k[\frac{d}{dt}]$-module M, the following two properties are equivalent:
   - M is torsion,
   - the dimension $\dim_k M$ of M as a k-vector space is finite.
3. Any submodule of a finitely generated (free) $k[\frac{d}{dt}]$-module is a finitely generated (free) $k[\frac{d}{dt}]$-module. Any quotient module of a finitely generated $k[\frac{d}{dt}]$-module is a finitely generated $k[\frac{d}{dt}]$-module.
4. For a finitely generated $k[\frac{d}{dt}]$-module M, the following two properties are equivalent:
   - M is torsion-free,
   - M is free.

Remark 1.1. All modules considered in the sequel will be finitely generated $k[\frac{d}{dt}]$-modules.

Let M be a module. The derivation $\frac{d}{dt}$ defines an endomorphism of the torsion submodule tM, which may be viewed as a k-linear endomorphism τ of the finite-dimensional k-vector space tM. A *Smith zero* [3] of M is an eigenvalue of τ over the algebraic closure $\overline{k}$ of k.

Notation. Write $[S]$ the submodule spanned by a subset S of M.

1.2 Systems

A *k-linear system* Λ is a module. A *k-linear dynamics* is a k-linear system Λ with an *input*, i.e., with a finite subset $u = (u_1, \ldots, u_m)$ such that the quotient module $\Lambda/[u]$ is torsion. The input u is assumed to be free, i.e., the submodule $[u]$ is free of rank m. Then, the rank of Λ is equal to m. A *k-linear input-output system* is a k-linear dynamics Λ with an *output*, i.e., with a finite subset $y = (y_1, \ldots, y_p)$ of Λ.

There exists a short exact sequence
$$0 \to \mathcal{N} \to \mathcal{F} \to \Lambda \to 0$$
The module $\mathcal{F}$ is free. The free module $\mathcal{N}$, which is called sometimes the *module of relations*, should be viewed as a *system of equations* defining Λ.

Example 1.1. Consider the system of equations

$$\sum_{\kappa=1}^{\mu} a_{\iota\kappa} w_\kappa = 0 \tag{11.4}$$

where $a_{\iota\kappa} \in k[\frac{d}{dt}]$, $\iota = 1, \ldots, \nu$. The unknowns are $w_1, \ldots, w_\mu$. Let $\mathcal{F}$ be the free module spanned by $f_1, \ldots, f_\mu$. Let $\mathcal{N} \subseteq \mathcal{F}$ be the submodule spanned by $\sum_{\kappa=1}^{\mu} a_{\iota\kappa} f_\kappa$. The module corresponding to (11.4) is $\mathcal{F}/\mathcal{N}$.

1.3 Controllability and observability

A k-linear system Λ is called *controllable* if, and only if, the module Λ is free[4]. Any basis of Λ, which may be viewed as a fictitious output, is called a *flat*, or *basic, output*.

Example 1.2. Consider the classic state-variable representation

$$\frac{d}{dt}\begin{pmatrix} x_1 \\ \vdots \\ x_n \end{pmatrix} = A \begin{pmatrix} x_1 \\ \vdots \\ x_n \end{pmatrix} + B \begin{pmatrix} u_1 \\ \vdots \\ u_m \end{pmatrix} \tag{11.5}$$

where $A \in k^{n\times n}$, $B \in k^{n\times m}$. The control variables $u = (u_1, \ldots, u_m)$ are assumed to be independent. It follows from [13] that (11.5) is controllable, i.e., that $\mathrm{rk}(B, AB, \ldots, A^{n-1}B) = n$, if, and only if, the corresponding module Λ is free. As a matter of fact, the torsion submodule $t\Lambda$ in the decomposition (11.3) corresponds to the Kalman uncontrollable subspace.

Example 1.3. Assume that (11.5) is controllable. There exists a static state feedback which transforms it into the famous *Brunovský canonical form* (see, e.g., [27]; see also [16, 18] for a module-theoretic derivation which comprises the time-varying case) which reads $z^{(\nu_i)} = v_i$, $i = 1, \ldots, m$, where the v_i's are the new control variables and the ν_i's the *controllability*, or *Kronecker, indices*.

The next property is clear.

Proposition 1.1. *The set $z = (z_1, \ldots, z_m)$ is a flat output.*

Example 1.4. Consider the input-output system

$$\mathcal{A}\begin{pmatrix} y_1 \\ \vdots \\ y_p \end{pmatrix} = \mathcal{B}\begin{pmatrix} u_1 \\ \vdots \\ u_m \end{pmatrix} \tag{11.6}$$

where $\mathcal{A} \in k[\frac{d}{dt}]^{p\times p}$, $\det \mathcal{A} \neq 0$, $\mathcal{B} \in k[\frac{d}{dt}]^{p\times m}$. It is known that (11.6) is controllable if, and only if, $\mathcal{A}$ and $\mathcal{B}$ are left coprime (see, e.g., [25] and [3, 17, 18]). Denote by Λ its corresponding module.

[4] See [15] for the connection with Willems' behavioral approach [51].

Proposition 1.2. *The output $y = (y_1, \ldots, y_p)$ is flat if, and only if, the following two conditions are satisfied:*
1) the matrices $\mathcal{A}$ and $\mathcal{B}$ are left coprime;
2) the system is square, i.e., $m = p$, and the matrix $\mathcal{B}$ is unimodular.

Proof. The first condition ensures the controllability of (11.6) and is equivalent to the freeness of Λ. The second condition is equivalent to $u_i \in [y]$, $i = 1, \ldots, m$.

A k-linear system Λ with input $u = (u_1, \ldots, u_m)$ and output $y = (y_1, \ldots, y_p)$ is called *observable* if, and only if, $[u, y] = \Lambda$ (see [13], where it is shown that this intrinsic definition coincides with the classic one for systems given by the Kalman state-variable representation).

1.4 Equivalence and endogenous feedback

Two systems Λ_ι, $\iota = 1, 2$, are said to be *equivalent* if, and only if, the modules Λ_1 and Λ_2 are isomorphic. In other words any variable of one system may be expressed as a $k[\frac{d}{dt}]$-linear combination of the variables of the other: those expressions define an *endogenous feedback* (compare with [19, 22]).

Example 1.5. The controllable dynamics $x_\iota^{(n_\iota)} = u_\iota$ corresponds to the rank 1 free module Λ_ι with basis x_ι, $\iota = 1, 2$. The isomorphism $\Lambda_1 \to \Lambda_2$, $x_1 \longmapsto x_2$, shows the equivalence of the two dynamics. Note that, if $n_1 \neq n_2$, u_2 is not the image of u_1 under this isomorphism.

1.5 Transmission zeros

Take a system Λ with input $u = (u_1, \ldots, u_m)$ and output $y = (y_1, \ldots, y_p)$. Let $\varpi : \Lambda \to \Lambda/t\Lambda$ be the canonical epimorphism (the free module $\Lambda/t\Lambda$ is the *transfer module* [45]). Set $\overline{u} = (\overline{u}_1, \ldots, \overline{u}_m)$, $\overline{y} = (\overline{y}_1, \ldots, \overline{y}_p)$, where $\overline{u}_i = \varpi u_i$, $i = 1, \ldots, m$, $\overline{y}_j = \varpi y_j$, $j = 1, \ldots, p$.

Lemma 1.1. *The restriction of ι to $[u]$ induces an isomorphism $[u] \to [\overline{u}]$.*

Proof. It follows at once from $[u] \cap t\Lambda = \{0\}$ and $\ker \varpi = t\Lambda$.

Let $\mathcal{T}$ be the quotient module $[\overline{u}, \overline{y}]/[\overline{y}]$. The *transmission zeros* of Λ (see [3]) are the Smith zeros of $\mathcal{T}$.

When k is a subfield of the field $\mathbf{C}$ of complex numbers, the system Λ is said to be *minimum phase* if, and only if, the real parts of the transmission zeros are strictly negative.

Remark 1.2. Assume that the input-output system Λ is controllable, i.e., that the module Λ is free. Then $\mathcal{T} = [u, y]/[y]$.

Example 1.6. Take again (11.6), which is assumed to be controllable, i.e., $\mathcal{A}$ and $\mathcal{B}$ are left coprime. The transmission zeros of (11.6) are the zeros of the invariant factors of $\mathcal{B}(s)$, where s stands for $\frac{d}{dt}$.

The notion of equivalence is independent of any distinction between the system variables. We now introduce an equivalence for input-output systems, which preserves the output. Consider a system Λ with input u and output y and another system Σ with the same output y and a new input v. These two systems are said to be *equivalent* (*by endogenous feedback*) if, and only if, the following conditions are satisfied:

1. there exists an isomorphism $\alpha : \Lambda \to \Sigma$,
2. the restriction of α to $[u, y]$ defines an isomorphism between $[u, y]$ and $[v, y]$.

The next result follows at once.

Proposition 1.3. *The transmission zeros are invariant under equivalence.*

Remark 1.3. Two input-output systems which are equivalent by static state feedback are not necessarily equivalent in the above sense as static state feedback does not necessarily preserves observability. This is why transmission zeros are not preserved by static state feedback equivalence (see, e.g., [11]).

1.6 Controllable and observable SISO systems

To a controllable and observable system Λ with a single input u and a single output y corresponds a rank 1 free module Λ such that $\Lambda = [u, y]$. Let z be a basis of Λ (any other basis is of the form αz, $\alpha \in k, \alpha \neq 0$). Set $u = Qz, y = Pz; P, Q \in k[\frac{d}{dt}]$.

Proposition 1.4. *The polynomials P and Q are coprime. The transmission zeros of Λ are the zeros of $P(s)$ over $\overline{k}$, where the variable s stands for $\frac{d}{dt}$.*

Proof. Bézout's theorem shows that $[u, y] = [z]$ $(= \Lambda)$ if, and only if, P and Q are coprime. From $Qy = Pu$ and from the property of the transmission zeros of (11.6), we see that the transmission zeros of Λ are the zeros of $P(s)$ over $\overline{k}$.

The transfer function of the input-output system Λ is $\frac{P(s)}{Q(s)}$ (see [17] for a module-theoretic approach to the Laplace transform and to transfer matrices, which comprises the time-varying case). The next property is clear.

Proposition 1.5. *The system output y is a flat output if, and only if, the numerator of the transfer function is a non-zero constant.*

2. Application to a linear SISO system

2.1 Description of the System

The following transfer function [8, 9], which has been obtained by identification,

$$\frac{0.27s^2 - 0.1187s + 6.1041}{s^4 + 0.2765s^3 + 6.1041s^2}$$

describes an experimental flexible structure used for studying elastic vibrations. It consists of two freely rotating disks connected by a thin shaft as shown in Figure 11.1.

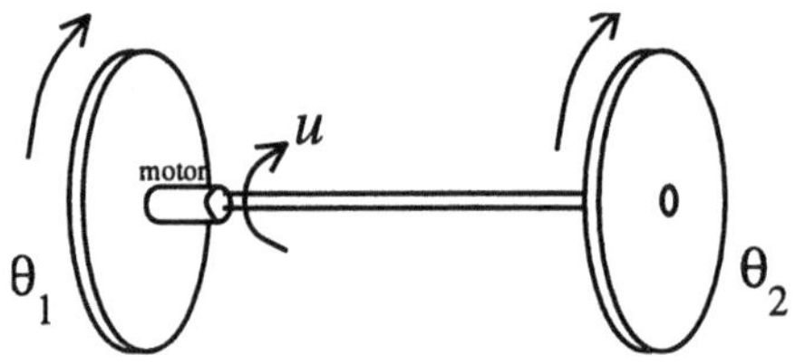

Fig. 11.1. Flexible system example: Schematic diagram

The control is the torque u provided by a dc motor. The output $y = \theta_2$ is the farthest away angular rotation, which is measured by potentiometers. The transmission zeros $0.2198 \pm 4.7497\sqrt{-1}$ are unstable and y is non-minimum phase.

2.2 Open loop strategy

According to (11.1) and (11.2), the flat output z satisfies the following relations

$$\begin{aligned} y &= \left[a_1 \frac{d^2}{dt^2} + a_2 \frac{d}{dt} + a_3\right] z \\ u &= \left[\frac{d^4}{dt^4} + b_1 \frac{d^3}{dt^3} + b_2 \frac{d^2}{dt^2}\right] z \\ z &= \left[c_1 \frac{d}{dt} + c_2\right] u + \left[c_3 \frac{d^3}{dt^3} c_4 \frac{d^2}{dt^2} + c_5 \frac{d}{dt} + c_6\right] y \end{aligned}$$

(where $a_1 = 0.27, a_2 = -0.1187, a_3 = 6.1041, b_1 = 0.2765, b_2 = 6.1041, c_1 = 0.0001635, c_2 = 0.002497, c_3 = -0.0006058, c_4 = -0.00968, c_5 = 0.00319, c_6 = 0.16382$)

We are driving y from an operating point $y(T_1) = \overline{y}_1$ to another one $y(T_2) = \overline{y}_2$, for $T_1 < T_2$. Set $z_d(T_i) = \bar{y}_i/a_3$, $i = 1, 2$. The time function $z_d(t)$, $T_1 \leq t \leq T_2$ is defined as a suitable polynomial in t, of degree 9. It yields functions $y(t)$, $u(t)$, $T_1 \leq t \leq T_2$, which are, at least, continuous. The behavior of y, u and z is illustrated in Figure 11.2.

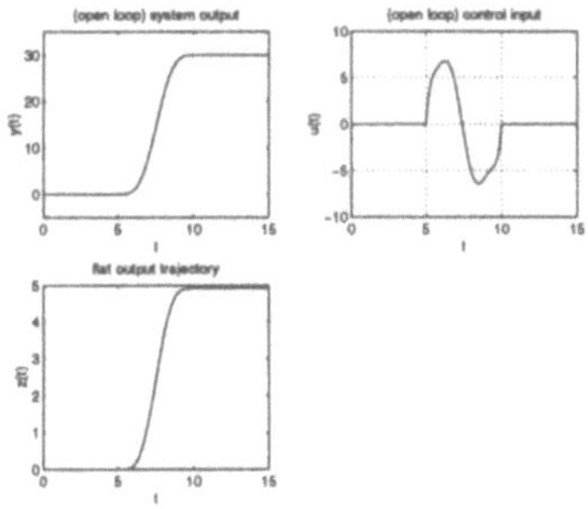

Fig. 11.2. Open loop control

2.3 Time scaling

Apply, as already often done for flat systems (see, e.g., [19, 20] and [1]), a time-scaling which is here linear, i.e., $t = \frac{\tau}{\alpha}$, where $\alpha > 0$ is a constant. This results in

$$\begin{aligned} u &= \left[\alpha^4 \frac{d^4}{d\tau^4} + \alpha^3 b_1 \frac{d^3}{d\tau^3} + \alpha^2 b_2 \frac{d^2}{d\tau^2}\right] z \\ y &= \left[\alpha^2 a_1 \frac{d^2}{d\tau^2} + \alpha a_2 \frac{d}{d\tau} + a_3\right] z \end{aligned}$$

For $0 < \alpha < 1$, we can reduce the magnitude of u if necessary (compare with [8, 9]). Figure 11.3 shows y, u and z for $\alpha = 0.5$.

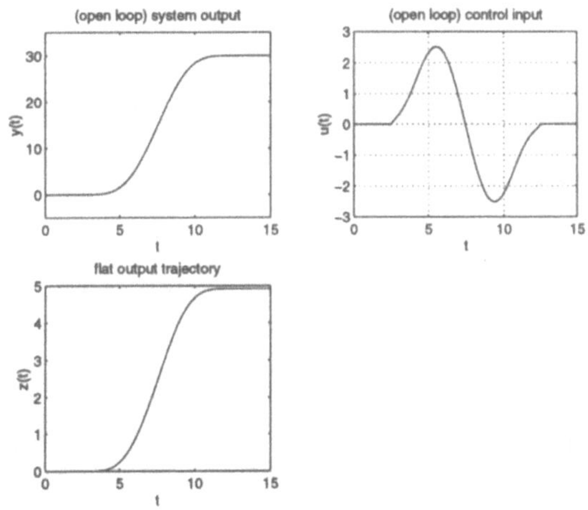

Fig. 11.3. Open loop control under time scaling of the reference trajectory

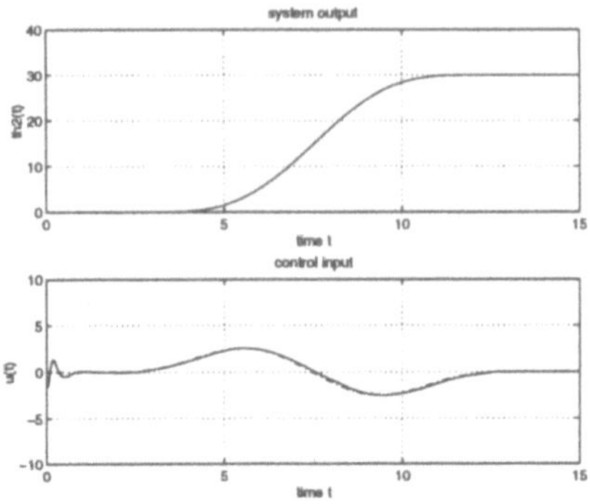

Fig. 11.4. Close loop control

2.4 Feedback stabilization

In Figure 11.4 we present simulations of a closed-loop control strategy using a standard observer-controller scheme for a controllable and observable state space representation:

$$\begin{aligned}
\dot{x}_1 &= x_2 \\
\dot{x}_2 &= x_3 \\
\dot{x}_3 &= x_4 \\
\dot{x}_4 &= -b_2 x_3 - b_1 x_4 \\
y &= a_3 x_1 + a_2 x_2 + a_1 x_3
\end{aligned}$$

3. Nonlinear systems

3.1 Differential fields

All fields have characteristic zero. An (*ordinary*) *differential field* K (see, e.g. [4, 28])[5] is a field which is equipped with a single derivation $\frac{d}{dt} =$ "$\cdot$" such that

$$\forall a \in K, \ \frac{da}{dt} = \dot{a} \in K$$
$$\forall a, b \in K, \ \frac{d}{dt}(a+b) = \dot{a} + \dot{b}, \ \frac{d}{dt}(ab) = \dot{a}b + a\dot{b}$$

A differential field extension L/K is given by two differential fields K, L such that $K \subseteq L$. An element in L is *differentially algebraic* over K if, and only if, it satisfies an algebraic differential equation with coefficients in K. An element in L is *differentially transcendental* over K if, and only if, it is not differentially algebraic over K. The extension L/K is differentially algebraic if, and only if, any element of L is differentially algebraic over K; L/K is *differentially transcendental* if, and only if, it is not differentially algebraic.

[5] See [14, 19] for a detailed review.

A set $\xi = \{\xi_i \mid i \in I\}$ of elements in L is said to be *differentially algebraically (in)dependent* over K if, and only if, the set of derivatives of arbitrary orders $\{\xi_i^{(\nu_i)} \mid i \in I, \nu_i = 0, 1, 2, \dots\}$ is algebraically (in)dependent over K. An independent set which is maximal with respect to inclusion is called a *differential transcendence basis* of L/K. Two such bases have the same cardinality which is the *differential transcendence degree* of L/K; it is written diff tr d^0 L/K.

Notation. The differential subfield of L generated by K and the set ξ is denoted $K\langle\xi\rangle$.

A differentially transcendental field extension L/K is said to be *pure* if, and only if, there exists a differential transcendence basis ξ such that $L = K\langle\xi\rangle$.

Remark 3.1. From now on all differential field extensions will be finitely generated.

3.2 Systems

Let k be a given differential ground field. A k-system is a differential field extension K/k [12, 14, 19]. A k-dynamics is a k-system K/k equipped with an *input*, i.e., with a finite set $u = (u_1, \dots, u_m)$ such that $K/k\langle u\rangle$ is differentially algebraic. From now on the input is assumed to be *independent*, i.e., $k\langle u\rangle/k$ is a pure differentially transcendental extension and diff tr d^0 $k\langle u\rangle = m$. An *output* is a finite set $y = (y_1, \dots, y_p)$ of elements in K/k.

3.3 Equivalence, endogenous feedback, differential flatness

Two systems K_ι/k, $\iota = 1, 2$, are *k-equivalent* [19] if, and only if, there exist differential fields M_ι such that M_ι/K_ι is algebraic and the two differential extensions M_1/k and M_2/k are differentially k-isomorphic. In other words, any variable of one of the systems may be expressed as an algebraic function of the variables of the other one and of their derivatives up to some finite order. Those expressions define an *endogenous feedback* between the two systems [19].

A system K/k is (*differentially*) *flat* [19] if, and only if, K/k is k-equivalent to a pure differentially transcendental extension of k. A differential transcendence basis $z = (z_1, \dots, z_m)$ of K/k such that $K/k\langle z\rangle$ is (non-differentially) algebraic, which may be viewed as a fictitious output, is called a *flat*, or *linearizing*, *output*. Recall that a flat system is equivalent to a controllable linear system.

3.4 Residual dynamics

Consider a system K/k with input u and output y. Its *residual dynamics* is the differential field extension $K/k\langle y\rangle$; its *input residual dynamics* is $k\langle u, y\rangle/k\langle y\rangle$

(compare with the notion of zero dynamics [26, 36]). When the system K/k is flat, it might also be useful to consider the extension $k\langle u, z\rangle/k\langle z\rangle$, where z is a flat output.

Take two systems K/k and L/k with respective inputs u and v but with the output y. They are said to be *k-equivalent* (*by endogenous feedback*) if, and only if, the next two conditions are satisfied:

1. There exists a differential k-isomorphism $\alpha : \mathcal{M}/k \to \mathcal{N}/k$, where the differential fields $\mathcal{M}$ and $\mathcal{N}$ are algebraic extensions of K and L.
2. There exists differential fields M and N, which are finite algebraic extensions of $k\langle u, y\rangle$ and $k\langle v, y\rangle$, such that the restriction of α to M induces a differential k-isomorphism between M/k and N/k.

The next result is clear.

Proposition 3.1. *Take two k-equivalent input-output systems with the same output y. Then their (input) residual dynamics are $k\langle y\rangle$-equivalent.*

4. A dc-to-dc power converter

4.1 Description and flatness

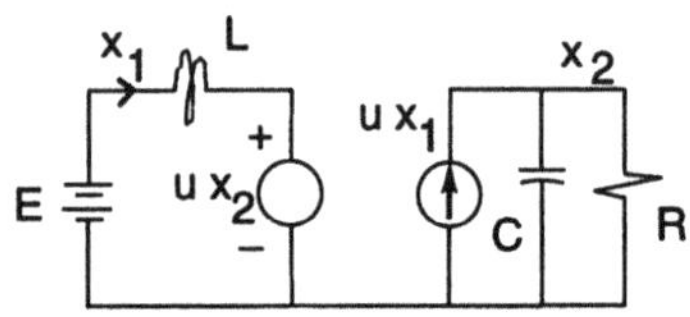

Fig. 11.5. "Boost" converter circuit

The *average* behavior of a pulse-width-modulated "boost" converter circuit (see Figure 11.5) may be, according to [35], defined by

$$\begin{aligned} \dot{x}_1 &= -\frac{1}{L}ux_2 + \frac{E}{L} \\ \dot{x}_2 &= \frac{1}{C}ux_1 - \frac{1}{RC}x_2 \end{aligned} \tag{11.7}$$

where x_1 is the average inductor current, x_2 is the average output capacitor voltage and u is the *duty ratio function* taking values in the closed interval $[0, 1]$ and acting as an input variable. It is known [47] that the output x_1 is minimum phase, while x_2 is not.

System (11.7) is static state feedback linearizable [46] and, therefore, differentially flat (see [19, 22]). A flat output y is given by the average stored energy

$$y = \frac{1}{2}Lx_1^2 + \frac{1}{2}Cx_2^2$$

From $x_1, x_2 > 0$, we obtain

$$\begin{aligned}
x_1 &= -\frac{ERC}{2L} + \left[\left(\frac{ERC}{2L}\right)^2 + \left(\frac{2y + RC\dot{y}}{L}\right)\right]^{\frac{1}{2}} \\
x_2 &= \left\{-R\dot{y} - \frac{R^2CE^2}{2L} + ER\left[\left(\frac{ERC}{2L}\right)^2 + \left(\frac{2y + RC\dot{y}}{L}\right)\right]^{\frac{1}{2}}\right\}^{\frac{1}{2}} \\
u &= \frac{LRC}{[2Lx_1(y,\dot{y}) + ERC]\, x_2(y,\dot{y})}\left[\frac{E^2}{L} + \frac{2}{R^2C}x_2^2 - \ddot{y}\right] \qquad (11.8)
\end{aligned}$$

4.2 Trajectory planning

It is desired to transfer the non-minimum phase output x_2 from a given equilibrium value $x_2(T_1) = V_{d_1}$ towards a second equilibrium value $x_2(T_2) = V_{d_2}, T_2 > T_1$. Those two equilibrium values yield, from (11.8), constant values $\overline{y}_1$, $\overline{y}_2$ for y. An open loop trajectory $y^*(t)$ for the flat output y may be specified in terms of a suitable polynomial

$$y^*(t) = \overline{y}_1 + \left[21\left(\frac{t-T_1}{T_2-T_1}\right)^5 - 35\left(\frac{t-T_1}{T_2-T_1}\right)^6 + 15\left(\frac{t-T_1}{T_2-T_1}\right)^7\right](\overline{y}_2 - \overline{y}_1)$$

where $y^*(T_i) = \overline{y}_i$, $i = 1, 2$, $y^{*(\nu_1)}(T_1) = 0$, $\nu_1 = 1, \ldots, 4$, $y^{*(\nu_2)}(T_2) = 0$, $\nu_2 = 1, 2$.

4.3 Stabilizing feedback

One proceeds to impose on the flat output tracking error $e(t) = y(t) - y^*(t)$ the following asymptotically stable behavior $\ddot{e}(t) + 2\xi\omega_n e(t) + \omega_n^2 e(t) = 0$. It yields

$$\ddot{y} = \ddot{y}^*(t) - 2\xi\omega_n\left(\dot{y} - \dot{y}^*(t)\right) - \omega_n^2\left(y - y^*(t)\right)$$

A stabilizing feedback loop around the reference trajectory is thus obtained by utilizing (11.8):

$$\begin{aligned}
u &= \frac{LRC}{[2Lx_1 + ERC]\, x_2}\left[\frac{E^2}{L} + \frac{2}{R^2C}x_2^2 - \ddot{y}^*(t) + 2\xi\omega_n\left(Ex_1 - \frac{x_2^2}{R} - \dot{y}^*(t)\right)\right. \\
&\quad \left. + \omega_n^2\left(\frac{1}{2}Lx_1^2 + \frac{1}{2}Cx_2^2 - y^*(t)\right)\right] \qquad (11.9)
\end{aligned}$$

Taking as an output $\zeta = x_1$, we propose the following observer for the output feedback implementation of the control law (11.9).

$$\begin{array}{rcl} \dot{\hat{x}}_1 & = & -\frac{1}{L}u\hat{x}_2 + \frac{E}{L} + \frac{R_1}{L}(\zeta - \hat{\zeta}) \\ \dot{\hat{x}}_2 & = & \frac{1}{C}u\hat{x}_1 - \frac{1}{RC}\hat{x}_2 \\ \hat{\zeta} & = & \hat{x}_1 \end{array}$$

with $R_1 > 0$ being an observer design parameter.

The state estimation error components, $\epsilon_1 = \zeta - \hat{\zeta} = x_1 - \hat{x}_1$, $\epsilon_2 = x_2 - \hat{x}_2$, evolve according to the following *controlled* dynamics

$$\begin{array}{rcl} \dot{\epsilon}_1 & = & -\frac{1}{L}u\epsilon_2 - \frac{R_1}{L}\epsilon_1 \\ \dot{\epsilon}_2 & = & \frac{1}{C}u\epsilon_1 - \frac{1}{RC}\epsilon_2 \end{array}$$

If one takes as a Lyapunov function candidate the positive definite function $V(\epsilon) = \frac{1}{2}(L\epsilon_1^2 + C\epsilon_2^2)$, one easily verifies that the time derivative of $V(\epsilon)$, along the controlled trajectories of the state estimation error vector ϵ, is not only independent of u, but it is also negative definite,

$$\dot{V}(\epsilon) = -R_1\epsilon_1^2 - \frac{1}{R}\epsilon_2^2 \leq -\alpha V(\epsilon)$$

for $\alpha = min\{R_1, 1/R\}/max\{L, C\} > 0$. i.e., the quadratic function $V(\epsilon)$ *exponentially* converges to zero. As a consequence, the components of the estimation error vector e also exponentially converge to zero.

The output feedback controller is now given by

$$\begin{array}{rcl} u & = & \frac{LRC}{[2Lx_1 + ERC]\,\hat{x}_2}\left[\frac{E^2}{L} + \frac{2}{R^2C}\hat{x}_2^2 - \ddot{y}^*(t) + 2\xi\omega_n\left(Ex_1 - \frac{\hat{x}_2^2}{R} - \dot{y}^*(t)\right)\right. \\ & & \left. +\omega_n^2\left(\frac{1}{2}Lx_1^2 + \frac{1}{2}C\hat{x}_2^2 - y^*(t)\right)\right] \end{array}$$

4.4 Simulation results

Two voltage transitions were performed. First, from arbitrary initial conditions, it was set as a control objective the stabilization of the voltage x_2 around a constant equilibrium value $x_2 = V_{d1} = 30$ V. From the reached equilibrium, a second transfer, starting at time $T_1 = 0.07$ s., was enforced to reach, at time $T_2 = 0.12$ s., a new equilibrium value given by, $x_2(T_2) = V_{d2} = 60$ V. The observer design parameter was chosen to be $R_1 = 0.01$. The simulation results shown in Figure 11.6 correspond to the following set of converter parameter values,

$$L = 20 \text{ mH} \;\; ; \;\; C = 20\ \mu\text{F} \;\; ; \;\; R = 30\ \Omega \;\; ; \;\; E = 15 \text{ V}$$

The controller design parameters ξ and ω_n were chosen to be $\xi = 0.9$, $\omega_n = 200$.

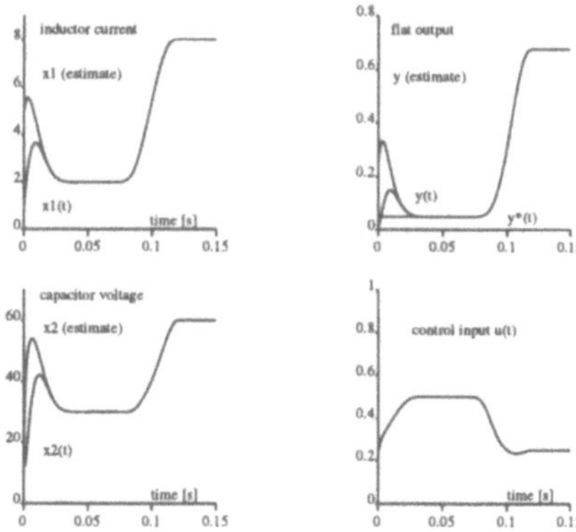

Fig. 11.6. Simulation of output feedback controlled dc-to-dc power converter

5. A PVTOL aircraft

5.1 Description and flatness

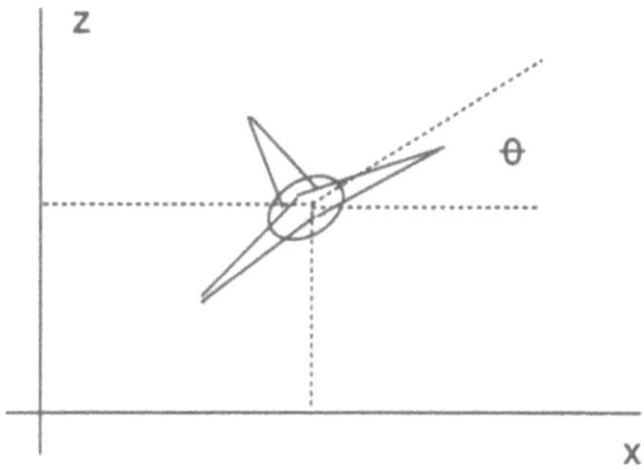

Fig. 11.7. Planar Vertical Take-Off and Landing Aircraft System

Consider, with [24, 32, 33] and [22], the following simplified description of a PVTOL aircraft (see Figure 11.7)

$$\begin{aligned} \ddot{x} &= -u_1 \sin\theta + \epsilon u_2 \cos\theta \\ \ddot{z} &= u_1 \cos\theta + \epsilon u_2 \sin\theta - g \\ \ddot{\theta} &= u_2 \end{aligned} \tag{11.10}$$

where x and z are the horizontal and vertical coordinates of the center of mass of the aircraft, respectively measured along an orthonormal set of fixed horizontal and vertical coordinates. The angle θ is the aircraft's longitudinal axis angular rotation as measured with respect to the fixed horizontal coordinate axis. The controls u_1 and u_2 represent normalized quantities related to the vertical thrust and the angular torque applied around the longitudinal axis of the aircraft respectively. The constant g is the gravity acceleration and ϵ is a fixed constant related to the geometry of the aircraft. The outputs x and z are known to be non-minimum phase [24].

System (11.10) is not static state-feedback linearizable, but differentially flat [22, 32, 33]. A physically motivated flat output is $y = (F, L)$ with $F = x - \epsilon \sin\theta$; $L = z + \epsilon \cos\theta$. Thus, state variables are given by

$$\begin{aligned} x &= F + \epsilon \frac{\ddot{F}}{\sqrt{(\ddot{F})^2 + (\ddot{L}+g)^2}} \\ z &= L - \epsilon \frac{(\ddot{L}+g)}{\sqrt{(\ddot{F})^2 + (\ddot{L}+g)^2}} \\ \theta &= \arctan\left(\frac{\ddot{F}}{\ddot{L}+g}\right) \end{aligned}$$

By setting

$$\varsigma = -\frac{\left(\ddot{F}\right)^2}{\left(\ddot{F}\right)^2 + \left(\ddot{L}+g\right)^2}$$

we obtain the following (linearizing) endogenous feedback

$$\begin{aligned} u_1 &= \varsigma + \epsilon\left(\dot{\theta}\right)^2 \\ u_2 &= \frac{1}{\varsigma}\left(-v_1 \cos\theta - v_2 \sin\theta - 2\dot{\varsigma}\dot{\theta}\right) \\ \ddot{\varsigma} &= -v_1 \sin\theta + v_2 \cos\theta + \varsigma\left(\dot{\theta}\right)^2 \end{aligned}$$

where $v_1 = F^{(4)}$, $v_2 = L^{(4)}$ are new control variables.

5.2 Trajectory planning

It is desired to transfer, in a finite amount of time $\Delta T > 0$, the aircraft position in the x-z plane, from a given fixed initial position, specified by a given set of constant horizontal and constant vertical coordinate values, $\overline{x}_{in}$ and $\overline{z}_{in}$, towards a second constant position represented by the set of coordinates $\overline{x}_f$ and $\overline{z}_f$ with the angular coordinate θ changing from an initial value $\overline{\theta}_{in} = 0$ towards a final value $\overline{\theta}_f = 0$. In [32] the same problem is solved by constructing a bounded trajectory for the internal dynamics, represented by the angular displacement θ, on the basis of the solutions of a sequence of linear ordinary differential equations with suitable initial conditions. This trajectory is in turn translated into a state space trajectory which is then tracked in a conventional manner.

The control objective of regulating the outputs x and z from given constant equilibrium values $x(T_1) = \overline{x}_{in}$ and $z(T_1) = \overline{z}_{in}$, towards a given second equilibrium value $x(T_2) = \overline{x}_f$ and $z(T_2) = \overline{z}_f$, in a prespecified amount of time $\Delta T = T_2 - T_1 > 0$, can be now translated into a corresponding transfer of the minimum phase flat outputs F and L from the initial equilibrium values $F(T_1) = \overline{F}_{in} = \overline{x}_{in}$ and $L(T_1) = \overline{L}_{in} = \overline{z}_{in} + \epsilon$ towards the final equilibrium value $F(T_2) = \overline{F}_f = \overline{x}_f$ and $L(T_2) = \overline{L}_f = \overline{z}_f + \epsilon$. A set of open loop trajectories $F^*(t)$ and $L^*(t)$ for the flat outputs (F, L) achieving a transfer between two equilibrium points $(\overline{F}_{in}, \overline{L}_{in})$ and $(\overline{F}_f, \overline{L}_f)$, may be specified in terms of suitable polynomials, as follows,

$$
\begin{aligned}
F^*(t) = \overline{F}_{in} + \Big[& 252 \left(\tfrac{t-T_1}{\Delta T}\right)^5 - 1050 \left(\tfrac{t-T_1}{\Delta T}\right)^6 + 1800 \left(\tfrac{t-T_1}{\Delta T}\right)^7 \\
& -1575 \left(\tfrac{t-T_1}{\Delta T}\right)^8 + 700 \left(\tfrac{t-T_1}{\Delta T}\right)^9 - 126 \left(\tfrac{t-T_1}{\Delta T}\right)^{10} \Big] (\overline{F}_f - \overline{F}_{in}) \\
L^*(t) = \overline{L}_{in} + \Big[& 252 \left(\tfrac{t-T_1}{\Delta T}\right)^5 - 1050 \left(\tfrac{t-T_1}{\Delta T}\right)^6 + 1800 \left(\tfrac{t-T_1}{\Delta T}\right)^7 \\
& -1575 \left(\tfrac{t-T_1}{\Delta T}\right)^8 + 700 \left(\tfrac{t-T_1}{\Delta T}\right)^9 - 126 \left(\tfrac{t-T_1}{\Delta T}\right)^{10} \Big] (\overline{L}_f - \overline{L}_{in})
\end{aligned}
$$

This particular choice of trajectories for the flat outputs, F and L, guarantees that at time T_1, the first four time derivatives of $F^*(t)$ and $L^*(t)$ are all zero, while at time T_2 the first five time derivatives of the planned flat outputs are also zero, thus avoiding noticeable discontinuities in the dynamically generated control inputs u_1 and u_2 as well as on the auxiliary inputs v_1 and v_2.

5.3 Stabilizing feedback

One proceeds to impose on the flat output tracking errors $e_F(t) = F(t) - F^*(t)$ and $e_L(t) = L(t) - L^*(t)$ the following asymptotically stable behaviors

$$
\begin{aligned}
e_F^{(4)}(t) + a_3 e_F^{(3)}(t) + a_2 \ddot{e}_F(t) + a_1 \dot{e}_F(t) + a_0 e_F(t) &= 0 \\
e_L^{(4)}(t) + b_3 e_L^{(3)}(t) + b_2 \ddot{e}_L(t) + b_1 \dot{e}_L(t) + b_0 e_L(t) &= 0
\end{aligned}
$$

where the sets of coefficients a_3, a_2, a_1, a_0 and b_3, b_2, b_1, b_0 are chosen so that the corresponding polynomials

$$
p_F(s) = s^4 + a_3 s^3 + a_2 s^2 + a_1 s + a_0 \ ; \ \ p_L(s) = s^4 + b_3 s^3 + b_2 s^2 + b_1 s + b_0
$$

are both Hurwitz, i.e., with all their roots having strictly negative real parts. It results in the following feedback controller explicitly based on the open loop specification of the flat outputs

$$
\begin{aligned}
v_1 &= F^{(4)} = F^{*(4)}(t) + a_3\left(F^{(3)}(t) - F^{*(3)}(t)\right) \\
&\quad + a_2\left(\ddot{F}(t) - \ddot{F}^*(t)\right) + a_1\left(\dot{F}(t) - \dot{F}^*(t)\right) + a_0\left(F(t) - F^*(t)\right) \\
&= F^{*(4)}(t) + a_3\left(-\dot{\varsigma}\sin\theta - \varsigma\dot{\theta}\cos\theta - F^{*(3)}(t)\right) + a_2\left(-\varsigma\sin(\theta) - \ddot{F}^*(t)\right) \\
&\quad + a_1\left(\dot{x} - \epsilon\dot{\theta}\cos\theta - \dot{F}^*(t)\right) + a_0\left(x - \epsilon\sin\theta - F^*(t)\right) \\
v_2 &= L^{(4)} = L^{*(4)}(t) + b_3\left(L^{(3)}(t) - L^{*(3)}(t)\right) \\
&\quad + b_2\left(\ddot{L}(t) - \ddot{L}^*(t)\right) + b_1\left(\dot{L}(t) - \dot{L}^*(t)\right) + b_0\left(L(t) - L^*(t)\right) \\
&= L^{*(4)}(t) + b_3\left(\dot{\varsigma}\cos(\theta) - \varsigma\dot{\theta}\sin(\theta) - L^{*(3)}(t)\right) + b_2\left(\varsigma\cos(\theta) - g - \ddot{L}^*(t)\right) \\
&\quad + b_1\left(\dot{z} + \epsilon\dot{\theta}\sin\theta - \dot{L}^*(t)\right) + b_0\left(z + \epsilon\cos\theta - L^*(t)\right)
\end{aligned}
$$

5.4 Simulation results

A maneuver transfering the PVTOL aircraft center of mass outputs (x, z) from a given initial equilibrium position towards a prescribed second equilibrium position was performed. The initial equilibrium point was set at $(\overline{x}_{in}, \overline{z}_{in}) = (0, 0)$ while the second equilibrium position for the center of mass was set to be located at $(\overline{x}_f, \overline{z}_f) = (1, 1)$. The maneuver was set to smoothly begin at $T_1 = 6$ time units, and it was prescribed to be completed at $T_2 = 14$ time units. The simulation results shown in Figure 11.8 correspond to the following set of (normalized) system parameter values $\epsilon = 0.5$, $g = 1$. The controller design parameters were chosen so that the polynomials $p_F(s)$ and $p_L(s)$ each had four roots located at the point -2, i.e, $a_3 = 8$, $a_2 = 24$, $a_1 = 32$, $a_0 = 16$, $b_3 = 8$, $b_2 = 24$, $b_1 = 32$, $b_0 = 16$.

Conclusion

This communication might be considered as a contribution to predictive control, which until now has been mainly developed for linear systems (see, e.g., [2, 41, 49]). It demonstrates, once again, the power of flatness, which is quite easy to teach to engineers (see, e.g., [31, 42, 44]), for dealing with concrete topics, concerning the motion planning and the stabilization of nonholonomic mechanical systems [19, 20], magnetic bearings [31], chemical reactors [42, 43], electric motors [6, 34], windshield wipers [1], tracking observers [23], etc. Other connections may be found in [33].

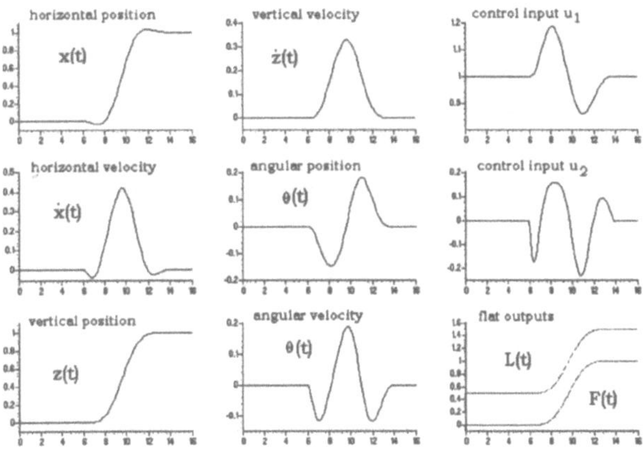

Fig. 11.8. State variables, control inputs and flat outputs in position transfer maneuver for a PVTOL example

References

1. L. Bitauld, M. Fliess and J. Lévine, A flatness based control synthesis of linear systems with an application to windshield wipers, *Proc. 4th Europ. Control Conf.*, Brussels, 1997.
2. P. Boucher and D. Dumur, *La commande prédictive*, Technip, Paris, 1996.
3. H. Bourlès and M. Fliess, Finite poles and zeros of linear systems: an intrinsic approach, *Internat. J. Control*, **68**, 1997, 897–922.
4. A. Buium, *Differential Algebra and Diophantine Geometry*, Hermann, Paris, 1994.
5. C.I. Byrnes and A. Isidori, Local stabilization of minimum phase nonlinear systems, *Systems Control Lett.*, **11**, 1988, 32–36.
6. A. Chelouah, E. Delaleau, P. Martin and P. Rouchon, Differential flatness and control of induction motors, *Proc. Symp. Control Optim. Supervision Computational Engineering in Systems Applications IMACS Multiconference*, Lille, 1996, pp. 80–85.
7. E. Delaleau, Suivi de trajectoires pour les systèmes linéaires, *Actes Coll. Cetsis-Eea*, Orsay, 1997, pp. 151–154.
8. S. Devasia, Optimal output trajectory redesign for invertible systems, *J. Guidance Control Dynamics*, **19**, 1996, 1189–1191.
9. J.S. Dewey and S. Devasia, Experimental and theoretical results in output-trajectory redesign for flexible structures, *Proc. 35th IEEE Control Decision Conf.*, Kobe,1996, pp. 4210–4215.

10. F.J Doyle, III, F. Allgöwer and M. Morari, A normal form approach to approximate input-output linearization for maximum phase nonlinear SISO systems, *IEEE Trans. Automat. Control*, **41**, 1996, 305–309.
11. F. Fallside, *Control System Design by Pole Zero Assignment*, Academic Press, New York, 1977.
12. M. Fliess, Automatique et corps différentiels, *Forum Math.*, **1**, 1989, 227–238.
13. M. Fliess, Some basic structural properties of generalized linear systems, *Systems Control Lett.*, **15**, 1990, 391–396.
14. M. Fliess, Generalized controller canonical forms for linear and nonlinear dynamics, *IEEE Trans. Automat. Control*, **35**, 1990, 994–1001.
15. M. Fliess, A remark on Willems'trajectory characterization of linear controllability, *Systems Control Lett.*, **19**, 1992, p. 43–45.
16. M. Fliess, Some remarks on the Brunovskỳ canonical form, *Kybernetika*, **29**, 1993, 417–422.
17. M. Fliess, Une interprétation algébrique de la transformation de Laplace et des matrices de transfert, *Linear Algebra Appl.*, **203-204**, 1994, 429–442.
18. M. Fliess, *Notes de Cours de D.E.A.*, Gif-sur-Yvette, 1997.
19. M. Fliess, J. Lévine, P. Martin and P. Rouchon, Flatness and defect of nonlinear systems: introductory theory and examples, *Internat. J. Control*, **61**, 1995, 1327–1361.
20. M. Fliess, J. Lévine, P. Martin and P. Rouchon, Design of trajectory stabilizing feedback for driftless flat systems, *Proc.* 3rd *European Control Conf.*, Rome, 1995 pp. 1882–1887.
21. M. Fliess, J. Lévine, P. Martin and P. Rouchon, Deux applications de la géométrie locale des diffiétés, *Ann. Inst. H. Poincaré Phys. Théor.*, **66**, 1997, 275–292.
22. M. Fliess, J. Lévine, P. Martin and P. Rouchon, A Lie-Bäcklund approach to equivalence and flatness of nonlinear systems, *IEEE Trans. Automat. Control*, to appear.
23. M. Fliess and J. Rudolph, Corps de Hardy et observateurs asymptotiques locaux pour systèmes différentiellement plats, *C.R. Acad. Sci. Paris*, **II-324**, 1997, 513–519.
24. J. Hauser, S. Sastry and G. Meyer, Nonlinear control design for slightly nonminimum phase systems: application to a V/STOL aircraft, *Automatica*, **28**, 1992, 665–679.
25. A. Ilchmann, Time-varying linear systems and invariants of system equivalence, *Internat. J. Control*, **42**, 1985, 759–790.
26. A. Isidori, *Nonlinear Control Systems*, 3rd ed., Springer, New York, 1995.
27. T. Kailath, *Linear Systems*, Prentice Hall, Englewood Cliffs, NJ, 1980.
28. E. Kolchin, *Differential Algebra and Algebraic Groups*, Academic Press, New York, 1973.
29. C. Kravaris and P. Daoutidis, Nonlinear state feedback control of second-order nonminimum-phase nonlinear systems, *Comput. Eng. Eng.*, **14**, 1990, 439–449.
30. S. Lang, *Algebra*, 3rd ed., Addison-Wesley, Reading, MA, 1993.
31. J. Lévine, J. Lottin and J.C. Ponsart, A nonlinear approach to the control of magnetic bearings, *IEEE Trans. Control Syst. Techn.*, **4**, 1996, 524–544.
32. P. Martin, S. Devasia and B. Paden, A different look at output tracking: control of a VTOL aircraft, *Automatica*, **32**, 1996, 101–107.
33. P. Martin, R.M. Murray and P. Rouchon, Flat systems, *Plenary Lectures and Mini-Courses ECC 97*, G. Bastin and M. Gevers Eds, Brussels, 1997, pp. 211–264.
34. P. Martin and P. Rouchon, Flatness and sampling control of induction motors, *Proc IFAC World Cong.*, San Francisco, 1996.

35. R. D. Middlebrook and S. Cúk, A general unified approach to modeling switching-converter power stages, *Proc. IEEE Power Electronics Specialists Conference*, 1976, pp. 18–34.
36. H. Nijmeijer and A.J. van der Schaft, *Nonlinear Dynamical Control Systems*, Springer, New York, 1990.
37. M. van Nieuwstadt, Trajectory Generation for Nonlinear Control Systems, *Ph.D. Thesis*, Caltech, Pasadena, 1996.
38. M. van Nieuwstadt, M. Ratinam and R.M. Murray, Differential flatness and absolute equivalence, *Proc.* 33rd *IEEE Conf. Decision Control*, Lake Buena Vista, FL, 1994, pp. 326–332.
39. J.-B. Pomet, A differential geometric setting for dynamic equivalence and dynamic linearization, in *Geometry in Nonlinear Control and Differential Inclusions*, B. Jakubczyk, W. Respondek and T. Rzeżuchowski Eds, Banach Center Publications, Warsaw, 1995, pp. 319–339.
40. L. Qiu and E.J. Davison, Performance limitations of non-minimum phase systems in the servomechanism problem, *Automatica*, **29**, 1993, pp. 337–349.
41. J. Richalet, *Pratique de la commande prédictive*, Hermès, Paris, 1993.
42. R. Rothfuß, *Anwendung der flachheitsbasierten Analyse und Regelung nichtlinearer Mehrgrößensysteme*, VDI, Düsseldorf, 1997.
43. R. Rothfuss, J. Rudolph and M. Zeitz, Flatness based control of a nonlinear chemical reactor model, *Automatica*, **32**, 1996, 1433–1439.
44. R. Rothfuß, J. Rudolph and M. Zeitz, Flachheit: Ein neuer Zugang zur Steuerung und Regelung nichtlinearer Systeme, *Automatisierungstechnik*, **45**, 1997, 517–525.
45. J. Rudolph, Viewing input-output system equivalence from differential algebra, *J. Math. Systems Estim. Control*, **4**, 1994, 353–384.
46. H. Sira-Ramírez and M. Ilic-Spong, Exact linearization in dc-to-dc power converters, *Internat. J. Control*, **50**, 1989, 511–524.
47. H. Sira-Ramírez and P. Lischinsky-Arenas, The differential algebraic approach in nonlinear dynamical compensator design for dc-to-dc power converters, *Internat. J. Control*, **54**, 1991, 111–134.
48. H. Sira-Ramírez, R. Pérez-Moreno, R. Ortega and M. García-Esteban, Passivity based controllers for the stabilization of dc-to-dc power converters, *Automatica*, **33**, 1997, 499–513.
49. R. Soeterboek, *Predictive Control*, Prentice Hall, New York, 1992.
50. A. Tornanbe, Output feedback stabilization of a class of non-minimum phase nonlinear systems, *Systems Control Lett.*, **19**, 1992, 193–204.
51. J.C. Willems, Paradigms and puzzles in the theory of dynamical systems, *IEEE Trans. Automat. Control*, **36**, 1991, 259–294.
52. R.A. Wright and C. Kravaris, Nonminimum-phase compensation for nonlinear processes, *AIChE J.*, **38**, 1992, 26–40.

State and storage

J.C. Willems

Department of Mathematics and Computing Science, P.O. Box 800, 9700 AV Groningen, The Netherlands, e-mail: J.C.Willems@math.rug.nl

Summary. [1] In this paper we explain in what sense, for dissipative linear time-invariant dynamical systems with quadratic supply rates, a storage function is a memoryless quadratic function of the state of the system obtained from combining the dynamics of the dissipative system with the dynamical aspects of the supply rate.

Keywords: dissipative systems, state, storage functions, quadratic differential forms.

1. Introduction

It is a pleasure for me to contribute this paper to the volume dedicated to Ioan Landau at the occasion of his 60-th birthday.

I first learned of Ioan's work at the occasion of a visit that he made to MIT almost 30 years ago. Our basic interests showed much overlap: we were both pursuing passivity, a subject that had been introduced in system theory by V.M. Popov, Ioan's teacher, and which has turned out to be one of the central notions in the field. Ioan applied these ideas mainly to adaptive control and identification [10]. My own interests dealt with stability issues and system theoretic aspects. This led to the theory of *dissipative systems* [19], a circle of ideas that has been a very useful one since, especially recently in the context of robust control. In developing this concept, I was very much influenced by the work of Brockett [4], Kalman [9], and Zames [25]. Also the techniques for

[1] I acknowledge the contributions of H.L. Trentelman, with whom I collaborated on the topics discussed in this paper.

electrical network synthesis using state space ideas, developed by Anderson [1] and others, played an important role in providing the required intuition.

In this paper I want to explain our most recent insights regarding the definition of a dissipative system. The definition which will be proposed does not involve the state, not even in the notion of a storage function. However, we will explain that the state enters nevertheless, indirectly, in that the storage function is necessarily a function of a state. For ease of exposition, and because of the status of the theoretical developments, we will deal only with controllable linear time-invariant systems and quadratic supply rates. For the proofs of the claims, and more background material, we refer to [16, 24, 22, 23, 17].

The theory of dissipative systems was set up in order to cope with a variety of special cases and related conditions that have emerged in stability theory, stemming from the small loop gain theorem and the positive operator theorem [25], conicity conditions [25, 14], and the theory of multipliers [18]. The theory of dissipative systems also allowed to unify Lyapunov-oriented approaches [3] and input/output approaches [25, 18] to stability analysis. The key to this was the role played by the storage function of a dissipative system as a candidate Lyapunov function, and of the required supply and the available storage as universal storage functions that could be constructed using optimal control techniques.

The basic idea of a dissipative system as formulated in [19] is as follows. In keeping with the central viewpoint in systems theory, we view a system as an entity in interaction with its environment. Often, through this interaction, a physical quantity of interest (other than information) is exchanged between the system and its environment, while some of it is stored in the system, and some of it is dissipated in the system. The exchange is the integral of the *supply rate*, which in [19] is assumed to be a function of the external variables (classically the input and the output). The amount stored is accounted for by the *storage function*, which in [19] is assumed to be an instantaneous function of the state of the system under consideration. The system is called dissipative if the rate of increase in the storage does not exceed the suppy rate: part of the supply and the storage is dissipated.

The notion of a dissipative system is one of the ideas stemming from the analysis of physical systems that has come to play an important role in systems theory and control. Dissipative dynamical systems can be understood as models for physical phenomena which exchange energy (or entropy) with their environment, as for example electrical circuits, in which part of the electric and magnetic energy is dissipated by the resistors in the form of heat, and visco-elastic mechanical systems in which friction causes a loss of mechanical energy in the from of heat. Since its formalization in [19], this notion has been the source of many generalizations [8, 15] and applications, for example in the area of adaptive [10] and even in robust nonlinear control. It is fair to add, however, that the idea of a dissipative system is to some extent present in the early state space work on electrical network synthesis [2, 9, 1], and the system theory work by [12]. The storage function of a dissipative system can be viewed as a natural generalization, to systems in interaction with their environment, of a Lyapunov function in autonomous sytems. As such its crucial importance cannot come as a surprise.

It is reasonable to expect that the value of the storage at a particular time depends on the past of the system's time-trajectory through the memory of the system. The memory in the time-trajectory is captured by the *state*. It is therefore natural to *require* that a storage function is a function of the state of the system. This is in fact what was done in [19]. In this paper, however, we will explain that it is not necessary to assume this instantaneous dependence on the state, since one can actually *prove* that for linear dynamical systems with quadratic supply rates, any storage function is a memoryless function of the state, more precisely, that the storage function can expressed as a memoryless quadratic function of any state variable of the linear dynamical system obtained by combining the dynamics of the original system with the dynamics of the supply rate.

In order to avoid issues of smoothness and convergence, we will assume infinite differentiability and compact support properties of trajectories whenever convenient, even though the results are valid without these assumptions; $\mathfrak{C}^\infty(\mathbb{R}, \mathbb{R}^q)$ denotes the set of all infinitely often differentiable functions from $\mathbb{R}$ to $\mathbb{R}^q$, and $\mathfrak{D}(\mathbb{R}, \mathbb{R}^q)$ denotes the subset of those elements$\in \mathfrak{C}^\infty(\mathbb{R}, \mathbb{R}^q)$ that have compact support. Given two column vectors x and y, the column vector obtained by stacking x over y is denoted by $\text{col}(x, y)$. Likewise, for given matrices A and B with the same number of columns, $\text{col}(A, B)$ denotes the matrix obtained by stacking A over B.

2. Behavioral systems

We first review some basic facts from the behavioral approach to linear dynamical systems. For more details we refer to [22, 20, 21, 22]. The basic philosophy of the behavioral approach is to view a dynamical system as a family of trajectories. This is a much more natural starting point than the usual input/output approach, particularly when one considers physical systems which are not naturally endowed with a signal flow direction. So this approach is particularly apt to treat things as flow of energy and dissipation in physical systems.

We consider dynamical systems described by differential equations, in particular by systems of linear constant coefficient differential equations of the form

$$R(\frac{d}{dt})w = 0, \tag{12.1}$$

in the real variables $w_1, w_2, \ldots, w_q$, arranged as the column vector w; R is a real polynomial matrix with, of course, q columns. This is denoted as $R \in \mathbb{R}^{\bullet \times q}[\xi]$, where ξ denotes the indeterminate. Hence if $R(\xi) = R_0 + R_1\xi + \cdots + R_N\xi^N$, then (12.1) equals the system of differential equations

$$R_0 w + R_1 \frac{dw}{dt} + \cdots + R_N \frac{d^N w}{dt^N} = 0.$$

Formally, (12.1) defines the dynamical system $\Sigma = (\mathbb{R}, \mathbb{R}^q, \mathfrak{B})$, with $\mathbb{R}$ the time-axis, $\mathbb{R}^q$ the signal space, and $\mathfrak{B}$ the *behavior,* i.e., the solution set of (12.1):

$$\mathfrak{B} = \{w \in \mathfrak{C}^\infty(\mathbb{R}, \mathbb{R}^q) \mid R(\frac{d}{dt})w = 0\}.$$

The family of dynamical systems Σ obtained this way is denoted by $\mathfrak{L}^q$. Instead of writing $\Sigma \in \mathfrak{L}^q$, we often write $\mathfrak{B} \in \mathfrak{L}^q$, since throughout the time-axis is $\mathbb{R}$ and the signal space $\mathbb{R}^q$. For obvious reasons, we refer to (12.1) as a *kernel representation* of $\mathfrak{B}$.

We will meet other representations of a given $\mathfrak{B} \in \mathfrak{L}^q$, in particular *latent variable representations* and *image representations.* We now briefly explain what these are. The system of differential equations

$$R(\frac{d}{dt})w = M(\frac{d}{dt})\ell \tag{12.2}$$

is said to be a *latent variable model,* with w the *manifest* variable vector, (the variables that the model aims at), and ℓ the *latent* variable vector (auxiliary variables introduced in the modelling process, or for the purposes of analysis). In (12.2), we assume that there are q manifest and d latent variables, and that R and M are polynomial matrices of appropriate dimension. Of course (12.2), being a differential system as (12.1), defines the behavior

$$\mathfrak{B}_f = \{(w, \ell) \in \mathfrak{C}^\infty(\mathbb{R}, \mathbb{R}^q) \mid (12.2) \text{ holds}\}.$$

$\mathfrak{B}_f$ will be called the *full behavior* of the latent variable model, in order to distinguish it from the *manifest behavior* which is the projection of $\mathfrak{B}_f$ on the manifest variable space, i.e., the set

$$\{w \in \mathfrak{C}^\infty(\mathbb{R}, \mathbb{R}^q) \mid \text{ there exists } \ell \in \mathfrak{C}^\infty(\mathbb{R}, \mathbb{R}^d) \text{ such that } (w, \ell) \in \mathfrak{B}_f\}. \tag{12.3}$$

This set is called the *manifest behavior* of (12.2). If, for a given $\mathfrak{B} \in \mathfrak{L}^q$, the manifest behavior (12.3) of (12.2) equals $\mathfrak{B}$, then (12.2) is called a *latent variable representation* of $\mathfrak{B}$. The latent variable representation is called *observable* if the latent variable is uniquely determined by the manifest variable, i.e., if $(w, \ell_1), (w, \ell_2) \in \mathfrak{B}_f$ implies that $\ell_1 = \ell_2$. It can be shown that (12.2) is observable if and only if $\text{rank}(M(\lambda)) = d$ for all $\lambda \in \mathbb{C}$.

The above explains the notion of observability in a behavioral setting. However, also controllability enters this framework in a novel and convincing way. A system $\mathfrak{B} \in \mathfrak{L}^q$ is said to be *controllable* if for each $w_1, w_2 \in \mathfrak{B}$ there exists a $w \in \mathfrak{B}$ and a $t' \geq 0$ such that $w(t) = w_1(t)$ for $t < 0$ and $w(t) = w_2(t - t')$ for $t \geq t'$. Controllability tests in terms of kernel representations are readily obtained. Thus, it can be shown that $\mathfrak{B}$ is controllable if and only if its kernel representation satisfies $\text{rank}(R(\lambda)) = \text{rank}(R)$ for all $\lambda \in \mathbb{C}$. Controllability can also be expressed in terms of the existence of a suitable representation using latent variables. Thus, controllable systems are exactly those that admit an *image representation* (such representations are very much related to the *flat systems* which have been studied in [7] and elsewhere). More concretely, $\mathfrak{B} \in \mathfrak{L}^q$ is controllable if and only if there exists an $M \in \mathbb{R}^{q \times \bullet}[\xi]$ such that $\mathfrak{B}$ is the manifest behavior of a latent variable model of the form

$$w = M(\frac{d}{dt})\ell. \tag{12.4}$$

For obvious reasons again, (12.4) is called an *image representation* of $\mathfrak{B}$. An image representation is called *observable* if it is observable as a latent variable representation. Hence, the image representation (12.4) is observable if and only if $\text{rank}(M(\lambda)) = d$ for all $\lambda \in \mathbb{C}$. A controllable system always has an observable image representation. It is evident that these notions are natural, far-reaching generalizations of their state space counterparts.

An important addition that we have recently added to the machinery of linear time-invariant systems are quadratic differential forms and two-variable polynomial matrices. These are studied extensively in [23]. We now give a brief review of the elements needed to understand the result of this paper. We denote by $\mathbb{R}^{q \times q}[\zeta, \eta]$ the set of square real polynomial matrices in the indeterminates ζ and η, i.e., expressions of the form

$$\Phi(\zeta, \eta) = \sum_{k,\ell} \Phi_{k\ell} \zeta^k \eta^\ell.$$

The above sum runs over the non-negative integers, and is assumed to be finite, and the matrices $\Phi_{k\ell} \in \mathbb{R}^{q \times q}$. The two-variable polynomial matrix Φ induces a *quadratic differential form* (QDF) $Q_\Phi : \mathfrak{C}^\infty(\mathbb{R}, \mathbb{R}^q) \to \mathfrak{C}^\infty(\mathbb{R}, \mathbb{R})$ by

$$Q_\Phi(w)(t) := \sum_{k,\ell} \left(\frac{d^k w}{dt^k}(t) \right)^{\mathrm{T}} \Phi_{k\ell} \left(\frac{d^\ell w}{dt^\ell}(t) \right).$$

If $\Phi \in \mathbb{R}^{q \times q}[\zeta, \eta]$ satisfies $\Phi(\zeta, \eta) = \Phi^*(\zeta, \eta) := \Phi(\eta, \zeta)^{\mathrm{T}}$, then Φ is called *symmetric*. The symmetric elements of $\mathbb{R}^{q \times q}[\zeta, \eta]$ are denoted by $\mathbb{R}_s^{q \times q}[\zeta, \eta]$. Associated with $\Phi \in \mathbb{R}_s^{q \times q}[\zeta, \eta]$ we form the symmetric matrix

$$\tilde{\Phi} = \begin{pmatrix} \Phi_{00} & \Phi_{01} & \cdots & \cdot & \cdots \\ \Phi_{10} & \Phi_{11} & \cdots & \cdot & \cdots \\ \vdots & \vdots & & \vdots & \\ \cdot & \cdot & \cdots & \Phi_{k\ell} & \cdots \\ \vdots & \vdots & & \vdots & \end{pmatrix}.$$

Note that, although the matrix $\tilde{\Phi}$ is infinite, only a finite number of its elements are non-zero. We can obviously factor $\tilde{\Phi}$ as $\tilde{\Phi} = \tilde{M}^{\mathrm{T}} \Sigma_M \tilde{M}$, with $\tilde{M}$ an infinite matrix having a finite number of rows and all but a finite number of elements equal to zero, and Σ_M a signature matrix, i.e., a matrix of the form

$$\Sigma_M = \begin{pmatrix} I_{r_+} & 0 \\ 0 & -I_{r_-} \end{pmatrix}.$$

This factorization leads, after pre-multiplication by $(I_q \quad I_q\zeta \quad I_q\zeta^2 \cdots)$ and post-multiplication by $\mathrm{col}(I_q \quad I_q\eta \quad I_q\eta^2 \cdots)$, to a factorization of Φ as $\Phi(\zeta, \eta) = M^{\mathrm{T}}(\zeta)\Sigma_M M(\eta)$. If we take $\tilde{M}$ of full row rank, then Σ_M is unique. We denote this Σ_M as Σ_Φ. The resulting r_+ equals the number of positive eigenvalues and r_- the number of negative eigenvalues of $\tilde{\Phi}$. Any factorization $\Phi(\zeta, \eta) = M^{\mathrm{T}}(\zeta)\Sigma_\Phi M(\eta)$ is called a *canonical factorization* of Φ. In a canonical factorization, the rows of $M(\xi)$ are linearly independent over $\mathbb{R}$. A canonical factorization is not unique, and all of them can be obtained from one by replacing $M(\xi)$ by $UM(\xi)$ with U a unimodular matrix such that $U^{\mathrm{T}}\Sigma_\Phi U = \Sigma_\Phi$. Note that if $\Phi(\zeta, \eta) = M^{\mathrm{T}}(\zeta)\Sigma_\Phi M(\eta)$ is a canonical factorization, and $\Phi(\zeta, \eta) = M_1^{\mathrm{T}}(\zeta)\Sigma_{M_1} M_1(\eta)$ is an arbitrary factorization, then there exists a real constant matrix H such that $M(\xi) = HM_1(\xi)$.

QDF's identified with two-variable polynomial matrices allow a very convenient calculus. One example of this is differentiation. If Q_Φ is a QDF, so is $\frac{d}{dt}Q_\Phi$. Indeed, $\frac{d}{dt}Q_\Phi = Q_{\overset{\bullet}{\Phi}}$ with $\overset{\bullet}{\Phi}(\zeta, \eta) := (\zeta + \eta)\Phi(\zeta, \eta)$. Suppose now that $\Phi \in \mathbb{R}_s^{q \times q}[\zeta, \eta]$ is given. An important question is: does there exist $\Psi \in \mathbb{R}_s^{q \times q}[\zeta, \eta]$ such that $\overset{\bullet}{\Psi} = \Phi$, equivalently $\frac{d}{dt}Q_\Psi = Q_\Phi$? Such a Ψ exists if and only if Φ contains a factor $(\zeta + \eta)$. Then $\Psi(\zeta, \eta) = \frac{1}{\zeta+\eta}\Phi(\zeta, \eta)$. It is shown in [23] that Φ contains a factor $(\zeta + \eta)$ if and only if $\partial\Phi = 0$, where $\partial\Phi$ is the one-variable polynomial matrix defined by $\partial\Phi(\xi) := \Phi(-\xi, \xi)$. It is also proven that $\partial\Phi = 0$ if and only if $\int_{-\infty}^{\infty} Q_\Phi(w)dt = 0$ for all $w \in \mathfrak{D}(\mathbb{R}, \mathbb{R}^q)$.

If $\Phi \in \mathbb{R}_s^{q \times q}[\zeta, \eta]$, we write $\Phi \geq 0$ if $Q_\Phi(w) \geq 0$ for all $w \in \mathfrak{C}^\infty(\mathbb{R}, \mathbb{R}^q)$. It is shown in [23] that $\Phi \geq 0$ if and only if there exists $D \in \mathbb{R}^{\bullet \times q}[\xi]$ such that $\Phi(\zeta, \eta) = D^{\mathrm{T}}(\zeta)D(\eta)$, equivalently $Q_\Phi(w) = \|D(\frac{d}{dt})w\|^2$ for all $w \in \mathfrak{C}^\infty(\mathbb{R}, \mathbb{R}^q)$. In addition, we need the concept of *average non-negativity*: we will say $\int Q_\Phi \geq 0$ if $\int_{-\infty}^{\infty} Q_\Phi(w)dt \geq 0$ for all $w \in \mathfrak{D}(\mathbb{R}, \mathbb{R}^q)$. Again, it was shown in [23] that $\int Q_\Phi \geq 0$ if and only if $(\partial\Phi)(i\omega) \geq 0$ for all $\omega \in \mathbb{R}$. In turn, this condition is equivalent to the existence of polynomial spectral factorizations of $\partial\Phi$: $(\partial\Phi)(i\omega) \geq 0$ if and only if there exists $D \in \mathbb{R}^{q \times q}[\xi]$ such that $(\partial\Phi)(\xi) = D^{\mathrm{T}}(-\xi)D(\xi)$ (see [5, 6]).

3. Supply, storage, and dissipation

In this section, we introduce the notions of supply rate, storage, and dissipation in the context of linear time-invariant systems with quadratic differential functionals. Let $\mathfrak{B} \in \mathfrak{L}^q$ be controllable, with $R(\frac{d}{dt})w = 0$ and $w = M(\frac{d}{dt})\ell$ respectively a kernel and an observable image representation of it. Consider also the QDF $Q_\Phi : \mathfrak{C}^\infty(\mathbb{R}, \mathbb{R}^q) \to \mathfrak{C}^\infty(\mathbb{R}, \mathbb{R})$ induced by the symmetric Φ; Q_Φ is called the *supply rate.* Intuitively, think of $Q_\Phi(w)$ as the power going into the physical device described by $\mathfrak{B}$. In many concrete situations, the power delivered is indeed a quadratic expression involving the system variables and their high order derivatives, whence the formalization of the supply rate as a QDF. For instance, in mechanical systems, the power equals $\sum_k F_k \frac{dq_k}{dt}$ with F_k the external force acting on the k-th pointmass and q_k its position; in electrical circuits it equals $\sum_k V_k I_k$ with V_k the potential and I_k the current into the circuit at the k-th port or terminal. The system $\mathfrak{B}$ is called *dissipative* with respect to the supply rate Q_Φ if along trajectories that start at rest and bring the system back to rest (or equivalently, along periodic trajectories, but we will not pursue this definition in this paper), the total amount of energy flowing into the device is non-negative: in other words, if the system dissipates energy. The supply rate will, even in a dissipative system, sometimes be positive, and sometimes be negative, due to the internal storage in the system. The difference between the rate of increase of the storage and the supply rate is the dissipation rate.

All this is formalized as follows.

Definition: The pair $(\mathfrak{B}, Q_\Phi)$ is called *dissipative* if $\int_{-\infty}^{\infty} Q_\Phi(w)dt \geq 0$ for all $w \in \mathfrak{B} \cap \mathfrak{D}(\mathbb{R}, \mathbb{R}^q)$. The QDF Q_Ψ induced by $\Psi \in \mathbb{R}_s^{q\times q}[\zeta, \eta]$ is called a *storage function* for $(\mathfrak{B}, Q_\Phi)$ if

$$\frac{d}{dt}Q_\Psi(w) \leq Q_\Phi(w) \text{ for all } w \in \mathfrak{B} \cap \mathfrak{C}^\infty(\mathbb{R}, \mathbb{R}^q).$$

The QDF Q_Δ induced by $\Delta \in \mathbb{R}_s^{q\times q}[\zeta, \eta]$ is called a *dissipation rate* for $(\mathfrak{B}, Q_\Phi)$ if $Q_\Delta(w) \geq 0$ for all $w \in \mathfrak{B} \cap \mathfrak{C}^\infty(\mathbb{R}, \mathbb{R}^q)$ and

$$\int_{\infty}^{\infty} Q_\Phi(w)dt = \int_{\infty}^{\infty} Q_\Delta(w)dt \text{ for all } w \in \mathfrak{B} \cap \mathfrak{D}(\mathbb{R}, \mathbb{R}^q).$$

If the supply rate Q_Φ, the dissipation function Q_Δ, and the storage function Q_Ψ are related by

$$\frac{d}{dt}Q_\Psi(w) = Q_\Phi(w) - Q_\Delta(w) \text{ for all } w \in \mathfrak{B} \cap \mathfrak{C}^\infty(\mathbb{R}, \mathbb{R}^q),$$

then we call the triple $(Q_\Phi, Q_\Psi, Q_\Delta)$ *matched on* $\mathfrak{B}$.

Dissipativeness is immediately related to the existence of a storage function and a dissipation rate. Indeed, the following turn out to be equivalent:

1. $(\mathfrak{B}, Q_\Phi)$ defines a dissipative system;

2. $M(-i\omega)^{\mathrm{T}}\Phi(-i\omega, i\omega)M(i\omega) \geq 0$ for all $\omega \in \mathbb{R}$;
3. $(\mathfrak{B}, Q_\Phi)$ admits a storage function;
4. $(\mathfrak{B}, Q_\Phi)$ admits a dissipation rate.

Furthermore, corresponding to any dissipation rate Q_Δ there exists a storage function Q_Ψ, and corresponding to any storage function Q_Ψ there exists a dissipation rate Q_Δ such that $(Q_\Phi, Q_\Psi, Q_\Delta)$ is matched on $\mathfrak{B}$.

For a proof of these statements, see [16].

4. The state

A state model is a special type of latent variable model, one in which the latent variable (called the state, and, in keeping with tradition, by x) separates the past and the future: given the present value of the state, any past leading to it is compatible, that is, concatenable, with any future emanating from it. Thus, a latent variable model $R(\frac{d}{dt})w = M(\frac{d}{dt})x$ is said to be a *state model* if whenever (w_1, x_1) and (w_2, x_2) are both elements of its full behavior $\mathfrak{B}_f$, and if $x_1(0) = x_2(0)$, then the concatenation $(w, x) := (w_1, x_1) \wedge (w_2, x_2)$ also satisfies the differential equation $R(\frac{d}{dt})w = M(\frac{d}{dt})x$. However, in general this concatenation needs not be $\mathfrak{C}^\infty$, and therefore one should interpret this solution as a weak solution, that is, as a solution to the differential equation in the sense of distributions.

Let $\mathfrak{B} \in \mathfrak{L}^q$. A latent variable representation of $\mathfrak{B}$ is called a *state representation* of $\mathfrak{B}$ if it is a state model. Thus, when we have a state model, in order to decide how a future continuation of a trajectory in $\mathfrak{B}$ looks like, it is not necessary to analyze the whole past of this trajectory, rather only the *present* value of the state is relevant. As such, the state x parametrizes the memory of the system.

The state property can be related to a special equation structure. State models lead to latent variable kernel representations of the form

$$Gw + Fx + E\frac{dx}{dt} = 0. \tag{12.5}$$

Here, E, F, and G are real constant matrices. Equation (12.5) is an (implicit) first order differential equation containing derivatives of at most first-order in x and zero-th order in w. It is easy to see (see, for example, [13]) that any latent variable model of the form (12.5) is in fact a state space model. Conversely, every state model $R'(\frac{d}{dt})w = M(\frac{d}{dt})x$ is equivalent (in the sense that the two have the same full behavior) to a representation of the above first-order form. The number of state variables, i.e., the size of x, is called the *dynamic order* of the state model, and is denoted by n.

Let $\mathfrak{B}$ be the manifest behavior of any (not necessarily observable) state representation (12.5). It turns out that there exists an *observable* state representation of $\mathfrak{B}$, and that this state representation has smallest possible dynamic order among all state representations of $\mathfrak{B}$. We call such state representations *minimal*. Moreover, the state variable of the not-necessarily observable specimen is related to the state of an observable state representation by a linear map. Formally, let $\mathfrak{B} \in \mathfrak{L}^q$ be the manifest behavior of (12.5). Then there exists an observable state representation $G'w + F'x' + E'\frac{dx'}{dt} = 0$ of $\mathfrak{B}$ (its full behavior is denoted by $\mathfrak{B}'_f$) with dynamic order $n' \leq n$, and a linear map $L : \mathbb{R}^n \to \mathbb{R}^{n'}$ such that $(w, x) \in \mathfrak{B}_f$, $(w, x') \in \mathfrak{B}'_f$ implies $x' = Lx$.

For a proof of this result, see, for example, [16].

The above implies that any memoryless function of a minimal state is in fact a memoryless function of any state. The state representation $Gw + Fx + E\frac{dx}{dt} = 0$ of $\mathfrak{B}$ is observable (i.e., x is observable from w) if and only if there exists $X \in \mathbb{R}^{\bullet \times q}[\xi]$ such that for all $w \in \mathfrak{B}$ we have: $(w, x) \in \mathfrak{B}_f \Leftrightarrow x = X(\frac{d}{dt})w$. The operator $X(\frac{d}{dt})$ is called a *state map* for $\mathfrak{B}$. Hence, if $R(\frac{d}{dt})w = 0$ is a kernel representation of $\mathfrak{B}$, then $X(\frac{d}{dt})$ is called a state map for $\mathfrak{B}$ if

$$\begin{pmatrix} R(\frac{d}{dt}) \\ X(\frac{d}{dt}) \end{pmatrix} w = \begin{pmatrix} 0 \\ I \end{pmatrix} x$$

is a state representation of $\mathfrak{B}$. An analogous construction holds for image representations [13]. We will now explicitate this for controllable input/output representations. Let $\mathfrak{B}$ be controllable, and let $w = M(\frac{d}{dt})\ell$ be an observable image representation of it. Let Π be a permutation matrix such that $\Pi M = \text{col}(U, Y)$, with YU^{-1} a matrix of proper rational functions (a permutation matrix Π that achieves this always exists, see [22]). This corresponds to permuting the components of w as $\Pi w = \text{col}(u, y)$, with $u = U(\frac{d}{dt})\ell$ and $y = Y(\frac{d}{dt})\ell$, such that u is an input and y is an output, in the sense these terms are normally used, see for example [22]. The number of input components of $\mathfrak{B}$, i.e., the size of u, is denoted by m.

The construction of a state map can be carried out in terms of this input/output partition as follows. Consider the set of real polynomial row vectors

$$\mathfrak{F} := \{f \in \mathbb{R}^{1 \times q}[\xi] \mid fU^{-1} \text{ is strictly proper}\}.$$

It can be shown that the set of row vectors of polynomials $\mathfrak{F}$ actually forms a finite-dimensional vector space over $\mathbb{R}$. Its dimension equals in fact the minimal dynamic order over the state representations of $\mathfrak{B}$. Let $X \in \mathbb{R}^{\bullet \times q}[\xi]$. It was shown in [13] that $X(\frac{d}{dt})$ is a state map for $\mathfrak{B}$ if and only if the rows of the polynomial matrix XM span $\mathfrak{F}$, i.e., every element of $\mathfrak{F}$ is a real linear combination of the rows of XM.

Suppose now that we have a system $\mathfrak{B} \in \mathfrak{L}^q$, and suppose a state representation of this system is given, with state variable x. Assume that to the manifest variable w we add an extra component, say f, i.e., we consider a new system $\mathfrak{B}_{\text{ext}}$ with the property that $w \in \mathfrak{B}$ if and only if there exists f such that $\text{col}(w, f) \in \mathfrak{B}_{\text{ext}}$. We now establish conditions under which this new variable f can be written as a linear function of the state variable of the original system $\mathfrak{B}$, or as a linear function of the state variable and input variable of $\mathfrak{B}$.

Let $\mathfrak{B} \in \mathfrak{L}^q$, and let

$$\begin{pmatrix} u \\ y \end{pmatrix} = \begin{pmatrix} U(\frac{d}{dt}) \\ Y(\frac{d}{dt}) \end{pmatrix} \ell$$

be an observable image representation with YU^{-1} a matrix of proper rational functions. Let

$$G\begin{pmatrix} u \\ y \end{pmatrix} + Fx + E\frac{dx}{dt} = 0$$

be an observable state representation of $\mathfrak{B}$ (with full behavior denoted by $\mathfrak{B}_f$). Let $F \in \mathbb{R}^{\bullet \times d}[\xi]$ and let $\mathfrak{B}_{\text{ext}}$ be the system with image representation

$$\begin{pmatrix} u \\ y \\ f \end{pmatrix} = \begin{pmatrix} D(\frac{d}{dt}) \\ N(\frac{d}{dt}) \\ F(\frac{d}{dt}) \end{pmatrix} \ell.$$

Then there exists a real constant matrix $H \in \mathbb{R}^{\bullet \times n}$ such that $f = Hx$ for all f and x for which there exists $\text{col}(u, y) \in \mathfrak{B}$ such that $\text{col}(u, y, f) \in \mathfrak{B}_{\text{ext}}$ and $\text{col}(u, y, x) \in \mathfrak{B}_f$, if and only if FD^{-1} is a matrix of strictly proper rational functions. There exist real constant matrices $H \in \mathbb{R}^{\bullet \times n}$ and $J \in \mathbb{R}^{\bullet \times m}$ such that $f = Hx + Ju$ for all f, x and u for which there exists y such that $\text{col}(u, y, f) \in \mathfrak{B}_{\text{ext}}$ and $\text{col}(u, y, x) \in \mathfrak{B}_f$, if and only if FD^{-1} is a matrix of proper rational function.

For a proof of this result, see [16].

5. Every storage function is a state function

In this section we state the result of this paper. It states that a quadratic storage function for a dissipative system can always be represented as a memoryless quadratic function of any state variable, and that a quadratic dissipation rate can always be represented as memoryless quadratic function of any state variable jointly with the manifest variable of a given system.

We will first treat the case that all the dynamics are captured by the supply rate, i.e., when $\mathfrak{B} = \mathfrak{C}^\infty(\mathbb{R}, \mathbb{R}^q)$. Let $\Phi \in \mathbb{R}_s^{q \times q}[\zeta, \eta]$. This Φ induces the QDF that defines the supply rate. Note that since it contains derivatives, is has dynamics. These dynamics lead, as we shall see, to a state. Assume that $(\mathfrak{C}^\infty(\mathbb{R}, \mathbb{R}^q), Q_\Phi)$ is dissipative. It turns out that every storage function is a quadratic function of any state variable of a particular system $\mathfrak{B}_\Phi$ obtained from the dynamics of Φ. Also, every dissipation rate is a quadratic function of any state variable, jointly with the manifest variable of this system $\mathfrak{B}_\Phi$. We now explain how the system $\mathfrak{B}_\Phi$ is obtained. Let $\Phi(\zeta,\eta) = M^{\mathrm{T}}(\zeta)\Sigma_\Phi M(\eta)$ be a canonical factorization of Φ, with $\Sigma_\Phi \in \mathbb{R}^{r\times r}$. Consider the system $\mathfrak{B}_\Phi \in \mathfrak{L}^r$ (with manifest variable $v \in \mathbb{R}^r$) with image representation

$$v = M(\frac{d}{dt})w$$

Theorem: Let $Gv + Fx + E\frac{dx}{dt} = 0$ be a state representation of $\mathfrak{B}_\Phi$, with full behavior $\mathfrak{B}_f$. Let Q_Ψ be a storage function for $(\mathfrak{C}^\infty(\mathbb{R},\mathbb{R}^q), Q_\Phi)$. Then there exists a matrix $K = K^{\mathrm{T}} \in \mathbb{R}^{n\times n}$ such that $\mathrm{col}(M(\frac{d}{dt})w, x) \in \mathfrak{B}_f$ implies $Q_\Psi(w) = x^{\mathrm{T}}Kx$. Furthermore, if Q_Δ is a dissipation rate for $(\mathfrak{C}^\infty(\mathbb{R},\mathbb{R}^q), Q_\Phi)$, then there exists $L = L^{\mathrm{T}} \in \mathbb{R}^{(n+q)\times(n+q)}$ such that $\mathrm{col}(M(\frac{d}{dt})w, x) \in \mathfrak{B}_f$ implies

$$Q_\Delta(w) = \begin{pmatrix} x \\ v \end{pmatrix}^{\mathrm{T}} L \begin{pmatrix} x \\ v \end{pmatrix}.$$

Proof: See [16].

Next, we treat the general case, in which there are dynamics both in the system and in the supply rate. Let $\mathfrak{B} \in \mathfrak{L}^q$ be an arbitrary controllable system. Let $\Phi \in \mathbb{R}_s^{q\times q}[\xi]$ induce the QDF that defines the supply rate. Now dynamics originate both from $\mathfrak{B}$ and from Φ. Assume that $(\mathfrak{B}, Q_\Phi)$ is dissipative. Also in this case, every storage function turns out to be a memoryless quadratic function of any state variable, and every dissipation rate turns out to be a memoryless quadratic function of any state variable jointly with the manifest variable of a system $\mathfrak{B}_\Phi$. This time, however, the system $\mathfrak{B}_\Phi$ is obtained by combining the dynamics of $\mathfrak{B}$ and Φ, as follows. Again, let $\Phi(\zeta,\eta) = M^{\mathrm{T}}(\zeta)\Sigma_\Phi M(\eta)$ be a canonical factorization of Φ, with $\Sigma_\Phi \in \mathbb{R}^{r\times r}$. Now, consider the system $\mathfrak{B}_\Phi \in \mathfrak{L}^r$ (with manifest variable $v \in \mathbb{R}^r$) represented by

$$v = M(\frac{d}{dt})w, \quad w \in \mathfrak{B}.$$

Let $Gv + Fx + E\frac{dx}{dt} = 0$ be a state representation of $\mathfrak{B}_\Phi$, with full behavior $\mathfrak{B}_f$. Let Q_Ψ be a storage function for $(\mathfrak{B}, Q_\Phi)$. Then there exists $K = K^{\mathrm{T}} \in \mathbb{R}^{n\times n}$ such that $w \in \mathfrak{B}$ and $\mathrm{col}(M(\frac{d}{dt})w, x) \in \mathfrak{B}_f$ implies $Q_\Psi(w) = x^{\mathrm{T}}Kx$. Further, if Q_Δ is a dissipation rate for $(\mathfrak{B}, Q_\Phi)$, then there exists $L = L^{\mathrm{T}} \in \mathbb{R}^{(n+d)\times(n+d)}$ such that $w \in \mathfrak{B}$ and $\mathrm{col}(M(\frac{d}{dt})w, x) \in \mathfrak{B}_f$ implies

$$Q_\Delta(w) = \begin{pmatrix} x \\ v \end{pmatrix}^{\mathrm{T}} L \begin{pmatrix} x \\ v \end{pmatrix}.$$

Proof: See [16].

Finally, we discuss the special case that the supply rate Q_Φ is first order in w, i.e., $Q_\Phi(w) = w^{\mathrm{T}} P w$, with $P = P^{\mathrm{T}} \in \mathbb{R}^{q \times q}$. Let $\mathfrak{B} \in \mathfrak{L}^q$ be controllable, and assume that $(\mathfrak{B}, Q_\Phi)$ is dissipative. In this case every storage function is simply a quadratic function of any state variable of $\mathfrak{B}$, and every dissipation rate is a quadratic function of any state variable of $\mathfrak{B}$, jointly with the manifest variable of $\mathfrak{B}$.

Corolarry: Let $Gw + Fx + E\frac{dx}{dt} = 0$ be a state representation of $\mathfrak{B}$, with full behavior $\mathfrak{B}_f$. Let Q_Ψ be a storage function for $(\mathfrak{B}, Q_\Phi)$. Then there exists $K = K^{\mathrm{T}} \in \mathbb{R}^{n \times n}$ such that $\mathrm{col}(w, x) \in \mathfrak{B}_f$ implies $Q_\Psi(w) = x^{\mathrm{T}} K x$. If Q_Δ is a dissipation rate for $(\mathfrak{B}, Q_\Phi)$, then there there exists $L = L^{\mathrm{T}} \in \mathbb{R}^{(n+d)\times(n+d)}$ such that $\mathrm{col}(w, x) \in \mathfrak{B}_f$ implies

$$Q_\Delta(w) = \begin{pmatrix} x \\ w \end{pmatrix}^{\mathrm{T}} L \begin{pmatrix} x \\ w \end{pmatrix}.$$

Proof: Follows immediately from the previous theorem.

For examples relating these results to mechanical systems and electrical circuits, we refer to [16].

References

1. B.D.O. Anderson and S. Vongpanitlerd, *Network Analysis and Synthesis,* Prentice Hall, 1973.
2. V. Belevich, *Classical Network Theory,* Holden Day, 1968.
3. R.W. Brockett and J.L. Willems, Frequency domain stability criteria, Parts I and II, *IEEE Transactions on Automatic Control,* volume 10, pages 255-261 and 407-413, 1965.
4. R.W. Brockett, Path integrals, Lyapunov functions, and quadratic minimization, *Proceedings of the 4-th Allerton Conference on Circuit and System Theory,* pages 685-698, 1966.
5. F.M. Callier, On polynomial spectral factorization by symmetric extraction, *IEEE Transactions on Automatic Control*, volume 30, pages 453-464, 1985.
6. W.A. Coppel, *Linear Systems*, Notes in Pure Mathematics, Australian National University, volume 6, 1972.
7. M. Fliess and S.T. Glad, An algebraic approach to linear and nonlinear control, pages 223-267 of *Essays on Control: Perspectives in the Theory and Its Applications*, edited by H.L. Trentelman and J.C. Willems, Birkhäuser, 1993.
8. D.J. Hill and P.J. Moylan, Stability of nonlinear dissipative systems, *IEEE Transactions on Automatic Control*, volume 21, pages 708-711, 1976.
9. R.E. Kalman, On a new characterization of linear passive systems, *Proceedings of the 1st Allerton Conference*, pages 456-470, 1963.
10. I.D. Landau, *Adaptive Control - The Model Reference Approach*, Marcel Dekker, 1979.
11. J.W. Polderman and J.C. Willems, *Introduction to Mathematical Systems Theory: A Behavioral Approach,* Springer Verlag, 1998.
12. V.M. Popov, Hyperstability and optimality of automatic systems with several control functions, *Revue Roumaine Sci. Tech. Electrotech. et Energ.*, volume 9, pages 629-690, 1964.

13. P. Rapisarda and J.C. Willems, State maps for linear systems, *SIAM Journal on Control and Optimization,* volume 35, pages 1053-1091, 1997.
14. M.G. Safonov *Stability and Robustness of Multivariable Feedback Systems,* The MIT Press, 1980.
15. H.L. Trentelman and J.C. Willems, The dissipation inequality and the algebraic Riccati equation *The Riccati Equation* (Eds. S. Bittanti, A.J. Laub, and J.C. Willems) Springer Verlag, pages 197-242, 1991.
16. H.L. Trentelman and J.C. Willems, Every storage function is a state function, *Systems & Control Letters,* volume 32, pages 249-259, 1997.
17. H.L. Trentelman and J.C. Willems, H_∞ control in a behavioral context, Part 1: The full information case, submitted to the *IEEE Transactions on Automatic Control.*
18. J.C. Willems, *The Analysis of Feedback Systems,* The MIT Press, 1971.
19. J.C. Willems, Dissipative dynamical systems - Part I: General theory, Part II: Linear systems with quadratic supply rates, *Archive for Rational Mechanics and Analysis,* volume 45, pages 321-351, 352-393, 1972.
20. J.C. Willems, From time series to linear system. Part I: Finite dimensional linear time invariant systems ; Part II: Exact modelling, Part III: Approximate modelling, *Automatica,* volume 22, pages 561-580, 675-694, 1986, volume 23, pages 87-115, 1987.
21. J.C. Willems, Models for dynamics, *Dynamics Reported,* volume 2, pages 171-269, 1989.
22. J.C. Willems, Paradigms and puzzles in the theory of dynamical systems, *IEEE Transactions on Automatic Control,* volume 36, pages 259–294, 1991.
23. J.C. Willems and H.L. Trentelman, On quadratic differential forms, *SIAM Journal on Control and Optimization,* to appear.
24. J.C. Willems, Path integrals and stability, in *Essays in Mathematical Control Theory,* Edited by J. Baillieul and J.C. Willems, Springer Verlag, to appear.
25. G. Zames, On the input-output stability of time-varying nonlinear feedback systems. Part I: Conditions derived using concepts of loop gain, conicity, and positivity; Part II: Conditions involving circles in the frequency plane and sector nonlinearities, *IEEE Transactions on Automatic Control,* volume 11, pages 228-238 and 465-476, 1966.

Concepts of "State" in Introductory Control Education

D. M. Auslander

Mechanical Engineering Department, University of California, Berkeley, CA, 94720-1740, USA, e-mail: dma@me.berkeley.edu

Summary. The concept of "state" as used in control engineering is powerful and unifying. The thesis of this essay, however, is that contemporary control problems extend well beyond the scope of what is now traditional feedback control education and that state concepts should be more broadly interpreted at the introductory educational level. This would give practicing engineers much better tools and insight for dealing with the complexities of modern machine control.

1. Introduction: Machine Control and Mechatronics

Control system education has evolved over the last 50 years or so to cover a broad range of feedback control structures and design techniques. Approaches cover everything from intuitive design and parameterization for specific applications to mathematically based techniques of considerable sophistication. It is important to note, however, that this has been an evolutionary process and remains based on assumptions and technological constraints that were in place at the time this process started. The most important of these was that the feedback control system existed largely as an independent entity. Other aspects of machine coordination and control were handled mostly with mechanical means such as gears, linkages, cams, shafts, belts, etc. Electrical switching was normally handled by relays. This is reflected in the current educational organization in which control system courses deal only with the mathematics and application of feedback control.

All of this changed mid-way through this period with the invention of the microprocessor. Although microprocessor computing power was initially too weak for anything but replacement of relatively slow logic (the original microprocessors used a 4-bit internal bus), as the performance improved it became possible to include both feedback control and general machine coordination and sequencing. As a result, the general design rule for mechanical equipment has become: use mechanical components for elements that are needed for the delivery of power to the target system; consider use of software or electronics for all other coordinating elements. In practice this means, for example, that two shafts that are required to rotate at a fixed velocity ratio can be driven by separate motors. Software acts as the coordinating element between the two shafts. In the past it would have been necessary to use gears, a belt or some similar mechanical component to provide the coordination. When software is used, the ratio can be changed parametrically; the mechanical system has to be disassembled and reassembled to change ratio. This is a crucial advantage in today's world of "custom mass production" (i.e., short run production at costs that are close to what in the past could only be achieved with long runs).

These changes have important educational implications, particularly at introductory levels. While engineering students are now routinely exposed to classes in feedback control, they have little or no exposure to techniques for dealing with the level of organizational complexity common in modern mechanical systems. It is the thesis of this essay that concepts of " state," so elegantly woven into feedback control theory and instruction, could be used as a unifying principle for the broader subject of mechanical system control (mechatronics).

2. Context: Manufacturing Industry

The suggestions made here arise from observation of mechanical systems used in a variety of manufacturing industries. There has been a steady trend towards more complex machines, increasing dependence on control software for core machine functionality, higher performance and demands for increased reliability. Typical applications in manufacturing are (examples in parentheses): packaging (form, fill and seal), web processes (printing, fiber manufacturing), semiconductor manufacturing (wafer handling, planarization, process control), medical systems (therapeutic and diagnostic radiation systems), home furnishings (foam cutting, carpets), pipelines (automated welding), machine tools (milling machines, lathes), coordinate measuring machines, general purpose robotic manipulators. Less intense observation in other industrial sectors such as transportation leads to similar conclusions.

A machine used in any of these applications could have 5 to 20 (or more) servo-controlled axes (possibly with some stepping motor axes), 20 to 200 discrete and analog input/output points, a computer-based, graphical operator interface, an internal network connecting geographically distributed control elements, external communication with factory-wide networks. The control computation would be accomplished by one or a combination of integrated controllers, PLCs (programmable logic controllers), motion control subsystems, CNC (computer numerical control) units, PCs (personal computers), industrial or embedded computers.

Feedback control is present is almost all systems in these manufacturing applications. It appears commonly in two places: as motion control for servo-based axes and as process control for temperatures, flows, etc. as appropriate to the particular machine. Most feedback control problems encountered in this class of equipment can be solved with controllers of complexity corresponding to PID or lead-lag compensator type. Mechanical systems are normally designed to be stiff enough so that higher order controllers are not required (if compliance problems are detected the mechanical design is changed more often than the controller). Gain scheduling is used on occasion, for example, in controlling winding or unwinding in a web process.

Programming such systems should be a modest (but significant) part of an exercise in system design. Too often, however, it isn't. Rather, the software is dealt with as the entirety of the control system. It is suggested that an extended use of "state" could help correct these tendencies.

3. Hybrid Control

It might be argued at this point that the recent interest in the control theory community in hybrid control speaks to the issues raised here. It appears, however, that this subject is viewed more in the light of an extension to traditional concerns in feedback control system design (that is, stability, optimality, robustness) than as a means to highlight design problems in the types of machines described above. For example, of the papers in Control Using Logic-Based Switching (Morse, 1997) most look at problems in which switching and nonlinearity represent types of feedback control not dealt with by traditional control theory. Some representative titles of papers in that volume are: "Hybrid Control of Global Stabilization of Nonlinear Systems," "A Notion of Discontinuous Feedback," "Optimal Hybrid Control with Applications to Automotive Powertrain Systems," "Towards a General Complexity-Based Theory of Identification and Adaptive Control." Indeed, only one author in the volume bring up issues related to those presented here, Sangiovanni-Vincentelli (1997). Even he focuses on the feedback control problem as central to the embedded system design problem. In examining the design methodology he lists as the first step (p. 22), "Define the control goal, e.g., desired behavior, stability, robustness," and in examining the decomposition problem he states (p 23), "The design problem examined here is how to define the regions of operations of the plant and how to determine a control law based on an approximated model of the plant in each region." (Italics from the original.)

Hybrid control is an important area of control research. It does not, however, address the issues of concern here. In manufacturing systems of the types listed above, feedback control is used at the component level and usually does not represents a significant part of the design effort. It is crucial to proper system operation, but is used as a utility – a part of the system that requires some tuning but is otherwise expected to operate transparently.

It is also important to observe that the people involved in designing these systems (at least in the US) generally have Bachelors or Masters level educational background, most with little more than whatever control system courses were required in their curricula. Thus the focus here on the introductory control courses that people designing and implementing complex manufacturing systems are likely to have taken.

4. State

"The state of a system is a set of numbers such that the knowledge of these numbers and the input functions will, with the equations describing the dynamics, provide the future state and output of the system." (Dorf, 1980)

This definition, and others like it, fits the needs of feedback control system design very well. It focuses on prediction, which is needed to design algorithms for control of dynamic systems. In the kind of machine control described above, the feedback control is taken as a given and assumed to work properly. There is most often no model describing the behavior of the entire system so the "equations describing the dynamics... " do not exist. The lack of a model is less of a problem than it appears because with the underlying feedback control behaving predictably for those components needing it, stability, etc., is not the primary design problem. In the same sense, predictability of system behavior is available as long as feedback control error is kept very low, which it is through setpoint profiling. For example, too large a following error in a servo control will cause a system exception. Nonetheless, state information remains crucial so it can be used to define permissible activities – the key to maximizing productivity is always knowing what can, and can't, be done safely.

State information in machine control systems exists in several forms: as real numbers (or as close as computers can get to real numbers!) from analog sensors and from internal calculations, in logic form from discrete sensors, and in finite-state form from various control elements. All of this information taken together is needed to design appropriate sequences for achieving successful machine control.

5. Design and State

The fundamental design problem for control of complex machines is the specification of exactly what the machine is supposed to do – a functional specification. This must of necessity encompass a combination of simultaneous and sequential activities. Simultaneous because all of the parts of the machine operate at the same time. Sequential because the manufacturing process involves sequential steps in making a product. For continuous production machines such as web processes, sequence is involved in startup and shutdown.

One way to handle this specification is with a combination of tasks, representing simultaneous activity, with a finite-state machine detailing the internal, sequential activity of each task (Auslander, 1995, see also http://euler.Berkeley.EDU/~dma/res.html). This provides the necessary lev el of detail for implementing the design as well as an understandable medium of communication with others who could critique the design.

The task/finite-state machine model associates states with the control system software. Each task is represented by a transition logic system, which has a finite number of states. Thus, each task has an associated state variable which can take on a finite number of values representing the state of the finite-state machine.

The complete state of a system is thus represented by internal values of control components (positions, velocities, setpoints, etc.), measured values of external systems not under feedback control, and state variables of transition logic software elements.

6. State and Computing Media

A variety of computing media are used to implement control systems. While digital computing is at the heart of most of them, the specifics make a big difference in how much (if any) state information is available from one portion of the system to another.

6.1 Algorithmic Software

Algorithmically based programming languages (C, C++, Basic, etc.) are the most common ways of implementing the logic of a control system. They are characterized by being very broadly applicable but have very little depth. That is, because a language like C++ can be used for everything from word processors to machine tool controllers, there is essentially nothing in the syntax of the base language specific to control. Any context for "state" in such languages must be constructed outside the basic syntax. Without a mandated style for doing that (see Auslander, 1995) state information is usually buried in the code and becomes implicit rather than explicit. It is thus impossible for other tasks to use that information in a consistent way. On the other hand, if programs are constructed in such a way that state information is explicit, it would be readily available to other tasks.

6.2 Programmable Logic Controllers (PLCs)

Programmable logic controllers originally entered the market as replacements for relay logic systems. They were configured as industrially hardened computers programmed using "ladder logic," a language that is an analog for relay logic. Ladder logic has remained the dominant programming tool for PLCs (at least in the US) and is a major factor in their continuing popularity because the language is readily accessible to technicians as well as engineers and can be debugged and maintained on the factory floor.

Ladder logic, as implemented in a PLC, is actually a representation of synchronous sequential logic (for example, Dietmeyer, 1978). In contrast, relay logic, on which ladder logic is based, is a realization of asynchronous sequential logic, which is much more difficult to design than synchronous logic. Synchronous sequential logic design is normally based on state transition logic (finite-state machine) but the irony is that design of ladder logic programs rarely follow this model! A typical text on PLC programming (Johnson, 1987), for example, does not even have the word " state" in its index. Ladder logic programs can be based on the same state model used for software development (Auslander, 1996) but that or similar techniques do not seem to be used much in practice. The result is very much like that of algorithmic software: any information about state is buried in the program code and is not easily accessible.

PLC programming could be based on Grafcet, sequential functions charts (SFC) or Petri nets, but these are also not commonly used in the US.

PLCs usually offer external access to most internal information, so if state were made explicit it could be accessed by other system control components.

6.3 Hard-Wired Logic

Hard wired logic is also normally based on a synchronous sequential programming model and state transition diagrams. Thus state is inherent to the realization of logic. Programming for more complicated systems implemented as complex programmable logic devices (CPLD) or field-programmable gate arrays (FPGA) now resembles software more than it does traditional circuit diagrams of electronic hardware design. Design languages such as Abel and VHDL allow direct input of state transition logic using statements that look very much like standard programming languages.

State information from hardwired logic can be made available to other control components if the signals are brought outside the logic chips as digital logic signals. These would then need to be read by a digital I/O interface.

6.4 Robot Controllers

Robots are to the mechanical world what computers are to the rest of the world: general purpose devices that can be used to solve a variety of problems by appropriate programming. Robot controllers and their associated languages are almost all proprietary to the company selling the robot. Generally, these languages are command oriented and give no access to the robot's state information. Programming is commonly done with a combination of command scripts and teaching, that is, physically moving the robot to a desired location and then having the controller store that location for command access.

The lack of state information access for robots is not normally a problem when they are used stand-alone or only loosely connected to other machines. When used in close coordination with other machine elements, however, this can cause serious problems (see example, below).

6.5 Machine Tool Controllers (CNC)

Computer numerical controllers, CNCs, are designed for machining operations such as milling, turning, grinding, etc. They are most commonly programmed using G/M codes, a system originally developed for paper tape controlled machine tools. Little or no state information is available to other machine control elements. Like robots, this is not a problem when CNCs are used simply to control machine tools.

6.6 Integrated Controller

An integrated controller is a computer based machine controller that uses a single processor, or at least a single address space in the case of symmetric multiprocessing, to perform all of the functions of a machine control system. The advantage to such a configuration is that as long as proper programming conventions are followed, all tasks have immediate access to all state information. Disadvantages are that it must perform all of those functions as well as the specialized devices and that it may not support geographical distribution of computing as well.

7. State Information Availability

Although instant access to all state information is the ideal, it is rarely reached in practice. For the purpose of this discussion, "instant" is defined as available from RAM in a computer. Thus an integrated system could meet this ideal for a project small enough to be controlled by a single unit. However, system size and complexity, geographical constraints, or functional requirements often lead to systems with multiple processing units.

7.1 Task Information Needs

Once the functional specifications for all of the tasks have been established the information needs are the next step in the design process. In the design model being used here the task is considered to be the minimum computational unit, so information can be associated with tasks. This is an important consideration because information transfer among computing units (processes or threads) must be handled very carefully to assure data integrity.

Task use of its own state information is normally not a problem. What needs to be identified is its use of state information that originates outside of the task. For each task, then, state information needs must be listed according to the originating task for each piece of information along with the use to which the information will be put. Typical examples are mechanical position or position setpoint information to avoid collisions, temperatures to establish when processes can take place or identify dangerous conditions, state machine information for task coordination, etc.

This information provides an initial constraint on possible computational architectures for a control system. The computational components making up the system must be capable of supplying the needed information. For example, if detailed position information is required for a robot, a conventional robot controller may not be able to supply that information (see A Real World Example, below).

7.2 Update Rate

How often the state information needs to be updated and the constraints on latencies are the next crucial pieces of the design puzzle. This provides a second constraint on the computational architecture – how fast the interconnects among computational components must be in order to meet the update rate specifications. Latencies and update rate are closely related, but not exactly the same property. Update rate refers to how much data is required per unit time. For example, smooth profiling of an axis motion might require 500 data points per second. Latency refers to the elapsed time between the occurrence of an event and the completion of an action dependent on that event. For example, when a switch closure is the trigger for a motion, there could be a latency restriction that the motion must be completed within a specified time of the switch closure. Update rate depends primarily on the total data handling capability of the communication medium. Latency depends on communication data rate, but also can depend on how heavily loaded the communication channel is, what kinds of software and hardware drivers are used, etc.

7.3 Communication Modes

The amount of state information that can be exchanged among units is limited first by the bandwidth of the communication media connecting them. From the fastest to the slowest, common media used for control system interconnects are:

- Bus backplane (PCI, VME, etc.) 1 - 10 x 108 bits/sec
- High speed network (Ethernet, token ring, etc.) 1 - 10 x 107 bits/sec
- Serial (RS232 or RS422) 1 - 10 x 104 bits/sec
- Serial Buses (Bitbus, CAN and derivatives) 1 - 10 x 104 bits/sec

Bus/backplane systems can interconnect a small number of closely located processors. High speed networks can connect a much larger number of processors over a substantial distance (unlimited when used in wide-area network configurations). Serial (RS232, 422) are point-to-point only and usable over modest distances while serial buses can connect many nodes over a modest distance.

Of equal importance for reliable communication are the protocols for information transfer supported by each of these media. The bus/backplane and the serial connections have not established protocols while the high speed networks and the serial buses have well established protocols. These protocols establish packet-based transfer mechanisms that include basic error checking to assure data integrity. At the other extreme, reliable data transfer using a point-to-point serial connection is hard to guarantee unless some similar protocol is established on an ad hoc basis.

8. A Real-World Example: Wafer Handling

As the world of semiconductor manufacturing becomes more and more automated, handling wafers entering and leaving various process steps is becoming a standard part of many semiconductor manufacturing machines. With the shift in the industry from 200mm wafers to 300mm wafers automated wafer handling becomes even more imperative because of the weight of filled wafer cassettes and the value of partially processed wafers.

This particular wafer handling system is shown in schematic in 13.1. Wafer cassettes are delivered to the two front-opening unified pods (FOUPs). Robot #1 is used to remove wafers from the FOUPs and place them on the notch aligner. This device finds the notch on the wafer for correct orientation and also is used to make sure the wafer is centered. Robot #2 takes the wafer from the notch aligner and places it for processing. Processed wafers are handled in reverse – robot #2 removes them from the processing unit and places then on the notch aligner and robot #1 returns them to the FOUP.

The original design for this system included conventional robot controllers with serial connections for each of the robots. Vacuum grippers were used by the robots to hold on to the wafers. The FOUPs and notch aligners (there were two in the original design) also used serial connections. The notch aligners physically centered the wafer for robot pickup as well as aligning it correctly.

The wafer throughput of this system was never greater than one-third of the specified throughput.

With a deadline approaching, a redesign was launched to attempt to improve the system throughput. Because of the tight time schedule, it was

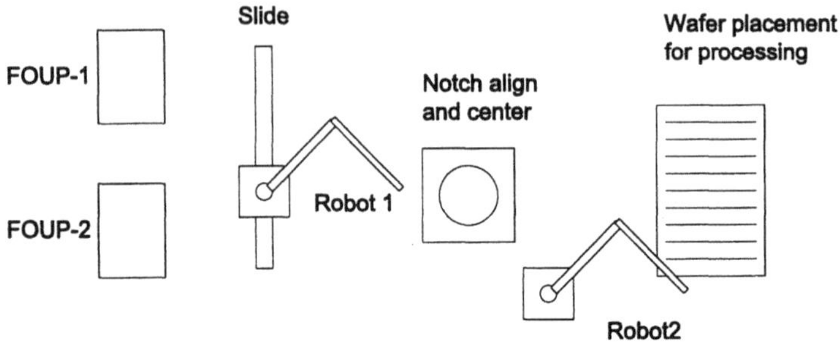

Fig. 13.1. Wafer Handling System

imperative that most of the changes be made in the control system rather than making major changes in the mechanical design.

Observations of the operation of the system with the original control scheme showed that three areas accounted for much of the delay in cycle time:

- Robot motions stopped between commanded operations.

- Overlapping of robot operations could not be accomplished because of potential collisions.

- Physical centering of the wafers on the notch aligners required a considerable amount of time.

With these in mind, a redesign was suggested which had as its major difference the use of an integrated controller in place of the robot controllers. The integrated controller had full access to all of the robot state information and was able to coordinate the robots to achieve highly overlapped operation. Because of the easy access to state information, robots could be safely moved in tandem even when they were in relatively close proximity. Because all of the details of the robot motions were known, the stops between commanded moves could also be eliminated and the overall robot acceleration and speed could be improved. Several other changes were also made:

- A servo controlled robot wrist was substituted for the original wrist which had more constrained motions. Control of this wrist was also done using the integrated controller. Wafer orientation motions could be made while other robot motions were going on. This capability is important in this system because there is a restriction that if there were to be loss of vacuum, and thus the potential of a dropped wafer, the falling wafer would never be in a position to hit other wafers in the system. Thus, the orientation had to be closely controlled with respect to instantaneous position of the end effector.

- The pair of notch aligners was replaced with a single notch aligner. The replacement unit computed the center position of the wafer, but did not physically recenter it, thus saving considerable time. Because the integrated controller was operating both of the robots, it could use the center position data to constantly adjust the robot pickup location thereby obviating the need for a physical centering step.

- Some configuration changes were made to smooth material flow in the unit and minimize required robot motions.

The work indicated here resulted in a nearly fourfold improvement in throughput. It was completed in approximately three months, including development of software for the integrated controller, design and manufacture of the new wrist, and integration of the new notch aligner. The overall cost of the new system was approximately the same as the cost of the original system.

9. Conclusion

The wafer handling example is a dramatic example of the necessity of basing a substantive part of the system design process on issues of state information distribution. While it is likely that not all systems will be able to see a quadrupling of productivity through changes in control system design and configuration, it is likely that many systems could see economically important gains in that manner. These results also open many conventional practices to questioning. Robot configurations based on serial communication, for example, are based on assumptions of stand-alone operation. But many current manufacturing systems use these robots in close coordination with other machinery. The same is true of other components. In the wafer handling example, tight time schedules did not permit examination of the effects of extending the integrated control to the notch aligner or FOUPs, but there could be some important savings there too. Finally, given the demonstrated importance of this material, students should be exposed to it as a regular part of control system curricula.

References

1. Auslander, D.M., A-C Huang, M. Lemkin, "A Design and Implementation Methodology for Real Time Control of Mechanical Systems," Mechatronics, Pergamon, 5, No. 7, 1995, pp 811-832. (Also see the downloadable file mechsoft.pdf at the Internet site: http://euler.Berkeley.EDU/~dma/res.html)
2. Auslander, D.M., C. Pawlowski, J. Ridgely, "Reconciling Programmable Logic Controllers (PLCs) with Mechatronics Control Software," IEEE International Conference on Control Applications, Detroit, MI, September, 1996. (Also available from the Internet site noted above.)
3. Dietmeyer, Donald L., Logic Design of Digital Systems, Allyn and Bacon, Inc., Boston, MA, 1978, p 304.
4. Dorf, Richard C., Modern Control Systems, Addison-Wesley, Reading, MA, 1980, p. 312.

5. Johnson, David G., Programmable Controllers for Factory Automation, Marcel Dekker, Inc., New York, 1987, 266 pages.
6. Morse, Stephen A., editor, Control Using Logic-Based Switching, Springer-Verlag, London, 1997, 276 pages.
7. Sangiovanni-Vincentelli, A., "Embedded System Design and Hybrid Simulation," in Morse (1997).

Discrete-time state representations, a new paradigm

S. Monaco[1] *and D. Normand-Cyrot*[2]

[1] Dipartimento di Informatica e Sistemistica, Università di Roma "La Sapienza", Via Eudossiana 18, 00184 Rome, Italy, e-mail: monaco@riscdis.ing.uniroma1.it

[2] Laboratoire des Signaux et Systèmes, CNRS-ESE, Plateau de Moulon, 91190 Gif sur Yvette, France, e-mail: cyrot@lss.supelec.fr

Dedicated to Ioan Doré Landau
on his 60th birthday

Summary. Continuing the study of discrete-time dynamical systems, a new state representation of discrete-time dynamics is proposed, whose development has been largely inspired by pioneering works of I. D. Landau. Breaking with the traditional idea of modelling discrete-time processes as difference equations we here propose to represent discrete-time dynamics as coupled difference and differential equations. The difference equation models, through jumps, the free evolution or a nominal one associated with some constant control value while the differential equation models the influence of the control. The differential structure is at the basis of a geometric study which enables us to stress some intriguing analogies with continuous-time, in particular regarding the control dependency of the dynamics compared to the time dependency. The present paper aims at convincing the reader that such a differential modelling might be the proper way to define and represent forced discrete-time behaviours. This would completely renew the understanding and the methodologies used when discussing discrete-time dynamics in modelling or control contexts.

1. Introduction

In some recent works ([10] [11],[13]) the authors show that under quite general assumptions, discrete-time dynamics admit exponential representations. Such a representation enables us to unify the study of discrete-time and sampled dynamics and provide the natural geometrical framework for setting concepts and results which generalize in the discrete-time context the continuous-time ones. With reference to a particular class of dynamics, it is shown in [11], that the exponential representation proposed in [10] is nothing else but the nonlinear flow associated with a certain differential equation assigned to the control variables.

This observation stimulated a reflection on the existence of an alternative mathematical representation for modelling forced discrete-time behaviours. Following this idea we here propose to represent discrete-time dynamics as coupled differential and difference equations. The differential equation models the control action, around an equilibrium point, or a nominal control value, while the difference equation models the free evolution. The forced evolution is analytically control-dependent while the free dynamics evolves along time instants as a difference equation through jumps. The present paper aims at convincing the reader that such a kind of hybrid representation as mixed differential and difference equations might be directly issued from modelling a forced discrete-time behaviour. This would completely renew the understanding and methodologies used when discussing discrete-time dynamics in modelling or control contexts.

In ([11], [12], [3], [4]) the benefits of these exponential and differential state representations have already been discussed with reference to particular cases and illustrated through the results obtained concerning invariance, accessibility, feedback linearization, decoupling or passivity conditions.

Hereafter this approach is developed in a general context. Defining discrete-time dynamics as mixed forced differential and free difference equations, we show that their integration along the control variables for suitable initialization along the free evolution restores usual difference equations. This is related to the exponential representation of the flow associated with the solutions of such differential equations.

Arguing thus, one finally recovers the analogy between small time and small control variations in continuous-time and discrete-time contexts, respectively. As a matter of fact analytic time-varying continuous-time dynamics admit the same formal exponential representations as discrete-time dynamics so that an extension of various control results set for stationary dynamics to time-varying systems appear to be possible thanks to the results obtained in discrete time.

The paper is organized as follows. In Section 2 we introduce our definition of forced discrete-time dynamics as coupled difference and differential state equations and motivate it with the example of modelling a controlled population growth. It is further shown in Section 3 how to compute the usual discrete time model from its differential representation. In Section 4, several specific situations are classified thus introducing first and higher-order discrete-time dynamics referring to the finite number of vector fields characterizing the control dependency as well as drift invertible and driftless dynamics.

2. Discrete-time dynamics as coupled difference and differential state equations

Definition 2.1. *(Analytically parametrized single-input discrete-time dynamics)*

Let $x \in M$, be an analytic manifold, $u \in U$, a neighbourhood of zero in $\mathbb{R}$ and assume the maps analytic in their arguments. Given a function $F_0 : M \to M$ and a parametrized vector field G^0 on M, an analytically parametrized single-input discrete-time dynamics is defined by the two following equations

$$x^+ = F_0(x) \tag{14.1}$$

$$\frac{d}{du}(x^+(u)) = G^0(x^+(u), u); \quad x^+(0) = x^+. \tag{14.2}$$

As a matter of fact, a nonlinear difference equation of the usual form

$$x_{k+1} = F(x_k, u_k) \tag{14.3}$$

can be recovered by integrating (14.2) between 0 and u_k for the initial condition $x_k^+(0)$ computed from (14.1) setting $x_k^+ = x_k^+(0) := F_0(x_k)$. One obtains,

$$x_{k+1} := x_k^+(u_k) = x^+(0) + \int_0^{u_k} G^0(x^+(v), v)dv$$

where $F : M \times U \to M$ is analytic and such that $F(., 0) = F_0(.)$.

According to Definition (2.1) the state at time $(k+1)$ is computed by mixing a difference equation generating a jump "after" time (k), together with a differential equation which models the control effect.

It should be pointed out that F_0 and G^0 may be assumed parametrized by $\bar{u}$, say $F_{\bar{u}}(x)$ and $G^0(\bar{u})(x, u)$, to describe a discrete dynamics around a nominal one assigned through the constant control value $\bar{u}$.

A controlled population model

The following example, from [14], shows that our approach can be usefully applied to model real phenomena. In [14] the authors make reference to a classical Leslie model describing the natural growth of age-structured populations adapted to the case of controlled populations. The starting point of the modelling procedure is the uncontrolled evolution described by the linear drift

$$F_0(x(k)) = Ax(k)$$

in which the i-th component $x_i(k)$ of the n-dimensional vector $x(k)$ is the number of individuals of age i at the beginning of year k. The $(n \times n)$ *Leslie matrix* A takes the form

$$A = \begin{pmatrix} s_0 f_1 & s_0 f_2 & \cdots & s_0 f_{n-1} & s_0 f_n \\ s_1 & 0 & \cdots & 0 & 0 \\ 0 & s_2 & & & 0 \\ \vdots & & \ddots & & \vdots \\ 0 & \cdots & 0 & s_{n-1} & 0 \end{pmatrix}$$

where $s_i < 1, i = 0, \ldots, n-1$, are strictly positive *survival coefficients* and $f_i, i = 0, \ldots, n$, are nonnegative *fertility coefficients.* The growth is controlled by removing individuals (forest exploitation, harvesting, hunting, ...). Thus the removing process must be modelled together whith the growing process in order to obtain a complete description of the process. It is quite intuitive to assume that the removing season starts when the reproduction is over and that the *harvesting effort, u,* is constant, the rate of change of the population due to harvesting is proportional to the number of individuals, which is equivalent, in our context, to assuming

$$G^0(x, u) = -Qx$$

where Q is the matrix of the *catchability coefficients* and can be assumed diagonal

$$Q = \begin{pmatrix} q_0 & & & 0 \\ & q_1 & & 0 \\ & & \ddots & \\ 0 & & & q_{n-1} \end{pmatrix}$$

With this in mind and following our approach, one sets,

$$x^+ = Ax \tag{14.4}$$

$$\frac{d}{du}(x^+(u)) = -Qx^+(u); \quad x^+(0) = x^+. \tag{14.5}$$

We will afterwards show that the model (14.4-14.5) restores the one computed in [14].

3. The discrete-time evolution from its differential representation

In the following we study the relations between $F(x, u)$ in (14.3) and $G^0(x, u)$ in (14.2). To do so, we refer to the expansions

$$G^0(x, u) = G_1^0(x) + \sum_{i \geq 1} \frac{u^i}{i!} G_{i+1}^0(x) \tag{14.6}$$

$$F(x,u) := F_0(x) + \sum_{i\geq 1} \frac{u^i}{i!} F_i(x) \tag{14.7}$$

Let us define the operator $\mathcal{D}_u := \frac{\partial}{\partial u} + L_{G^0(x,u)}$, where L_G stands for the Lie derivative associated with G (i.e. $L_G(F) := J_x[F] \times G$), where $J_x[.]$ denotes the Jacobian with respect to x of the function in between the brackets.

Lemma 3.1. *Given the discrete-time dynamics (14.1-14.2), $F : M \times U \to M$ in (14.3) satisfies the power expansion (14.7) with*

$$F_i(x) = \mathcal{D}_u^{i-1}(G^0(x,u))\big|_{u=0}\big|_{x=F_0(x)} \tag{14.8}$$

Proof. Replacing $x^+(u)$ with $F(x,u)$, (14.2) can immediately be rewritten as

$$\frac{d}{du}(F(x,u)) = G^0(F(x,u),u). \tag{14.9}$$

from which successive derivatives w.r.t. u easily prove (14.8), i.e.,

$$\begin{aligned}
\frac{d}{du}(F(x,u)) &= G_1^0(F_0(x)) + \sum_{i\geq 1} \frac{u^i}{i!}(\mathcal{D}_v^i(G^0(x,v))\big|_{v=0})\big|_{x=F_0(x)} \\
&:= e^{u\mathcal{D}_v}(G^0(x,v))\big|_{v=0}\big|_{x=F_0(x)}
\end{aligned}$$

Specifying (14.8) one finds for the first terms

$$\begin{aligned}
F_2(x) &= [\frac{\partial G^0(.,v)}{\partial v} + L_{G^0(.,v)}(G^0(x,v))]\big|_{v=0}\big|_{x=F_0(x)} \\
&= G_2^0(x)\big|_{x=F_0(x)} + L_{G_1^0(.)}(G_1^0(x))\big|_{x=F_0(x)} \\
F_3(x) &= G_3^0(x)\big|_{x=F_0(x)} + 2L_{G_1^0(.)}(G_2^0(x))\big|_{x=F_0(x)} \\
&\quad + L_{G_2^0(.)}(G_1^0(x))\big|_{x=F_0(x)} + L_{G_1^0(.)} \circ L_{G_1^0(.)}(G_1^0(x))\big|_{x=F_0(x)} \\
&\dots
\end{aligned}$$

Exponential representation

The evolution over one step described by (14.3), takes an exponential form, namely the exponential representation of the flow, $\phi_u^{G^0}(x)$, of the differential equation (14.2). Hereafter, one specifies its expression in terms of the canonical vector fields $(G_i^0(.); i \geq 1)$'s.

Theorem 3.1. *The solution of the differential equation (14.2) initialized at (14.1) takes the form*

$$x^+(u) = e^{u\mathcal{G}^0(.,u)}(Id)\big|_{x^+(0)} \tag{14.10}$$

where $u\mathcal{G}^0(x,u) := M \to T_xM$ *is a Lie element in the* $(G_i^0(.) : i \geq 1)$*'s whose expansion is given by*

$$u\mathcal{G}^0(.,u) = \sum_{i\geq 1} \frac{u^i}{i!} B_i(G_1^0, \ldots, G_i^0) \tag{14.11}$$

where B_i *stands for a homogeneous Lie polynomial of degree* i *in its arguments when* $G_i^0(.)$ *is, by convention, said to be of degree* i.

The proof of Theorem **3.1**, detailed in [10], can be worked out by performing the logarithmic expansion of

$$\begin{aligned} F(x,u) &= e^{u\mathcal{D}_v}(Id)\big|_{v=0}\big|_{x=F_0(x)} \\ &= [x + uG_1^0(x) + \sum_{i\geq 2} \frac{u^i}{i!}(\mathcal{D}_v^{i-1}(G^0(x,v)))\big|_{v=0}]\big|_{x=F_0(x)} \end{aligned}$$

The expression of the (B_i)'s as homogeneous polynomials in the (G_i^0)'s are given in [10] from which one deduces their Lie polynomials expressions. For the first terms, one finds

$$u\mathcal{G}^0(.,u) = uB_1(G_1^0) + \frac{u^2}{2!}B_2(G_1^0, G_2^0) + \frac{u^3}{3!}B_3(G_1^0, G_2^0, G_3^0) + O(u^4)$$

with

$$\begin{aligned} B_1 &= G_1^0, B_2 = G_2^0, \\ B_3 &= G_3^0 + 1/2[G_1^0, G_2^0], B_4 = G_4^0 + [G_1^0, G_3^0], \end{aligned}$$

where $[.,.]$ indicates the Lie bracket of two vector fields.

Remark 3.1. This representation reveals why the Lie algebra $\mathcal{L}$, generated by vector fields $((G_i^0); i \geq 1)$'s and the associated Lie group structure enter in the controllability, accessibility, invariance, properties of discrete-time dynamics as analysed in ([8], [7], [2]) and further developed with reference to a particular class of systems in [11].

Starting from Definition (2.1), we shown how to build a difference equation in the usual sense (14.3) as the result of the integration of the given differential equation (14.2) with suitable initialization (14.1) deduced from the free evolution. Reciprocally, starting from an analytic difference equation $x_{k+1} = F(x_k, u_k)$, assuming the existence of a function G^0, analytic on M, parametrized by u and satisfying condition (14.9), we have shown in [13] how to interpret x_{k+1} as the result of the integration between 0 and u_k of (14.9) with initialization in $F_0(x_k)$. This comment reinforces the dual meaning of the present analysis in order to propose a new paradigm for discrete-time state representations. Let us finally note that the invertibility of $F_0(.)$ (or more generally of $F_{\bar{u}}(.) = F(., \bar{u})$ for a given control value) is a necessary and sufficient condition for obtaining two identical formulations since in such a case the existence and unicity of $G^0(x,u)$ are proven.

The exponential representation of the population model

We can now go back to our example and compute the exponential representation associated with the differential representation (14.4-14.5). One has

$$x(k+1) = F(x(k), u(k)) = e^{-u(k)Q}(Id)|_{x=Ax(k)}$$

The so computed state difference equation

$$x(k+1) = Ax(k) - u(k)QAx(k) + \frac{u(k)^2}{2}Q^2Ax(k) + \cdots = e^{-u(k)Q}Ax(k)$$

restores the model proposed in [14] thanks to quite complex and less intuitive arguments.

The associated flow

The flow associated with the solution of (14.2), denoted by $\phi_u^{G^0}$, can usually be defined by its asymptotic expansion

$$\begin{aligned}\phi_u^{G^0} &:= \overrightarrow{\exp} \int_0^u G^0(., v)dv \\ &:= I + \sum_{m\geq 1} \int_0^u \int_0^{v_1} \cdots \int_0^{v_{m-1}} G^0(., v_{m-1}) \circ \ldots \circ G^0(., v_1) dv_m \ldots dv_1.\end{aligned}$$

where I represents the identity operator. It satisfies (14.2), i.e.

$$\frac{d}{du}(\phi_u^{G^0}(.)) = G^0(\phi_u^{G^0}(.), u).$$

One has,

Lemma 3.2. *Given a discrete-time dynamics (14.1-14.2) then, for any x in M, there exists an open neighbourhood $\mathcal{U}$ of 0, depending on x and a smooth map $\phi_.^{G^0} := \mathcal{U} \times M \to M$ such that the solution of (14.2), initialized at (14.1) takes for $u \in \mathcal{U}$ the form*

$$x^+(u) := \phi_u^{G^0}(x^+(0)) = e^{u\mathcal{G}^0(.,u)}(Id)\big|_{x^+(0)}. \tag{14.12}$$

Lemma 3.2 simply clarifies that for any x in M the map $\mathcal{I}_x : \mathcal{U} \to M$ defined by $\mathcal{I}_x(u) := \phi_u^{G^0}(x)$, is the integral curve of $G^0(., u)$ passing through x; i.e. the u-dependent vector field $G^0(., u)$ is the infinitesimal generator of the group action associated with $\phi_u^{G^0}$.

Remark 3.2. It is interesting to note that the results of Theorem 3.1 and Lemma 3.2 apply to characterizing the flow associated with a nonlinear time-varying differential equation of the form

$$\frac{d}{dt}(x(t)) = f(x(t),t) := f_0(x(t)) + \sum_{i\geq 1} \frac{t^i}{i!} f_i(x(t)) \tag{14.13}$$

where for $i \geq 0$, the functions $f_i(.)$ are analytic from M to M and describe the expansion around 0 of $f(.,t)$. Similarly to the previous analysis, the solution $x(t)$ of (14.13), at time $t \in T$, a neigborhood of $t = 0$, is given by

$$x(t) = x(t_0) + \int_0^t f(x(s),s)ds$$

which admits an exponential representation similar to (14.10-14.11), i.e.,

$$x(t) = \phi_t^f(x(0)) = e^{t\mathcal{F}(.,t)}(Id)\big|_{x(0)}$$

with

$$t\mathcal{F}(.,t) = \sum_{i\geq 1} \frac{t^i}{i!} B_i(f_0, \dots, f_{i-1}),$$

thus characterizing the exponential representation of the associated flow, ϕ_t^f, usually defined by the asymptotic expansion

$$\overrightarrow{\exp} \int_0^t f(.,s)ds = I + \sum_{m\geq 1} \int_0^t \int_0^{t_1} .. \int_0^{t_{m-1}} f(.,t_{m-1}) \circ \dots \circ f(.,t_1) dt_m .. dt_1.$$

Input-output behaviour

Given an analytic output map, $H : M \to M$ and denoting by $y^+(u) = H(x^+(u))$ its evaluation at $x^+(u)$, one gets the following set of coupled difference and differential equations,

$$y^+ = H \circ F_0(x)$$

$$\frac{d}{du}(y^+(u)) = L_{G^0(.,u)}(y^+(u)), \quad y^+(0) = y^+$$

which define the input-output behaviour over one step, i.e.,

$$\begin{aligned} y_{k+1} &:= H(x^+(u_k)) = H \circ F_0(x_k) + \int_0^{u_k} L_{G^0(.,v)}(H)(x^+(v))dv \\ &= e^{u_k \mathcal{G}^0(.,u_k)}(H)\big|_{F_0(x_k)} \end{aligned} \tag{14.14}$$

4. Some specific cases

First order discrete-time dynamics

Let the assumption **A** : $G_i^0(x) = 0$ for all $i \geq 2$.

Under **A**, the vector field $G^0(x,u)$ does not depend on the control. Such a dynamics can be defined from (14.1-14.2) as follows:

Definition 4.1. *(First-order analytically parametrized single-input discrete-time dynamics)*
Let $u \in U$ and $x \in M$ and assume the maps analytic in their arguments. Given a function $F_0 : M \to M$ and a parametrized vector field G^0 on M, a first-order analytically parametrized single-input discrete-time dynamics is defined by the two equations,

$$x^+ = F_0(x) \tag{14.15}$$

$$\frac{d}{du}(x^+(u)) = G_1^0(x^+(u)); \quad x^+(0) = x^+. \tag{14.16}$$

At a first glance, assumption **A** may appear quite restrictive. However, if one accepts the paradigm to modelling discrete-time dynamics from computing the control directional derivative $G^0(.,u)$ of the control-dependent state evolution $x^+(u)$, it turns out to be rather natural to obtain $G^0(.,u)$ independent from u as a first approximation. In fact, **A** must be compared to assuming stationary dynamics rather than time-varying ones when modelling continuous-time processes. Arguing so, equations (14.15-14.16) will be referred to as first-order discrete-time dynamics, while higher-order dynamics will refer to polynomial approximations w.r.t. u of $G^0(.,u)$. In addition, let us note that the state extension $\frac{du}{du} = 1$ enables us to recover this simplest case from the general situation described by equations (14.1-14.2).

Let us further investigate such a class of systems discussed in [11] as the discrete-time analogues of affine continuous-time dynamics. The reason for such an analogy is that first-order discrete-time dynamics are described by only one forced vector field G_1^0 and the drift term F_0 as affine continuous-time dynamics are described by the well-identified f and g. Hereafter, specifying the results of section 3 we explain how the study can be simplified in this particular class when discussing criteria characterizing structural and control properties. The same reasoning can be extended to the muti-input case.

A first preliminary comment is that under **A**, one easily deduces the existence of a coordinates change $z = T(x)$ under which $G_1^0(x)$ is transformed into a constant vector field B.

Linear dynamics enter in this class assuming in addition the linearity of F_0. Given A and B, two matrices of suitable dimensions and setting $x^+(u) = Ax + Bu$, one gets $x^+ = Ax$ and $\frac{d(x^+(u))}{du} = B$. Starting from x_k for all $k \geq 0$, one first computes $x_k^+(0) = x_k^+ = Ax_k$ and then recovers

$$x_k^+(u_k) := x_{k+1} = Ax_k + \int_0^{u_k} B dv = Ax_k + Bu_k$$

after integration between 0 to u_k.

Secondly, for any $x_k \in M$ and $u_k \in U$, integrating (14.16) between 0 and u_k with intialization in $x_k^+(0) = F_0(x_k)$, one recovers a discrete-time dynamics in the usual sense,

$$x_{k+1} := x_k^+(u_k) = x^+(0) + \int_0^{u_k} G_1^0(x^+(v))dv := F(x_k, u_k) \tag{14.17}$$

where $F(x,u) : M \times U \to M$ is nonlinear in u but can easily be computed as

$$F(x,u) = (I + uL_{G_1^0} + \sum_{i\geq 2} \frac{u^i}{i!} L^i_{G_1^0})(Id)\big|_{x=F_0(x)} = e^{uL_{G_1^0}}(Id)\big|_{x=F_0(x)} \tag{14.18}$$

It is thus immediately verified that the flow associated with the solution (14.17) of (14.16), denoted by $\phi_.^{G_1^0} : U \times M \to M$ takes presently the simple form $\phi_u^{G_1^0} = e^{uG_1^0}$.

It follows that the group action associated with $\phi_u^{G_1^0}$ is the usual linear parameter group satisfying (14.16) when $x^+(u) = \phi_u^{G_1^0}(x)$. More precisely, denoting $\phi_u^{G^0}$ as $\phi_{u,0}^{G^0}$ to precise the integral bounds, one obtains $\phi_{u,u}^{G_1^0}(x) = x$, $(\phi_{u,\bar{u}}^{G_1^0})^{-1}(x) = e^{-(u-\bar{u})G_1^0}(x)$ and for any $\bar{u}, u \in U^2$,

$$x^+(u) = \phi_{u,0}^{G_1^0}(x^+(0)) = \phi_{u,\bar{u}}^{G_1^0}(\phi_{\bar{u},0}^{G_1^0}(x^+(0))),$$

which is nothing else but

$$e^{uG_1^0} = e^{\bar{u}G_1^0} {}_o e^{(u-\bar{u})G_1^0} = e^{\bar{u}G_1^0 + (u-\bar{u})G_1^0}.$$

Accessibility and observability properties can therefore easily be deduced. As an example, it is straightforward to see that the set of accessible states in one step is the integral curve of $G_1^0(.)$ passing through $F_0(x)$, i.e. $\mathcal{I}_x(u) := \phi_u^{G_1^0}(F_0(x))$.

When considering an analytic output map $H : M \to M$ one finds the integral representation

$$y_{k+1} = H(x^+(0)) + \int_0^{u_k} L_{G_1^0}(y^+(v))dv\big|_{x^+(0)} = e^{u_k L_{G_1^0}}(H)\big|_{F_0(x_k)}$$

In a Lyapunov design context, considering a positive real valued map $V : M \to \mathbb{R}$, one can express the first order difference by the integral form

$$V(x_{k+1}) - V(x_k) = V(F_0(x_k)) - V(x_k) + \int_0^{u_k} L_{G_1^0}(V(x^+(v)))\big|_{F_0(x_k)} dv$$

from which passivity and losslessness criteria can be deduced as in [12].

Let us note that these representations are at the basis of the definition of canonical normal forms in discrete time, which provide the suitable geometric tools for studying decoupling with stability [3], or adaptive control design.

Dynamics with invertible drift term

Assuming F_0 invertible enables us to describe the dynamics over several time-steps as composed exponential forms. Let $u_0, \ldots, u_p$ in U^{p+1} be any input sequence of lenght $p+1$ with $p \geq 0$.

Starting from (14.1-14.2) with F_0 invertible, one builds the transport of $G^0(., u)$ along the drift term $F_0(x)$ thus obtaining $G^1(x, u)$, i.e.,

$$G^1(., u) := Ad_{F_0} G^0(., u) := [\frac{\partial F_0(x)}{\partial x} \times G^0(., u)]|_{F_0^{-1}(.)}$$

Iteratively, one defines,

$$G^p(., u) := Ad_{F_0} \circ \ldots \circ Ad_{F_0} G^0(., u) = Ad^p_{F_0} G^0(., u) \tag{14.19}$$

and its expansion with respect to u

$$G^p(., u) = G^p_1(x) + \sum_{i \geq 1} \frac{u^i}{i!} G^p_{i+1}(x) \tag{14.20}$$

thus defining the $(G^p_i(x))$'s

$$G^p_i(x) := [\frac{\partial F^p_0(x)}{\partial x} \times G^0_i]|_{F_0^{-p}(x)} := Ad_{F_0^p} G^0_i(.) \tag{14.21}$$

Rewriting the composed dynamics over $p+1$ steps, $F(., u_p) \circ \ldots \circ F(., u_0)$, as

$$F(., u_p) \circ F_0^{-1} \circ F_0 \circ F(., u_{p-1}) \circ F_0^{-2} \circ \ldots \circ F(., u_0) \circ F_0^{-p+1} \circ F_0^{p+1}(.)$$

one immediately verifies that the state reached from x_0 in $p+1$ steps admits the exponential representation

$$\mathbf{x}^+_{p+1}(\mathbf{u}) = e^{u_0 \mathcal{G}^p(., u_0)} \circ \ldots \circ e^{u_p \mathcal{G}^0(., u_p)}(Id)|_{\mathbf{x}^+_{p+1}(\mathbf{0})} \tag{14.22}$$

with $\mathbf{x}^+_{p+1}(\mathbf{0}) := F_0^{p+1}(x_0)$.

From such an exponential representation, one easily deduces the Volterra type expansions [9] both for discrete-time input-state and input-output behaviours and computes the successive kernels in terms of compositions of vector fields. In [13], it is also shown that, by associating a differential equation to any of these so-defined vector fields, transported along the dynamics, one gets a set of triangular partial differential equations whose successive integration is possible thus recovering the flow which characterizes the dynamics over several steps. This is not presently explained in detail.

Driftless dynamics

Such a case is of interest for a detailed study. Since $F_0(x) = x$ in (14.1), the discrete-time dynamics is defined by the control-dependent differential equation (14.2) only.

In particular, the group action associated with the flow exactly corresponds to the operation of composition of analytic functions which comes out when computing the behaviour over several steps. It follows that the set of accessible states at any time instant belongs to the orbit of this group. For any $x_0 \in M$, and any sequence of inputs $u_0, \ldots, u_p$ one gets at the place of (14.22),

$$\mathbf{x}_{p+1}^{+}(\mathbf{u}) = e^{u_0 \mathcal{G}^0(.,u_0)} \circ \ldots \circ e^{u_p \mathcal{G}^0(.,u_p)}(Id)\Big|_{x_0}.$$

The Lie algebra generated by the vector fields in $span\{G_1^0, \ldots, G_i^0\}$, defines a foliation of the state space. From the previous expression and the Baker-Campbell-Hausdorff formula, it follows that the states reachable from x_0, after several steps, under small control values in order to keep the states in a neighbourhood of x_0, is a subset of the slice passing through x_0.

5. Conclusions

In this paper we proposed a representation of discrete-time dynamics as coupled difference and differential equations. The idea at the basis of this new paradigm is the modelling of the discrete-time behaviour around the free evolution as a control-derivative equation. We developed this thesis in the general case and considering nonlinearly u-parametrized differential equations but the simplest case, corresponding to a non u-parametrized differential equation, looks more adequate in practice. Arguing thus, we have drawn a parallel between time and control in continuous-time and discrete-time contexts, respectively. Pushing further the analogy, it comes to mind to compare the degrees of freedom associated with the control variables in continuous time with the possibility to represent discrete-time dynamics over several time steps. In fact controllability and observability properties in discrete-time differ wether or not one considers one or several time steps. This analogy becomes also clear when recalling the well-known benefits of multirate control design in discrete time.

References

1. A.A. Agracev and R.V. Gramkrelidge, *Exponential representations of flows and the chronological calculus*, Math. Sb., **107, 4**, 1978, pp. 467–532.
2. F. Albertini and E.D. Sontag, *Discrete-time transitivity and accessibility: analytic systems*, SIAM J. Cont. Opt., **33**, 1993, pp. 1599–1622.

3. C. Califano, S. Monaco and D. Normand-Cyrot, *Nonlinear decoupling with stability via static state feedback for a class of discrete-time systems*, Proc. IEEE-CDC, 1997.
4. C. Califano, S. Monaco and D. Normand-Cyrot, *A note on the discrete-time normal form*, IEEE Trans. on A.C., 1998.
5. M. Fliess, *Automatique en temps discret et algèbre aux différences*, Forum Math., **2**, 1990, pp. 213–232.
6. J.W. Grizzle, *A linear algebraic framework for the analysis of discrete time nonlinear systems*, SIAM J. on Cont. Opt., **31**, 1993, pp. 1026–44.
7. B. Jakubczyk and E.D. Sontag, *Controllability of nonlinear discrete time systems; a Lie algebraic approach*, SIAM J. Cont. and Opt., **28**, 1990.
8. S. Monaco and D. Normand Cyrot, *Invariant distributions for discrete time nonlinear systems*, Syst. and Cont. Lett., **5**, 1985 , pp. 191–196.
9. S. Monaco and D. Normand-Cyrot, *Functional expansions for nonlinear discrete time systems*, Math. Syst. Th., **21**, 1989, pp. 235–254.
10. S. Monaco and D. Normand-Cyrot, *A unifying representation for nonlinear discrete-time and sampled dynamics*, J. of Math. Syst., Est. and Cont., pp. 101–103 (summary) 1995, vol.7, **4**, 1997, 477-503.
11. S. Monaco and D. Normand-Cyrot, *Geometric properties of a class of nonlinear discrete-time dynamics*, Proc. ECC-97, **1**, TU-E-C5, Brussels, 1997.
12. S. Monaco and D. Normand-Cyrot, *On the conditions of passivity and losslessness in discrete-time*, Proc ECC-97, **3**, WE-A-J4, Brussels, 1997.
13. S. Monaco and D. Normand-Cyrot, *Differential representations with jumps for discrete-time dynamics*, Tech. Rep. 39-97, Dipt. Informatica e Sistemistica, University of Rome "La Sapienza", 1997.
14. S. Muratori and S. Rinaldi, *Structural properties of controlled population models*, Syst. and Cont. Lett., **10**, 1996 , pp. 147–153.
15. L.A.B. San Martin, *On global controllability of discrete-time control systems*, MCSS **8**, 1995, pp. 279–297.

Bias, Variance and Optimal Experiment Design: Some Comments on Closed Loop Identification

L. Ljung and U. Forssell

Division of Automatic Control Department of Electrical Engineering, Linköping University, S-581 83 Linköping, Sweden, e-mail: ljung,ufo@isy.liu.se

Summary. In this contribution we shall describe a rather unified way of expressing bias and variance in prediction error estimates. The emphasis is on systems operating in closed loop. We shall describe the identification criterion function in the frequency domain. The crucial entity is the joint spectrum of input and noise source. Different factorizations of this spectrum give different insights into the bias mechanisms of closed loop identification. It will be shown that so called *indirect identification* is the answer to the question of how to obtain consistent estimates of the dynamics part, even with an erroneous noise model. We also consider optimal design of experiments that seek to minimize the weighted variance of the dynamics estimate. It is shown that open loop experiments are optimal if the input power is constrained. However for any criteria that involve any kind of constraints on the output power, closed loop experiments will be optimal. The optimal regulator does not depend on the weighting function in the criterion to be minimized.

1. Introduction and Setup

Identification of systems operating in closed loop have long been of interest. The reasons are that many systems are not allowed to operate in open loop during an identification experiment. Adaptive control is another situation, where closed loop identification issues naturally arise. See among many references also [15]. The recent interest in so called identification for control has also spurred new methods and results, [18, 13, 4],[6] and [17].

See, among many general references on closed loop identification, [8], [16], [1], [9], [3], [2, 11, 5] and [12].

We shall consider identification of a linear system in a traditional prediction error setup. See [14] for all technical details. The true system is supposed to be described by

$$y(t) = G_0(q)u(t) + H_0(q)e(t) \tag{15.1}$$

where q is the delay operator, u is the input, y is the output and e is white noise with covariance matrix Λ_0.

The system is operating under arbitrary feedback, but we assume that all signals are quasi-stationary, so that the spectrum of

$$\zeta = \begin{bmatrix} u \\ e \end{bmatrix} \tag{15.2}$$

is well defined, and denoted by

$$\Phi_\zeta(\omega) = \begin{bmatrix} \Phi_u(\omega) & \Phi_{ue}(\omega) \\ \Phi_{eu}(\omega) & \Lambda_0 \end{bmatrix} \tag{15.3}$$

The system is identified within the model structure

$$y(t) = G(q,\theta)u(t) + H(q,\theta)e(t) \tag{15.4}$$

G will be called the dynamics model and H the noise model. The parameter θ is estimated by

$$\hat{\theta}_N = \arg \min_{\theta \in D_{\mathcal{M}}} V_N(\theta, Z^N) \tag{15.5}$$

$$V_N(\theta, Z^N) = \frac{1}{N} \sum_{t=1}^{N} \varepsilon^T(t,\theta)\Lambda^{-1}\varepsilon(t,\theta) \tag{15.6}$$

$$\varepsilon(t,\theta) = y(t) - \hat{y}(t|\theta) = H^{-1}(q,\theta)(y(t) - G(q,\theta)u(t)) \tag{15.7}$$

Here Λ is a symmetric, positive definite weighting matrix.

We shall discuss the asymptotic properties of $\hat{\theta}_N$ in the sequel.

2. Expressions for the Data Spectrum

The data spectrum Φ_ζ plays an important role in the analysis, and we shall therefore collect some results on it here. We shall introduce the following spectrum

$$\Phi_u^r = \Phi_u - \Phi_{ue}\Lambda_0^{-1}\Phi_{eu} \tag{15.8}$$

where we suppress the argument ω. This is the spectrum of that part of the input u, that cannot be estimated from e by a linear, time-invariant filter.

Similarly we introduce

$$\Phi_e^r = \Lambda_0 - \Phi_{eu}\Phi_u^{-1}\Phi_{ue} \tag{15.9}$$

The data spectrum Φ_ζ can now be written as

$$\begin{bmatrix} \Phi_u & \Phi_{ue} \\ \Phi_{eu} & \Lambda_0 \end{bmatrix} = \begin{bmatrix} I & \Phi_{ue}\Lambda_0^{-1} \\ 0 & I \end{bmatrix} \begin{bmatrix} \Phi_u^r & 0 \\ 0 & \Lambda_0 \end{bmatrix} \begin{bmatrix} I & 0 \\ \Lambda_0^{-1}\Phi_{eu} & I \end{bmatrix} \tag{15.10}$$

$$= \begin{bmatrix} I & 0 \\ \Phi_{eu}\Phi_u^{-1} & I \end{bmatrix} \begin{bmatrix} \Phi_u & 0 \\ 0 & \Phi_e^r \end{bmatrix} \begin{bmatrix} I & \Phi_u^{-1}\Phi_{ue} \\ 0 & I \end{bmatrix} \tag{15.11}$$

$$\tag{15.12}$$

From these factorization results we also find an expression for the inverse

$$\Phi_\zeta^{-1} = \begin{bmatrix} \Phi_u & \Phi_{ue} \\ \Phi_{eu} & \Lambda_0 \end{bmatrix}^{-1} = \begin{bmatrix} (\Phi_u^r)^{-1} & -(\Phi_u^r)^{-1}\Phi_{ue}\Lambda_0^{-1} \\ -\Lambda_0^{-1}\Phi_{eu}(\Phi_u^r)^{-1} & (\Phi_e^r)^{-1} \end{bmatrix} \tag{15.13}$$

In case the regulator is linear, time-invariant, we have

$$u(t) = r(t) - K(q)y(t) \tag{15.14}$$

where $K(q)$ is a linear regulator of appropriate dimensions and where the reference signal $\{r(t)\}$ is independent of $\{e(t)\}$. We then have the following expressions. Let S and S^i the the output and input sensitivity functions:

$$S_0 = (I + G_0K)^{-1}, \quad S_0^i = (I + KG_0)^{-1} \tag{15.15}$$

Then

$$\Phi_u = S_0^i\Phi_r(S_0^i)^* + KS_0\Phi_vS_0^*K^* \tag{15.16}$$

where Φ_r is the spectrum of the reference signal and $\Phi_v = H_0\Lambda_0H_0^*$ the noise spectrum. Superscript $*$ denotes complex conjugate transpose. We shall denote the two terms in (15.16)

$$\Phi_u^r = S_0^i\Phi_r(S_0^i)^* \tag{15.17}$$

and

$$\Phi_u^e = KS_0\Phi_vS_0^*K^* = S_0^iK\Phi_vK^*(S_0^i)^* \tag{15.18}$$

The cross spectrum between u and e is

$$\Phi_{ue} = -KS_0H_0\Lambda_0 = -S_0^iKH_0\Lambda_0 \tag{15.19}$$

3. The Main Expression

From standard asymptotic theory we know that

$$\hat{\theta}_N \to \arg\min_{\theta} \overline{V}(\theta) \quad \text{w.p.1 as} \quad N \to \infty \tag{15.20}$$

where

$$\begin{aligned}\overline{V}(\theta) &= \overline{E} V_N(\theta, Z^N) \\ &= \int_{-\pi}^{\pi} \operatorname{tr}\left\{ \Lambda^{-1} H_\theta^{-1} \begin{bmatrix} \Delta G_\theta & \Delta H_\theta \end{bmatrix} \Phi_\zeta \begin{bmatrix} \Delta G_\theta^* \\ \Delta H_\theta^* \end{bmatrix} (H_\theta^{-1})^* \right\} d\omega\end{aligned} \tag{15.21}$$

Here we have introduced the simplified notation

$$\begin{aligned}\Delta G_\theta &= G_0(e^{i\omega}) - G(e^{i\omega}, \theta) \\ \Delta H_\theta &= H_0(e^{i\omega}) - H(e^{i\omega}, \theta)\end{aligned}$$

To see (15.21) we rewrite (15.7) using (15.1) as

$$\begin{aligned}\varepsilon &= H_\theta^{-1}(y - G_\theta u) = H_\theta^{-1}(G_0 - G_\theta)u + H_\theta^{-1} H_0 e \\ &= H_\theta^{-1}(G_0 - G_\theta)u + (H_\theta^{-1} H_0 - I)e + e \\ &= H_\theta^{-1}[(G_0 - G_\theta)u + (H_0 - H_\theta)e] + e\end{aligned}$$

We then use that H_0 and H_θ are both monic (so that the difference starts with a delay) and that $e(t)$ is independent of $(G_0 - G_\theta)u(t)$, which is the case if either the regulator or the model/system contains a delay. Parseval's relationship then gives (15.21).

4. Identifiability, Bias and the Indirect Method

4.1 Consistency and Identifiability

Identifiability essentially means that the estimate is consistent and will converge to the true system, when the system is contained in the model structure. This is a joint requirement on the model structure and on the experiment condition, i.e., the data spectrum Φ_ζ.

The basic expression (15.20) with (15.21) shows that

> *consistency and identifiability follows when no model in the structure $\begin{bmatrix} \Delta G_\theta & \Delta H_\theta \end{bmatrix}$ lies in the null space of Φ_ζ.*

A sufficient condition for identifiability is thus that the data spectrum is positive definite (non-singular) almost everywhere. From (15.10) we see that this happens if and only if (we assume Λ_0 to be non-singular)

$$\Phi_u^r(\omega) > 0, a.e.\omega \tag{15.22}$$

In [14] this is termed that the experiment is *informative enough.* With the definition (15.8) this means that there should be a full rang part of u that cannot be estimated from e by a linear, time invariant filter. It essentially means that the regulator should not be linear, time-invariant with no extra signals (disturbances or setpoints). A persistently exciting setpoint (reference signal), a non-linear regulator or a time-varying regulator will basically secure (15.22).

Constraining the model structure may however secure identifiability even when (15.22) does not hold. Suppose, for example that the noise model H_θ is fixed to the true value H_0 ($\Delta H_\theta = 0$), and that the dynamics model structure G_θ contains the true system G_0. Then we see directly from (15.21) that consistency follows as soon as $\Phi_u(\omega) > 0$, $a.e.\,\omega$. However, it is essential that the noise model is the true one, otherwise the limit of G_θ will still be unique, but biased.

4.2 Bias Distribution

To more clearly see bias issues and lack of identifiability we can use the factorization (15.11) to rewrite the convergence expression as

$$\begin{aligned}\hat{\theta}_N \to \arg\min_{\theta\in D_{\mathcal{M}}} \int_{-\pi}^{\pi} \operatorname{tr}\big[\Lambda^{-1}\{H_\theta^{-1}[(G_0 + B_\theta - G_\theta)\Phi_u(G_0 + B_\theta - G_\theta)^* \\ + (H_0 - H_\theta)\Phi_e^r(H_0 - H_\theta)^*]H_\theta^{-*}\}\big]d\omega \quad \text{w. p. 1 as } N \to \infty\end{aligned} \tag{15.23}$$

where

$$B_\theta = (H_0 - H_\theta)\Phi_{eu}\Phi_u^{-1} \tag{15.24}$$

In the SISO, linear regulator case, we can characterize the "size" of B as

$$|B_\theta|^2 = \frac{\Lambda_0}{\Phi_u}\frac{\Phi_u^e}{\Phi_u}|H_0 - H_\theta|^2 \tag{15.25}$$

(see (15.16) and (15.19).) For a fixed noise model $H_\theta = H_*$, we see that the dynamics model will approximate the biased frequency function $G_0 + B_*$ in a frequency domain norm determined by H_*^{-1} and Φ_u (The ratio $\Phi_u/|H_*|^2$ in the SISO case). We obtain an approximation of the correct dynamics in case $B = 0$, which means that the noise model is correct $H_* = H_0$, or the system operates in open loop: $\Phi_{eu} = 0$.

4.3 The Indirect Method

We saw above that the noise model has to be correct, in order to ensure consistency of the dynamics part. Let us now ask the question:

Suppose the dynamics model structure G_θ contains the true dynamics G_0, and that we are not interested in the noise characteristics. Is is then possible to obtain a consistent estimate of G?

To answer that question, we rewrite the basic expression (15.21) using the alternative factorization (15.10):

$$\begin{aligned}\overline{V}(\theta) &= \int_{-\pi}^{\pi} \operatorname{tr}\left[\Lambda^{-1}H_\theta^{-1}(\Delta G_\theta \Phi_u^r(\Delta G_\theta)^* H_\theta^{-*}\right] d\omega \\ &+ \int_{-\pi}^{\pi} \operatorname{tr}\left[\Lambda^{-1}H_\theta^{-1}(\Delta G_\theta \Phi_{ue}\Lambda_0^{-1} + \Delta H_\theta)\Lambda_0 \right. \\ &\left. \times(\Delta G_\theta \Phi_{ue}\Lambda_0^{-1} + \Delta H_\theta)^* H_\theta^{-*}\right] d\omega\end{aligned} \tag{15.26}$$

To assure consistency even if the noise model is not correct, we should make the second term (the second integral) independent of ΔG_θ. We then expand the factor in the integral as follows (using the expression (15.19) for Φ_{ue})

$$\begin{aligned}&H_\theta^{-1}(G_\theta K S_0 H_0 - G_0 K S_0 H_0 + H_0) - I \\ &\quad = H_\theta^{-1}(G_\theta K - G_0 K + I + G_0 K) S_0 H_0 - I \\ &\quad = H_\theta^{-1}(I + G_\theta K) S_0 H_0 - I = \tilde{H}_\eta^{-1} S_0 H_0 - I\end{aligned}$$

where we introduced the noise model parameterization

$$H_\theta = (I + G_\theta K)\tilde{H}_\eta = S_\theta^{-1}\tilde{H}_\eta \tag{15.27}$$

where η is a parameterization, independent of θ. Here S_θ is the model sensitivity function, compare (15.15). With (15.27) we have thus achieved that the second integral of (15.26) is independent of the parameterization of G_θ. The first integral will thus determine to what the dynamics model converges: If $\tilde{H}$ is a fixed model we have

$$\hat{\theta}_N \to \arg\min_{\theta\in D_{\mathcal{M}}} \int_{-\pi}^{\pi} \operatorname{tr}\left[\Lambda^{-1}\left\{\tilde{H}^{-1}(S_\theta \Delta G_\theta)\Phi_u^r(S_\theta \Delta G_\theta)^* \tilde{H}^{-*}\right\}\right] d\omega \tag{15.28}$$

With (15.17) we find that

$$(S_\theta \Delta G_\theta)\Phi_u^r(S_\theta \Delta G_\theta)^* = (S_\theta \Delta G_\theta S_0^i)\Phi_r(S_\theta \Delta G_\theta S_0^i)^* \tag{15.29}$$

with

$$\begin{aligned}S_0 G_0 - S_\theta G_\theta &= G_0 S_0^i - S_\theta G_\theta \\ &= S_\theta((I + G_\theta K)G_0 - G_\theta(I + K G_0))S_0^i = S_\theta \Delta G_\theta S_0^i\end{aligned}$$

This shows that the noise model parameterization (15.27) will fit the closed loop model $S_\theta G_\theta$ to the closed loop system $S_0 G_0$ in a norm that is determined by the reference signal spectrum Φ_r and the fixed noise model $\tilde{H}$.

In fact the parameterization (15.27) corresponds to a well known method for dealing with closed loop identification data: The predictor for

$$y(t) = G(q,\theta)u(t) + H(q,\theta)e(t) \tag{15.30}$$

is

$$\hat{y}(t|\theta) = H^{-1}(q,\theta)G(q,\theta)u(t) + (I - H^{-1}(q,\theta))y(t) \tag{15.31}$$

Using $u = r - Ky$ and inserting (15.27) we get

$$\begin{aligned}\hat{y}(t|\theta) =& \tilde{H}^{-1}(q,\theta)(I + G(q,\theta)K(q))^{-1}G(q,\theta)r(t) \\ &+ (I - \tilde{H}^{-1}(q,\theta))y(t)\end{aligned} \tag{15.32}$$

But this is exactly the predictor also for the closed-loop model structure

$$y(t) = (I + G(q,\theta)K(q))^{-1}G(q,\theta)r(t) + \tilde{H}(q,\theta)e(t) \tag{15.33}$$

Identifying the closed loop, and then solving for the open loop dynamics is called *the indirect method.* We have here derived this approach as the answer to the question of how to obtain consistent dynamics models, without dealing with a noise model. Of course, in the open loop case ($K = 0$) (15.27) tells us that this is achieved by letting the noise model be parameterized independently from the dynamics.

5. Asymptotic Variance

5.1 Parameter and Transfer Function Covariance

The classical result on the asymptotic distribution of the parameter estimates (cf [14], Chapter 9) is as follows. Suppose that the true system is contained in the model structure, and that the experimental conditions are such that identifiability is secured. Let the true parameters be denoted by θ_0. In the multi-output case we also assume that $\Lambda = \Lambda_0$. Then $\sqrt{N}(\hat{\theta}_N - \theta_0)$ converges in distribution to the normal distribution with zero mean and a covariance matrix

$$P_\theta = [\overline{V}''(\theta_0)]^{-1} \tag{15.34}$$

where $\overline{V}$ is defined by (15.21) and prime denotes differentiation w.r.t θ.

Let us from now on concentrate on the SISO case for notational simplicity and denote

$$T_\theta = \begin{bmatrix} G_\theta & H_\theta \end{bmatrix}, \quad \hat{T}_N = \begin{bmatrix} \hat{G}_N & \hat{H}_N \end{bmatrix} \tag{15.35}$$

Then, using (15.21) we can write

$$P_\theta = \left[\frac{1}{2\pi} \int_{-\pi}^{\pi} \frac{1}{\Phi_v} T' \Phi_\zeta (T')^* d\omega \right]^{-1} \tag{15.36}$$

Here $\Phi_v = |H_0|^2 \Lambda_0$. From this expression and the factorizations (15.10) and (15.11) explicit expression can be found how Φ_r and K affect the parameter accuracy.

If we are interested in the covariance of $\hat{T}_N$, rather than the parameters, Gauss' approximation formula gives the expression

$$N \cdot \mathrm{Cov}\hat{T}_N(e^{i\omega}) = T'(e^{i\omega})^T$$
$$\times \left[\frac{1}{2\pi} \int_{-\pi}^{\pi} \frac{1}{\Phi_v(\xi)} T'(e^{i\xi}) \begin{bmatrix} \Phi_u(\xi) & \Phi_{ue}(\xi) \\ \Phi_{eu}(\xi) & \Lambda_0 \end{bmatrix} T'(e^{-i\xi})^T d\xi \right]^{-1} T'(e^{-i\omega}) \tag{15.37}$$

5.2 Asymptotic Black Box Expressions

The expression (15.37) shows an intriguing symmetry, with the factors T' in “cancelling positions”. In fact, suppose that parameterization of T has the following shift structure,

$$\theta = \begin{bmatrix} \theta_1 \\ \vdots \\ \theta_n \end{bmatrix}$$
$$\frac{d}{d\theta_k} T(q, \theta) = q^{-k+1} \frac{d}{d\theta_1} T(q, \theta)$$

which is satisfied by many black-box model parameterizations. Then, as n tends to infinity we have (see [14], Chapter 9):

$$\mathrm{Cov} \begin{bmatrix} \hat{G}_N(e^{i\omega}) & \hat{H}_N(e^{i\omega}) \end{bmatrix} \approx \frac{n}{N} \Phi_v(\omega) \begin{bmatrix} \Phi_u(\omega) & \Phi_{eu}(\omega) \\ \Phi_{ue}(\omega) & \Lambda_0 \end{bmatrix}^{-1} \tag{15.38}$$

n is here the “model order” and N is the number of data. From (15.13) we then find that

$$\mathrm{Cov} \left[\hat{G}_N(e^{i\omega}) \right] \approx \frac{n}{N} \frac{\Phi_v(\omega)}{\Phi_u^r(\omega)} \tag{15.39}$$

6. Optimal Experiment Design

In this section we will consider experiment design problems where the goal is to minimize a weighted norm of the covariance of G:

$$J = \int_{-\pi}^{\pi} \text{Cov}\hat{G}_N(e^{-i\omega})C(\omega)d\omega \tag{15.40}$$

The minimization shall be carried out with respect to the experiment design variables, which we take as K (the regulator) and Φ_r (the reference signal spectrum). Other equivalent choices are also possible, e.g., Φ_u and Φ_{ue} or K and Φ_u^r. To make the designs realistic we will also impose constraints on the input power or the output power, or both.

Consider the problem to minimize J given by (15.40) and using the asymptotic expression (15.39). The minimization is to be carried out under the constraint

$$\int_{-\pi}^{\pi} \{\alpha\Phi_u + (1-\alpha)\Phi_y\}d\omega \leq 1, \ \alpha \in [0,1] \tag{15.41}$$

with respect to the design variables K and Φ_r. The solution is to select the regulator $u = -Ky$ that solves the standard LQG problem

$$K^{opt} = \arg\min_K [\alpha Eu^2 + (1-\alpha)Ey^2], \quad y = G_0u + H_0e \tag{15.42}$$

The reference signal spectrum shall be chosen as

$$\Phi_r^{opt}(\omega) = \mu\sqrt{\Phi_v(\omega)C(\omega)}\frac{|1+G_0(e^{i\omega})K^{opt}(e^{i\omega})|^2}{\sqrt{\alpha+(1-\alpha)|G_0(e^{i\omega})|^2}} \tag{15.43}$$

where μ is a constant, adjusted so that

$$\int_{-\pi}^{\pi} \{\alpha\Phi_u + (1-\alpha)\Phi_y\}d\omega = 1 \tag{15.44}$$

This result can be proved as follows:

Proof. Replace the design variables K and Φ_r by the equivalent pair K and Φ_u^r. Then, by using expressions for the input and output spectra in terms of K and Φ_u^r we can rewrite the problem as

$$\min_{K,\Phi_u^r} \int_{-\pi}^{\pi} \frac{\Phi_v}{\Phi_u^r} C d\omega$$

under the contstraint

$$\int_{-\pi}^{\pi} \{(\alpha + (1-\alpha)|G_0|^2)\Phi_u^r + \frac{\alpha|K|^2 + (1-\alpha)}{|1+G_0K|^2}\Phi_v\}d\omega \leq 1 \tag{15.45}$$

$\alpha \in [0,1]$

The criterion function is independent of K hence the optimal controller K^{opt} can be found by solving the LQ problem

$$\min_K \int_{-\pi}^{\pi} \frac{\alpha|K|^2 + (1-\alpha)}{|1+G_0K|^2}\Phi_v d\omega \tag{15.46}$$

(Here it is implicitly assumed that $y(t) = G_0(q)u(t)+v(t)$, $u(t) = -K(q)y(t)$, and $\alpha \in [0,1]$.) This proves (15.42). Define the constant γ as

$$\gamma = 1 - \int_{-\pi}^{\pi} \frac{\alpha|K^{opt}|^2 + (1-\alpha)}{|1+G_0K^{opt}|^2}\Phi_v d\omega \tag{15.47}$$

Problem (15.45) now reads

$$\min_{\Phi_u^r}\{\int_{-\pi}^{\pi} \frac{\Phi_v}{\Phi_u^r} C d\omega : \int_{-\pi}^{\pi} (\alpha + (1-\alpha)|G_0|^2)\Phi_u^r d\omega \leq \gamma\} \tag{15.48}$$

This problem has the solution (cf. [14], p. 376)

$$\Phi_u^r = \mu\sqrt{\frac{\Phi_v C}{(\alpha + (1-\alpha)|G_0|^2)}} \tag{15.49}$$

where μ is a constant, adjusted so that

$$\int_{-\pi}^{\pi} ((\alpha + (1-\alpha)|G_0|^2)\Phi_u^r d\omega = \gamma \tag{15.50}$$

or in other words so that

$$\int_{-\pi}^{\pi} \{\alpha\Phi_u + (1-\alpha)\Phi_y\} d\omega = 1 \tag{15.51}$$

Consequently the optimal Φ_r is

$$\Phi_r^{opt} = \mu\sqrt{\Phi_v C}\frac{|1+G_0K^{opt}|^2}{\sqrt{\alpha + (1-\alpha)|G_0|^2}} \tag{15.52}$$

which ends the proof.

We stress that the optimal controller K^{opt} in (15.42) can easily be found by solving the indicated discrete-time LQ problem (if G_0 and Φ_v were known). Among other things this implies that the optimal controller K^{opt} is guaranteed to stabilize the closed-loop system and be linear, of the same order as G_0. This is a clear advantage over the results reported in, e.g., [10]. Furthermore, the optimal controller is independent of C which also is quite interesting and perhaps somewhat surprising. This means that whatever weighting C is used in the design criterion, it is always optimal to use the LQ regulator (15.42) in the identification experiment.

From the result we also see that closed-loop experiments are optimal as long as there is a constraint on the output power, i.e., as long as $\alpha \neq 1$. If $\alpha = 1$ then $K^{opt} = 0$ and the optimal input spectrum $\Phi_u^{opt} (= \Phi_r^{opt})$ becomes

$$\Phi_u^{opt} = \mu\sqrt{\Phi_v C} \tag{15.53}$$

If the constraint is on the output power only, Ey^2, the regulator K is the minimum variance controller. We then have the simple result, that any experiment that aims at minimizing the covariance of the dynamics, under output power constraint should be a minimum variance controller. To this we should add a reference signal with a power distribution that reflects the weighting function in the criterion. This result ties nicely with the special case treated in [7].

7. Conclusions

We have re-examined some basic results on bias and variance in closed loop identification. The common denominator in this analysis has been the data spectrum Φ_ζ. We have shown how different factorizations of this matrix give direct insights into identifiability, and bias distribution. It also gives a pragmatic "derivation" of the indirect method for closed loop identification.

This data spectrum also directly determines the variance of the estimated parameters and its inverse gives a simple and explicit expression for the asymptotic, black-box variance of the frequency functions. This in turn can be used to solve rather general experiment design problems, aiming at minimizing the covariance of the estimated dynamics under various realistic constraints.

References

1. B.D.O. Anderson and M. Gevers. Identifiability of linear stochastic systems operating under linear feedback. *Automatica*, 18(2):195–213, 1982.
2. K. J. Åström. Matching criteria for control and identification. In *Proceedings of the 2nd European Control Conference*, pages 248–251, Groningen, The Netherlands, 1993.
3. R. de Callafon and P. Van den Hof. Multivariable closed-loop identification: From indirect identification to Dual-Youla parametrization. In *Proceedings of the 35th Conference on Decision and Control*, pages 1397–1402, Kobe, Japan, 1996.
4. R. de Callafon, P. Van den Hof, and M. Steinbuch. Control relevant identification of a compact disc pick-up mechanism. In *Proceedings of the 32nd Conference on Decision and Control*, volume 3, pages 2050–2055, San Antonio,TX, 1993.

5. B. Egardt. On the role of noise models for approximate closed loop identification. In *Proceedings of the European Control Conference*, Brussels, Belgium, 1997.
6. M. Gevers. Towards a joint design of identification and control. In H. L. Trentelman and J. C. Willems, editors, *Essays on Control: Perspectives in the Theory and its Applications*, pages 111–151. Birkhäuser, 1993.
7. M. Gevers and L. Ljung. Optimal experiment design with respect to the intended model application. *Automatica*, 22:543–554, 1986.
8. I. Gustavsson, L. Ljung, and T. Söderström. Identification of processes in closed loop — Identifiability and accuracy aspects. *Automatica*, 13:59–75, 1977.
9. F. R. Hansen. *A fractional representation to closed-loop system identification and experiment design.* Phd thesis, Stanford University, Stanford, CA, USA, 1989.
10. H. Hjalmarsson, M. Gevers, and F. De Bruyne. For model-based control design,closed loop identification gives better performance. *Automatica*, 32, 1996.
11. I. D. Landau and K. Boumaiza. An output error recursive algorithm for identification in closed loop. In *Proceedings of the 13th IFAC World Congress*, volume I, pages 215–220, San Francisco, CA, 1996.
12. I. D. Landau and A. Karimi. Recursive algorithms for identification in closed loop: A unified approach and evaluation. *Automatica*, 33(8):1499–1523, 1997.
13. W. S. Lee, B. D. O. Anderson, I. M. Y. Mareels, and R. L. Kosut. On some key issues in the windsurfer approach to adaptive robust control. *Automatica*, 31(11):1619–1636, 1995.
14. L. Ljung. *System Identification: Theory for the User*. Prentice-Hall, 1987.
15. L. Ljung and I. D. Landau. Model-reference adaptive systems and self-tuning regulators: Some connections. In *Proc 7th IFAC World Congress*, pages 1899–1906, Helsinki, Finland, 1978. Paper no 46 A.2.
16. T. Söderström and P. Stoica. *System Identification*. Prentice-Hall International, 1989.
17. P. M. J. Van den Hof and R. J. P. Schrama. Identification and control — Closed-loop issues. *Automatica*, 31(12):1751–1770, 1995.
18. Z. Zang, R. R. Bitmead, and M. Gevers. Iterative weighted least-squares identification and weighted LQG control design. *Automatica*, 31(11):1577–1594, 1995.

Diagnostics and Monitoring as tightly coupled to Identification

A. Benveniste[1], *B. Delyon*[1], *and M. Basseville*[2]

[1] IRISA-INRIA, Campus de Beaulieu, 35042 Rennes cedex, France; e-mail: name@irisa.fr
[2] IRISA-CNRS, Campus de Beaulieu, 35042 Rennes cedex, France; e-mail: name@irisa.fr

Summary. In this paper we give an overview of the apporach we have developed for diagnostics and model validation, based on the so-called "asymptotic local" approach in mathematical statistics. All algorithms we propose for diagnostics and model validation are tightly related to corresponding algorithms for system identification. In a first part of this paper, we analyze this relationship from the theoretical point of view: in particular we report new results to show how central limit theorems and diagnostics algorithms based on the local approach are related to each other. The second part is devoted to reporting how these results can be practically applied for real applications.

Keywords : Diagnostics, identification, model validation, testing, local asymptotic normality, dependent processes and mixing.

1. The overall picture

Our general picture is shown in the figure below:

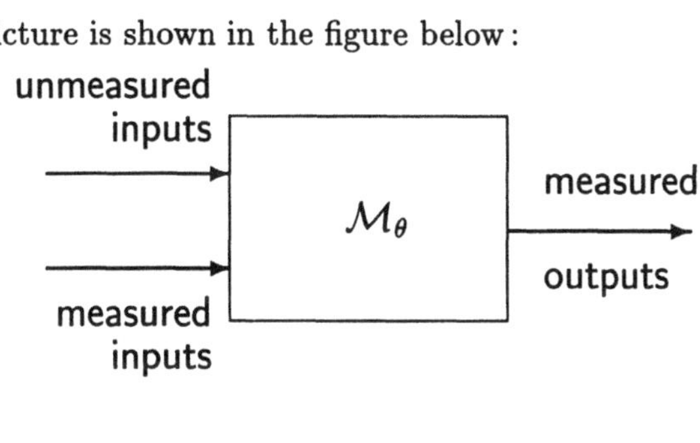

We consider a system with unmeasured inputs (disturbances, both random and nonrandom), and measured inputs and outputs, or, more generally, measurements. And we assume a parameterized model $\mathcal{M}_\theta$ for this system is at hand. We do not require that the system should belong to the model class, i.e., there may not exist a "true" value $\theta_\star$ such that $\mathcal{M}_{\theta_\star}$ is an exact model for the system.

Nevertheless we wish:

1. To fit a "best" θ from measured data, call $\theta_\star$ such a best parameter fit, and also
2. To be able to compare any given model parameter, generically denoted θ_o in the sequel, with a collected data set from the system or plant. In addition, we wish
3. That items 1 and 2 above match together in some way we shall make later precise.

The practical motivation for this is manyfold:

- It is a recommended practice that identified models should be *validated* against validation data, in order to assess how good the obtained fit actually was.
- Also, quite often, it happens that the system design phase included the prior development of a *design model* in order to minimize the "guess-and-trial" nature of plant design. This typically would result in a design model such as our $\mathcal{M}_{\theta_o}$ above. Of course, it is unlikely that the reality satisfactorily matches this design model. Thus it is desirable to confront $\mathcal{M}_{\theta_o}$ to data collected from the actual plant. Understanding whether $\mathcal{M}_{\theta_o}$ is good enough to capture the real systems behaviour, or alternatively which components of θ_o are most reponsible for the mismatch, helps in *updating this design model.*
- Finally, *monitoring and diagnosing gradual changes* in the behaviour of the plant can be performed by means of comparing some baseline model $\mathcal{M}_{\theta_o}$ to fresh collected sample data. In this case, the baseline parameter value θ_o would be typically be taken equal to the identified one, $\theta_\star$, using data collected from the safe plant.

In this paper we shall discuss these items. We shall report on both recent mathematical results that we obtained (section 2.), as well as our practical experience (section 3.).

2. From theory ...

Yes: theory is sometimes useful for practice [13].

2.1 Some heuristic discussion

Our mathematical background consists in investigating the relationships between the consistency of M-estimators, the local asymptotic normality of statistics and the asymptotic normality of estimators.

M-estimators have been introduced in the 60s by Huber (cf [10]). We consider an observation process $(Z_1, Z_2, ...)$ and try to estimate a parameter θ of its distribution as a solution $\widehat{\theta}_n$ of $\sum_{k=1}^{n} H(\theta, Z_k) = 0$:

$$\widehat{\theta}_n \quad = \quad \arg_\theta \ : \ \sum_{k=1}^{n} H(\theta, Z_k) = 0 \ , \tag{16.1}$$

where $H(\theta, z)$ is a so-called *estimating function.*

Of course this is based on our knowledge that the best fit parameter $\theta_\star$ satisfies

$$E[H(\theta_\star, Z_k)] = 0. \tag{16.2}$$

The obtained estimator is called an M-estimator [11]. This estimator corresponds sometimes to a *minimum contrast* or *quasi-likelihood* estimate

$$\widehat{\theta}_n = \arg\min_\theta \sum_{k=1}^{n} G(\theta, Z_k) \tag{16.3}$$

where G is the quasi-likelihood [9]. Obviously, if G is smooth enough, the solution of (16.3) is a particular solution of (16.1) by taking[1]: $H = \nabla G$. However, in some cases equations like (16.1) are solved and H is not the gradient of any function G, e.g., Instrumental Variable or Subspace type of methods in system identification.

The concept of *local asymptotic normality* was introduced by Lucien LeCam in the 50s for comparing parameterized distributions P_θ over infinite sequences $(Z_1, Z_2, ...)$. The idea is that generically, under suitable conditions, although the P_θs for different θs are singular with respect to each other, the marginal density Λ_n of $P_{\theta+\widetilde{\theta}/\sqrt{n}}(Z_1, ..Z_n)$ with respect to $P_\theta(Z_1, ..Z_n)$ exists; furthermore it is a random variable $\Lambda_n(Z_1, ..Z_n)$ which converges in distribution, under P_θ, when n tends to infinity to some limit (in this context $\widetilde{\theta}$ is fixed and $\sqrt{n}$ happens to be the correct normalization for obtaining such limit theorems). This limit may be seen as a kind of differential which characterizes the way in which P_θ varies in the direction $\widetilde{\theta}$; LeCam's theory shows how the identification of this limit leads to various limit-theorems concerning general statistics based on samples drawn from P_θ, when θ varies in such a $O(n^{-1/2})$-rescaled neighborhood of a given value. We won't go this far since

[1] For $f(x, z)$ a function, we shall denote by $\nabla_x f$ the partial derivative of f w.r.t. x, when it exists. Also, the derivative of f will sometimes be denoted by ∇f.

we will only study the distribution under $P_{\theta+\widetilde{\theta}/\sqrt{n}}$ of a specific statistic of the data:

$$U_n = \frac{1}{\sqrt{n}} \sum_{k=1}^{n} H(\theta, Z_k)$$

which is asymptotically Gaussian and zero-mean under P_θ. We will show that under mild conditions, U_n is asymptotically Gaussian with the same covariance matrix and mean [2]

$$-E_\theta[\nabla_\theta H(\theta, Z)] \; . \; \widetilde{\theta}$$

under $P_{\theta+\widetilde{\theta}/\sqrt{n}}$. This remark is the basis for the testing methods built upon general M-estimators, as introduced in [5, 6]. The advantage of this direct approach is to lead to similar conclusions for the behavior of statistics in a much simpler framework, allowing to deal with dependent sequences (LeCam's theory for dependent processes is quite complicated, and difficult to use in practice since it is generally very difficult to have a formula for the score, and even for the information matrix; however interesting results appear in [12]).

Here follows a sketch of the main steps of our argument on the relationship between identification and testing in the asymptotic local framework. First, we need that the law of large numbers $1/n \sum_1^n H(\theta, Z_k) \to E[H(\theta, Z)]$ holds uniformly in θ, this is guaranteed under mild conditions, as stated in Section 2.2. This allows us to establish, in the same section, the consistency of the M-estimators for stationary dependent sequences. Section 2.3 focuses on testing under the local approach: the asymptotic behavior of U_n is given under $P_{\theta+\widetilde{\theta}/\sqrt{n}}$ distribution, using mild mixing conditions. The key step in enlightening the link between local tests and identification is given in section 2.4, it consists of the following elementary argument: first, make θ explicit in

$$U_n(\theta) \;=\; \frac{1}{\sqrt{n}} \sum_{k=1}^{n} H(\theta, Z_k) \; .$$

Assuming that the (Z_k) are drawn from distribution $P_{\theta_\star}$ (i.e., $\theta_\star$ is the "true" underlying system parameter), then, for $\theta - \theta_\star$ small, one has

$$U_n(\theta) \;\approx\; U_n(\theta_\star) + \frac{1}{n}\left(\sum_{1}^{n} \nabla_\theta H(\theta_\star, Z_k)\right) (\theta - \theta_\star)\sqrt{n} \qquad (16.4)$$

Since $\widehat{\theta}_n$ is characterized by the equation $U_n(\widehat{\theta}_n) = 0$, we get, using (16.4) and the law of large numbers,

$$\sqrt{n}(\widehat{\theta}_n - \theta_\star) \;\approx\; -E\left[\nabla_\theta H(\theta_\star, Z)\right]^{-1} U_n(\theta_\star) \qquad (16.5)$$

[2] E_θ shall denote expectation under law P_θ.

and we deduce immediately the Central Limit Theorem for $\widehat{\theta}_n$ from the asymptotic behavior of the statistics U_n. This argument is detailed in section 2.4. The use of these theorems is illustrated in [7] on the Instrumental Variable method for estimating the AR part of an ARMA process with time-varying MA parameters, a problem first investigated in [15] in a more difficult setting.

Besides its interest for its own, this relationship between $\widehat{\theta}_n$ and $U_n(\theta_\star)$ has a practical interest. While estimating or calculating the sensitivity matrix $E\left[\nabla_\theta H(\theta_\star, Z)\right]$ is most of the time easy, estimating or calculating the error covariance matrix of $\widehat{\theta}_n$ based on formula (16.13) of theorem 2.4 often leads to illconditioned estimates. In contrast formula (16.5) suggests that it is enough to estimate the covariance matrix of $U_n(\theta_\star)$, which can be safely achieved from drawing samples of this statistics.

2.2 Consistency of the quasi-likelihood procedure

We state here separately some Uniform Law of Large Numbers since this result is of major importance in the theory of M-estimators. We put all our efforts in proposing simple and "optimal" assumptions. There are essentially two assumptions: ergodicity (A2) and smoothness (A3). For the sake of simplicity, we consider the case where Θ is bounded. Our assumptions are:

(A1). Θ is a compact subset of the Euclidean space $\mathbb{R}^n$ and $H(\theta, z)$ is a vector-valued measurable function defined on $\Theta \times \mathbb{R}^m$.

(A2). $(Z_k)_{1\leq k}$ is a strictly stationary ergodic sequence of random variables and

$$E[\ \sup_{\theta\in\Theta} |H(\theta, Z_1)|\ ] \ < \ \infty.$$

(A3). For each θ the function $\theta \longrightarrow H(\theta, Z)$ is continuous at θ for almost all value of Z (i.e. this full probability set may depend on θ) [3].

Lemma 2.1 (Uniform Law of Large Numbers). *Set* $h(\theta) = E[H(\theta, Z)]$. *Then, under assumptions (A1,A2,A3),*

$$\limsup_n \sup_{\theta\in\Theta} \left| h(\theta) - \frac{1}{n}\sum_{k=1}^{n} H(\theta, Z_k) \right| = 0 \tag{16.6}$$

holds.

Proof: See [7]. ∎

[3] This assumption allows us to handle a $H(\theta, Z)$ involving quantizers, see [6], p. 33.

Theorem 2.1. *Let assumptions (A1,A2,A3) hold. Then any sequence* $(\widehat{\theta}_n)$ *of random vectors,* $\sigma(Z_1, ... Z_n)$*–measurable, such that*

$$\lim_{n \longrightarrow \infty} \frac{1}{n} \sum_{k=1}^{n} H(\widehat{\theta}_n, Z_k) = 0$$

satisfies

$$\lim_n h(\widehat{\theta}_n) = 0.$$

Note. Obviously, adding an identifiability assumption like (C2) below implies the consistency of $\widehat{\theta}_n$.
Proof: Theorem 2.1 is an immediate consequence of lemma 2.1. ∎

2.3 Local asymptotic normality and testing

In this section, we consider a family of probability measures P_θ on the space Ω of the sequences of m-dimensional vectors; these measures are indexed by $\theta \in \Theta$. We are interested in the asymptotic distribution of the statistics

$$U_n = \frac{1}{\sqrt{n}} \sum_{k=1}^{n} H(\theta_0, Z_{n,k})$$

when θ_0 is fixed (θ_0 represents the nominal value of the parameter) and $(Z_{n,k})_{1 \leq k}$ is drawn from the distribution P_{θ_n} where

$$\theta_n = \theta_0 + \frac{\widetilde{\theta}}{\sqrt{n}} ,$$

and $\widetilde{\theta}$ is fixed. The normalization $1/\sqrt{n}$ is necessary in order to get a non-trivial limit distribution. The obtained asymptotic variance allows us to build change tests for the parameter θ_0.

Let us state first some prerequisite result on central limit-theorem for arrays of variables.

Definition 2.1 (uniformly mixing array). *We shall say that a process* $(Z_{n,k})_{1 \leq k}$*,* $n \geq 1$*, is a uniformly mixing array if for some* $C, \varepsilon, \delta > 0$*, and all* $k, l, n \geq 1$

$$||Z_{n,k}||_{2+\varepsilon} < C \tag{16.7}$$

$$||E[Z_{n,k+l}|\mathcal{F}_{n,l}]||_2 \leq C \; k^{-\delta-1/2} \tag{16.8}$$

where $\mathcal{F}_{n,k} = \sigma(Z_{n,1}, ... Z_{n,k})$*. We shall say that* $(Z_k)_{1 \leq k}$*, is a mixing sequence if* $(Z_{n,k}) = (Z_k)$ *is a uniformly mixing array.*

The importance of this concept is explained by the following theorem

Theorem 2.2. *Let $(Z_{n,k})_{1 \le k \le n}$ be a strictly stationary uniformly mixing array and*

$$S_n = \frac{1}{\sqrt{n}} \sum_{j=1}^{n} Z_{n,k}.$$

If the limit

$$V = \lim_n E[S_n S_n^T] \tag{16.9}$$

exists, then

$$S_n \xrightarrow{d} \mathcal{N}(0, V). \tag{16.10}$$

Proof: This theorem is established in [14] in its one dimensional version with a more general setting (the correspondence in notations is: $X_{n,k} = Z_{n,k}/\sqrt{n}$, $k_n(t) = [nt]$, $\sigma_{n,k} = 1/n$, $\psi_k = k^{-\delta-1/2}$). If we apply this one dimensional version to the inner product $\langle S_n, u \rangle$ for any vector u, we obtain the convergence in distribution of $\langle S_n, u \rangle$ to $\mathcal{N}(0, u^T V u)$, which implies, using characteristic functions, the convergence of S_n to $\mathcal{N}(0, V)$. ∎

Examples of and discussion about this type of mixing property is found in [7].

Let us now introduce the assumptions we need for the local asymptotic normality of U_n.

(B1). Θ is an arbitrary subset of the Euclidean space $\mathbb{R}^n$ and $H(\theta, z)$ is a vector-valued measurable function defined on $\Theta \times \mathbb{R}^m$. Given any $\theta \in \Theta$, process (Z_k) drawn from P_θ is stationary; for any sequence θ_n converging to θ, and $(Z_{n,k})_{1 \le k}$, $n \ge 1$ processes drawn from the distributions P_{θ_n}, the array of variables $(Z_{n,k})_{1 \le k}$, $n \ge 1$ defined by:

$$Z_{n,k} = H(\theta_n, Z_{n,k}) - E_{\theta_n}[H(\theta_n, Z_{n,k})],$$

is uniformly mixing with $\delta > 1/2$ (cf equation (16.8)) .

(B2). The function

$$(\theta, \theta_\star) \longrightarrow h_{\theta_\star}(\theta) = E_{\theta_\star}[H(\theta, Z)]$$

is continuous on $\Theta \times \Theta$, it satisfies

$$h_\theta(\theta) = 0, \quad \forall \theta \in \Theta\ ,$$

and its gradient $\nabla_\theta h$ w.r.t. θ is a continuous function of $(\theta, \theta_\star)$. We shall denote by

$$h'(\theta) \quad = \quad \nabla_\theta h_{\theta_\star}(\theta) \quad \text{for } \theta_\star = \theta$$

the value of $\nabla_\theta h$ on the diagonal of $\Theta \times \Theta$.

(B3). The function $H(\theta, z)$ satisfies

$$\sup_{\theta_\star \in \Theta} E_{\theta_\star}[|H(\theta, Z)|^2] < \infty, \quad \forall \theta \in \Theta.$$

And for each θ the function $\theta \longrightarrow H(\theta, Z)$ is continuous at θ for almost all value of Z (i.e. this full probability set may depend on θ).

The condition $\delta > 1/2$ in assumption (B1) will allow us to check directly that condition (16.9) is satisfied for our processes. Since next theorem involves only values of θ close to some nominal value θ_0 these assumptions will actually need to be satisfied only for Θ being some neighborhood of θ_0.

Theorem 2.3 (Asymptotic local normality of statistics). *We assume that $(B1, B2, B3)$ are satisfied. Consider*

$$\theta_n = \theta_0 + \frac{\widetilde{\theta}}{\sqrt{n}}$$

for $\widetilde{\theta}$ fixed (we assume that $\theta_n \in \Theta$, $n > 0$), and

$$U_n = \frac{1}{\sqrt{n}} \sum_{k=1}^{n} H(\theta_0, Z_{n,k})$$

where $(Z_{n,k})$ is drawn from the distribution P_{θ_n}, and set

$$R_\theta(i) = \mathrm{Cov}_\theta(H(\theta_0, Z_1), H(\theta_0, Z_{i+1})) ,$$

where Cov_θ denotes the covariance under distribution P_θ. Then

$$U_n \xrightarrow{d} \mathcal{N}(-h'(\theta_0)\widetilde{\theta}, \Sigma(\theta_0)),$$

where

$$\Sigma(\theta) = R_\theta(0) + \sum_{i>0} \left(R_\theta(i) + R_\theta(i)^T\right) .$$

Proof: See Appendix 4.3. ∎

2.4 Asymptotic normality of M-estimators

In this section we prove a asymptotic normality result on the estimates of $\theta_\star$ from observations (Z_k) generated under a fixed probability measure P. We are thus back in the semi-parametric setting. Consider the following set of assumptions:

(C1). Θ is an arbitrary subset of the Euclidean space $\mathbb{R}^m$ and $H(\theta, z)$ is a vector-valued measurable function defined on $\Theta \times \mathbb{R}^m$.

For any $\theta \in \Theta$, the sequence of variables

$$Z_k = H(\theta, Z_k) - E[H(\theta, Z_k)],$$

where $(Z_k)_{1 \le k}$ are drawn from the distribution P, form a stationary mixing sequence (in the sense of definition 2.1).

(C2). Set $h(\theta) = E[H(\theta, Z)]$; the unique solution $\theta_\star$ of $h(\theta) = 0$ is an interior point of Θ. Moreover, for any sequence θ_n, $\lim_n h(\theta_n) = 0$ implies $\lim_n \theta_n = \theta_\star$.

(C3). For any compact subset $\mathcal{K}$ of Θ, the function $H(\theta, z)$ satisfy

$$E[|H(\theta, Z)|^2] < \infty, \quad \forall \theta \in \Theta \ ,$$

and $h(\theta) = E[H(\theta, Z)]$ is continuously differentiable. Moreover, for each θ, the function $H(\theta, z)$ is, for almost all value of y, continuously differentiable w.r.t. θ and its derivative $\nabla_\theta H(\theta, z)$ w.r.t. θ satisfies

$$E[\sup_\theta |\nabla_\theta H(\theta, Z)|] < \infty \ , \tag{16.11}$$

$$\text{and} \quad h'(\theta) = E[\nabla_\theta H(\theta, Z)] \ , \tag{16.12}$$

where h' denotes the derivative of h.

Assumption (C2) is clearly an identifiability condition. Note that we have taken the dimension of θ to be the same as that of $H(\theta, z)$ for reasons that will become clear in the theorem to follow.

Theorem 2.4 (Asymptotic normality of M-estimators). *We assume that (C1,C2,C3) are satisfied. Consider a sequence $(Z_k)_{1 \le k}$ drawn under P; then any sequence $(\widehat{\theta}_n)$ which is a solution of*

$$\sum_{k=1}^{n} H(\widehat{\theta}_n, Z_k) = 0$$

which remains a.s. in a compact subset of Θ, converges almost surely to $\theta_\star$. Furthermore, h is continuously differentiable and if $h'(\theta_\star)$ is invertible, the following limit in distribution holds :

$$\sqrt{n}\ (\widehat{\theta}_n - \theta_\star) \xrightarrow{d} \mathcal{N}(0, h'(\theta_\star)^{-T} \Sigma h'(\theta_\star)^{-1}) \ ,$$

where

$$\Sigma = R(0) + \sum_{i>0} R(i) + R(i)^T \ , \quad R(i) = \mathrm{Cov}(H(\theta_\star, Z_1), H(\theta_\star, Z_{i+1})) \ . \tag{16.13}$$

Proof: See Appendix 5.. ∎

3. ... To practice

... Practice, however, is our main concern.

3.1 The framework

Referring to our first figure, assume we have at hand a dynamical system, parameterized by θ, with measurements taken from plant as inputs, and producing as output some vector process Z_k satisfying suitable conditions we shall describe soon. Thus our new basic picture is the following one:

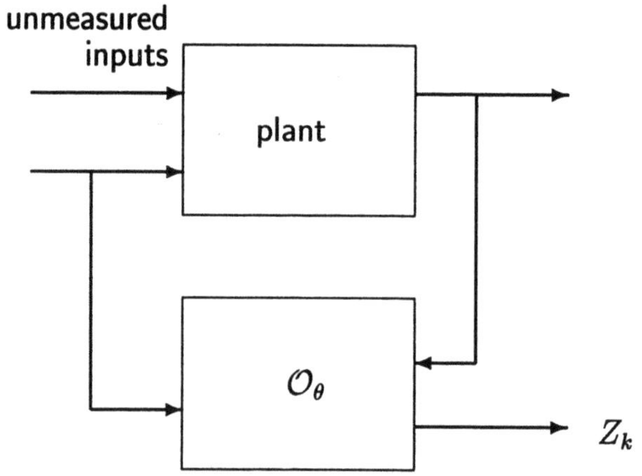

Typical choices for Z_k would be:

- Input/output pairs collected from plant. In this case, the $\mathcal{O}_\theta$ box does not depend on parameter θ. This is typically the case for plant models of the regression type, both linear or nonlinear:

$$y_k = f(\theta, \varphi_k) + \varepsilon_k$$

  where (y_k, φ_k) are measured. This is also the case for IV type of methods, or subspace methods.
- The same, plus, in addition, an observer of the internal state of the plant. There, the $\mathcal{O}_\theta$ box depends on parameter θ. This is typically the case for plant models that are (possibly nonlinear) dynamical systems:

$$\begin{cases} X_{k+1} &= f(\theta, X_k, u_k) \\ y_k &= g(\theta, X_k, u_k) \end{cases}$$

- More generally, z_k would collect some sort of "state" for our system, i.e., some suitable summary of measurement histories which enables us to predict the future behaviour of the system, using hypothesized parameter θ.

The next step is then to build a so-called *estimating function* $H(\theta, z)$ satisfying the following main conditions:

$$\begin{aligned} \mathbf{E}_\theta \, H(\theta_0, Z_k) &= 0 \;\text{ if }\; \theta = \theta_0 && (16.14) \\ \mathbf{E}_\theta \, H(\theta_0, Z_k) &\neq 0 \;\text{ if }\; \theta \neq \theta_0 \,, && (16.15) \end{aligned}$$

which is a rephrazing of condition **(B2)** of section 2.3. Conditions (16.14,16.15) express that, using *current* data from the plant, we can decide whether $\theta = \theta_0$ (plant $\equiv$ baseline parameter θ_o) or alternatively $\theta \neq \theta_0$ (plant $\not\equiv$ baseline parameter θ_o). As formulæ (16.14,16.15) suggest, we need that the referred expectations exist, which amounts to requiring some kind of asymptotic stationarity for our process Z_k. This point is extensively discussed in [2] from practical point of view.

Of course, designing such an estimating function is the heart of science for this method. Links with likelihood or quasi-likelihood, or contrast functions, or subspace algorithms are discussed in section 2.1 and in [2]. Systematic methods for constructing estimating functions are discussed in [9] from the theoretical standpoint, and in [2] from the practical standpoint.

3.2 The local approach as a way to decide upon possible changes

The next step is to decide as to whether $\theta = \theta_0$ (plant $\equiv$ baseline parameter θ_o), or alternatively $\theta \neq \theta_0$ (plant $\not\equiv$ baseline parameter θ_o) holds, based on some current sample data taken from the plant. Of course, deciding if some noise corrupted and uncertain estimate differs from zero amounts to correctly setting a threshold. This is where the mathematical analysis of section 2. comes to a rescue.

Let us thus consider an estimating function $H(\theta, Z_k)$, which we assume to be asymptotically stationary and to satisfy the local identifiability assumption (16.14)-(16.15). We define the so-called *quasi-score* U_N:

$$U_N(\theta_0) = \frac{1}{\sqrt{N}} \sum_{k=1}^{N} H(\theta_0, Z_k) \tag{16.16}$$

Following condition **(B2)**, we introduce the *sensitivity matrix*:

$$M(\theta_0) \triangleq - \mathbf{E}_{\theta_0} \frac{\partial}{\partial \theta} H(\theta, Z_k) \Big|_{\theta=\theta_0} \tag{16.17}$$

and the covariance matrix :

$$\begin{aligned} \Sigma(\theta_0) &\triangleq \lim_{N\to\infty} \Sigma_N(\theta_0), && (16.18) \\ \Sigma_N(\theta_0) &\triangleq \mathbf{E}_{\theta_0} \left(U_N(\theta_0) \, U_N^T(\theta_0) \right) \\ &= \frac{1}{N} \sum_{k=1}^{N} \sum_{j=1}^{N} \mathbf{E}_{\theta_0} \left(H(\theta_0, Z_k) \, H^T(\theta_0, Z_j) \right) \end{aligned}$$

It is of interest to note that, because of (16.14), Jacobian matrix (16.17) also writes :

$$M(\theta_0) \;=\; -\,\frac{\partial}{\partial\theta}\,\mathbf{E}_{\theta_0}\,H(\theta,Z_k)\Big|_{\theta=\theta_0} \;=\; +\,\frac{\partial}{\partial\theta}\,\mathbf{E}_{\theta}\,H(\theta_0,Z_k)\Big|_{\theta=\theta_0} \tag{16.19}$$

Both expressions are useful for implementation, depending on the situation. A theoretical analysis of this material was given in section 2., Theorem 2.3. However, for the sake of the comfort of the reader, we wish to present a heuristic derivation of the implementation.

First order Taylor expansion of the quasi-score. We now play with this improved residual the same game as before with the efficient score, namely first order Taylor expansion. We consider the local alternative hypothesis as Theorem 2.3 (we rephraze it in a more intuitive way here):

$$\theta \;=\; \theta_o + \frac{\widetilde{\theta}}{\sqrt{N}} \tag{16.20}$$

and we write the first order expansion of the quasi-score U_N in (16.16) as :

$$U_N(\theta) \approx U_N(\theta_0) + \sqrt{N}\;\underbrace{\frac{1}{N}\left(\sum_{k=1}^{N}\frac{\partial}{\partial\theta}\,H(\theta,Z_k)\Big|_{\theta=\theta_0}\right)}\;\frac{\widetilde{\theta}}{\sqrt{N}} \tag{16.21}$$

By the CLT under p_{θ_0}, the first term of the right hand-side is seen to be asymptotically Gaussian distributed with mean 0 and covariance matrix $\Sigma(\theta_0)$. By the law of large numbers under p_{θ_0}, the under-braced part of the second term of the right hand-side is seen to converge towards $\left(\mathbf{E}_{\theta_0}\,\frac{\partial}{\partial\theta}\,H(\theta,Z_k)\big|_{\theta=\theta_0}\right)$ which is exactly what we defined as $(-\,M(\theta_0))$. From this, the asymptotic Gaussianity of the quasi-score under both null ($\theta=\theta_o$) and local alternative ($\theta=\theta_o+\widetilde{\theta}/\sqrt{N}$) hypotheses can be shown.

Asymptotic Gaussianity of the quasi-score.. From the above Taylor expansion and provided that $\Sigma(\theta_0)$ is positive definite, the following CLT can be shown to hold (see Theorem 2.3 above) :

$$U_N(\theta_0) \;\to\; \begin{cases} \mathcal{N}(0\;,\;\Sigma(\theta_0)) & \text{under} \quad \mathbf{P}_{\theta_0} \\[2ex] \mathcal{N}(M(\theta_0)\,\widetilde{\theta}\;,\;\Sigma(\theta_0)) & \text{under} \quad \mathbf{P}_{\theta_0+\widetilde{\theta}/\sqrt{N}} \end{cases}$$

Again, this theorem means that a small deviation in the parameter vector is reflected into a change in the mean value of the quasi-score vector U_N, which switches from zero to a non-zero value. Also, this quasi-score is asymptotically Gaussian distributed with equal covariance matrices under both null and local alternative hypotheses.

Before investigating the design of the tests based on quasi-scores, we briefly describe another interesting byproduct of the above Taylor expansion.

Design of locally optimum tests. As for the log-likelihood function, the main use of the above CLT concerns the design of asymptotically optimum tests between composite (multi-valued) hypotheses [1, 5]. When Jacobian matrix M in (16.17) is also assumed to be full column rank, the quasi-score-based test between the hypotheses $\mathbf{H}_0$ and $\mathbf{H}_1$;

$$\begin{aligned} \mathbf{H}_0 &: \theta = \theta_o \\ \mathbf{H}_1 &: \theta = \theta_o + \frac{\widetilde{\theta}}{\sqrt{N}}, \ \widetilde{\theta} \text{ unknown} \end{aligned}$$

is :

$$U_N^T \, \Sigma_N^{-1} \, M \left(M^T \, \Sigma_N^{-1} \, M\right)^{-1} \, M^T \, \Sigma_N^{-1} \, U_N \tag{16.22}$$

where the dependence on θ_0 has been removed for simplicity. To derive this formula, simply compute the likelihood ratio test

$$-\log \max_{\widetilde{\theta}} \mathcal{N}_{M(\theta_0)\,\widetilde{\theta},\Sigma(\theta_0)}(U_N) \ + \log \mathcal{N}_{0,\Sigma(\theta_0)}(U_N)$$

where $\mathcal{N}_{\mu,\Sigma}$ denotes the normal distribution with mean μ and covariance Σ, and $\mathcal{N}_{\mu,\Sigma}(x)$ is the value of this distribution at point x. The non-centrality parameter of this chi-square test statistics under hypothesis $\mathbf{H}_1$ is :

$$\widetilde{\theta}^T \, M^T \, \Sigma^{-1} \, M \, \widetilde{\theta} \, .$$

As far as actual computation is concerned, we only mention that the estimation of $M(\theta)$ in (16.17) from a data sample is obtained by replacing the expectation by sample average. The estimation of Σ is more tricky, the discussion at the end of Section 2.1 provides some hints. When M is square and invertible, test (16.22) boils down to : $U_N^T \, \Sigma_N^{-1} \, U_N$.

We stress that this local approach, based on alternative estimating functions, reduces to the classical likelihood-based one only in particular cases. More precisely, test (16.22) is a classical LeCam-type local likelihood test when the estimating function H is associated with the minimization of the log-likelihood functional $(-\mathbf{E}_\theta \log p_\theta)$. As shown in [5, 3], the ELS and IV identification methods provide us with detectors which are *not* equivalent to the ML-based one. See [4] for an extensive study of subspace tests.

4. Some industrial case studies

4.1 In operation vibration monitoring: a case of eigenstructure monitoring for a linear dynamical system

The monitoring problem[4]. In-operation damage monitoring and predictive maintenance for complex mechanical structures and rotating machines is of

[4] Here we refer to a work which was supported by several grants from IFREMER (Institut Français de Recherche pour la Mer), Brest, France and EDF (Elec-

key importance in many industrial areas : power engineering (rotating machines, core and pipes of nuclear power plants), civil engineering (large buildings subject to hurricanes or earthquakes, bridges, dams, offshore structures), aeronautics (wings and other structures subject to strength), automobile, rail transportation. These systems are subject to both fast and unmeasured variations in their environment and small slow variations in their vibrating characteristics. Of course, only the latter are to be detected, using the available data (accelerometers). Thus we are in the general framework of our first figure, in which there are *no* measured inputs.

Modeling. We assume that the behavior of the mechanical system can be described by a stationary linear dynamic system, and that, in the frequency range of interest, the input forces can be modeled as a (possibly) non-stationary white noise. Consequently the relevant model is the following matrix differential equation :

$$\left\{ \begin{array}{rcl} M\ddot{\mathcal{X}}(t) + C\dot{\mathcal{X}}(t) + K\mathcal{X}(t) & = & \nu(t) \\ Y(t) & = & L\mathcal{X}(t) \end{array} \right. \tag{16.23}$$

where t denotes continuous time, M, C, K are the mass, damping and stiffness matrices respectively, the (high dimensional) vector $\mathcal{X}$ collects the displacements of the degrees of freedom of the structure; the external (non measured) force ν is modeled as a non-stationary white noise with unknown time-varying covariance matrix $Q_\nu(t)$, measurements are collected in the (low dimensional) vector Y, and matrix L states where the sensors are located. The mechanical characteristics (M, C, K) of the system cannot be recovered from the output measurements. Hence, identifiable modal characteristics of the system are introduced : the vibration modes or eigenfrequencies denoted generically by μ, and the modal shapes or eigenvectors denoted generically by ψ_μ. These quantities are solutions of the following equation :

$$(M\mu^2 + C\mu + K)\Psi_\mu = 0 \quad , \quad \psi_\mu = L\Psi_\mu \tag{16.24}$$

and the collection of pairs (μ, ψ_μ) constitute our parameter vector θ. Moving to state-space form and discretizing the above model yields a linear dynamical system, for which our parameter θ describes the eigenstructure.

4.2 Gas turbine monitoring: a case of a nonlinear static system

Detecting faults in the combustion set of a gas turbine[5] is a problem which can be addressed within the framework of system changes in a nonlinear

tricité de France), Direction des Etudes et Recherches, Clamart, France. Further work is being accomplished in the context of Eureka project no 1562 SINOPSYS (Model based Structural monitoring using in-operation system identification) coordinated by LMS, Leuven, Belgium.

[5] Here we refer to a work which was supported by several grants from Alcatel-Alsthom-Recherche, Marcoussis, France and European Gas Turbine S.A., Belfort, France.

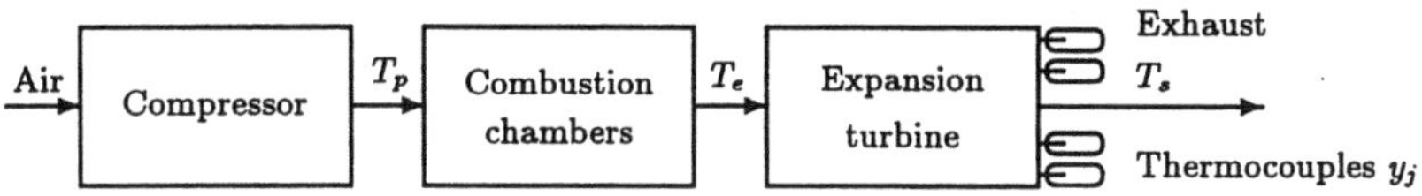

Fig. 16.1. Gas turbine system.

static model with noisy input and output measurements, and also unknown inputs.

The monitoring problem. Gas turbines are widely used as industrial motors, in electrical power generators and aircraft engines for example. As shown on figure 16.1, a gas turbine system is composed of a compressor, a set of combustion chambers, an expansion turbine and an exhauster. The problem here is to monitor the (ten) combustion chambers and the expansion turbine using the measurements given by the (eighteen) thermocouples at the output of the exhauster, to detect both sudden failures and gradual incipient faults, and to isolate which combustion chamber(s) failed.

This monitoring problem is complex for the following reasons. First, the knowledge about the physics of the involved phenomena (combustion, expansion, diffusion) is not complete and in some sense less mature than in the previous example of mechanical systems. There is thus a need for a semi-physical model, which provides us with a reasonable approximation of the measurements. Second, as explained below, this model has both noisy input and output measurements (errors-in-variable model). Third, one of the key difficulties here is that the considered gas turbines are used as secondary energy production devices in electricity power plants. Thus their functioning mode is changing quite often, and the effects of this non-stationarity on the system parameters are non negligible with respect to those of the faults.

Modeling. The two key ideas used for deriving a semi-physical model are to describe the output measurements (thermocouples at the output of the exhauster) as a linear combination of temperature profiles contributed by each combustion chamber, and to summarize complex phenomena (rotation and turbulence in the turbine and the exhauster) into three phase shift coefficients. As a result we end up with a nonlinear static system, in which the nonlinear part mainly follows from the presence of the phase shift coefficients.

4.3 Catalytic converter : a case of a nonlinear dynamic system

Detecting faults in the catalytic converter and the oxygen sensors of an automobile is a problem which can be addressed within the framework of changes in a parametric nonlinear dynamic model[6]. As obvious from figure 16.2, this system is still more tricky, because of a closed-loop which may blur some of the changes of interest, as do also varying operating conditions.

[6] Here we refer to a work which was supported by Renault, Direction de la Recherche, Rueil-Malmaison, France.

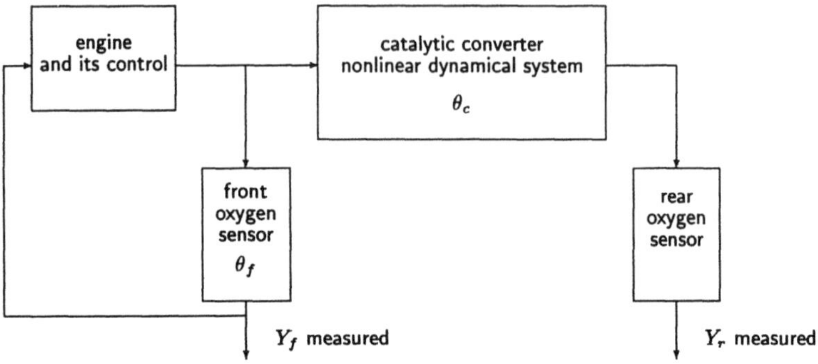

Fig. 16.2. Monitoring the catalytic converter and the front oxygen sensor of an automobile. The two parameters for monitoring are θ_c, θ_f.

The monitoring problem. On-board fault diagnosis is a crucial issue for the automotive industry. New on-board diagnostic system requirements (OBD II) have been introduced in California, which require monitoring vehicle emissions. Among the vehicle components to be monitored are the catalytic converter and the oxygen sensors.

A catalytic converter is aimed at processing the engine exhaust gases and reducing the quantity of pollutants. In particular, three way catalytic converters (TWC) oxidize the hydrocarbons HC and carbon monoxide CO and reduce nitrous oxides NO to produce water H_2O, carbon dioxide CO_2 and nitrogen N_2. A catalytic converter is characterized by its efficiency in converting HC, CO and NO_x. Thanks to a front oxygen sensor (located upstream from the catalytic converter), a closed-loop controller maintains the air-fuel ratio in the engine close to the stoichiometric value.

Two oxygen sensors – front and rear – measure the lack or excess of oxygen in the exhaust gases, in front of and after the catalytic converter respectively; see figure 16.2. We consider lambda sensors, which behave as follows. Their output voltage suddenly switches from zero to one when the gas mixture changes from lean to rich, and their response time is a function of temperature and air-fuel ratio, with a nearly bang-bang static characteristic. The response times for both lean-to-rich and rich-to-lean switches increase during the life cycle of a lambda sensor.

Due to the OBD recommendations, the problem is to monitor the catalytic converter and the front oxygen sensor.

Modeling issues. Models of all the elements of the de-pollution line (engine, catalytic converter and oxygen sensors) were available, see [2] for more details.

The engine is described by a nonlinear discrete-time dynamic model, whose inputs are the throttle position, the ignition advance and the fuel injection duration, and whose states and outputs are the engine speed, the intake manifold pressure, and the intake normalized air-fuel ratio. A PI regulator, based on the front oxygen sensor, makes the air-fuel ratio oscillate.

The catalytic converter is described by a nonlinear continuous-time state-space model, whose inputs are the front pollutants (hydrocarbons, carbon monoxide and nitrous oxides) and the oxygen, and whose states are the rear pollutants, the oxygen and the fraction of oxygen storage capacity filled with O_2. This model is parameterized with parameter θ_c. The front oxygen sensor is described by a sigmoid function fed with a first order transfer function, with the normalized air-fuel ratio as input and the sensor voltage as output. This model is parameterized with parameter θ_f.

Consequently, the two-dimensional (lambda sensors) measurements Y can be modeled with the aid of state-space model of the form :

$$\begin{cases} \dot{X} &= f(\theta, X, u) \\ Y &= h(\theta, X, u) \end{cases}$$

where functions f and h are strongly nonlinear, the parameter vector is : $\theta^T = \begin{pmatrix} \theta_c & \theta_f \end{pmatrix}^T$, and where, because of the closed-loop shown on figure 16.2, the input u depends on θ.

Appendix: proof of Theorem 2.3

Rewrite U_n as

$$\begin{aligned} U_n &= \frac{1}{\sqrt{n}} \sum_{k=1}^{n} (H(\theta_0, Z_{n,k}) - E_{\theta_n}[H(\theta_0, Z_{n,k})]) + \sqrt{n}\, E_{\theta_n}[H(\theta_0, Z_{n,k})] \\ &= S_n + T_1. \end{aligned}$$

The second term is

$$T_1 = \sqrt{n}\, h_{\theta_n}(\theta_0) = \sqrt{n}\, (h_{\theta_n}(\theta_0) - h_{\theta_n}(\theta_n))$$

which tends to $-h'(\theta_0)\widetilde{\theta}$ (assumption (B2)).

The mixing property implies, with theorem 2.2, that the first term converges in distribution to a zero-mean normal variable with the announced variance if we can prove that the covariance of S_n converges to $\Sigma(\theta_0)$. This will involve standard technicalities. One has, using the stationarity:

$$E[S_n S_n^T] = R_{\theta_n}(0) + \frac{1}{n} \sum_{i=1}^{n-1} (n-i)(R_{\theta_n}(i) + R_{\theta_n}(i)^T). \tag{16.25}$$

We have the following bound, uniform in θ on $R_\theta(i)$:

$$\begin{aligned} |R_\theta(i)| &= |\mathrm{Cov}_\theta(H(\theta_0, Z_1), H(\theta_0, Z_{i+1}))| \\ &\leq \|H(\theta_0, Z_{i+1}))\|_2 \; \|E[H(\theta_0, Z_i)|\mathcal{F}_1] - E[H(\theta_0, Z_i)]\|_2 \\ &\leq C i^{-\delta-1/2}. \end{aligned} \tag{16.26}$$

Equations (16.25) and (16.26) together imply that, for any n_0 and $n > n_0$

$$\begin{aligned}\left|E[S_nS_n^T] - R_{\theta_n}(0) + \sum_{i=1}^{n}(R_{\theta_n}(i) + R_{\theta_n}(i)^T)\right| &\leq 2\sum_{i=1}^{n}\frac{i}{n}|R_{\theta_n}(i)| \\ &\leq C'n^{-\delta+1/2}\end{aligned}$$

thus

$$\begin{aligned}\left|E[S_nS_n^T] - \Sigma(\theta_n)\right| &\leq C''n^{-\delta+1/2}\text{ , whence} \\ \lim_n\left|E[S_nS_n^T] - \Sigma(\theta_n)\right| &= 0.\end{aligned}$$

The bound (16.26) and the continuity w.r.t. θ of $R_i(\theta)$ (consequence of the Lebesgue theorem and assumption (B3)) together imply that $\Sigma(\theta)$ is a continuous function, and, finally,

$$\limsup_n\left|E[S_nS_n^T] - \Sigma(\theta_0)\right| = 0.$$

The proof is complete. ∎

5. Appendix: proof of Theorem 2.4

The convergence to $\theta_\star$ is a consequence of theorem 2.1 and (C2).

Set

$$U_n(\theta) = \frac{1}{\sqrt{n}}\sum_{k=1}^{n}H(\theta, Z_k).$$

Choose $\varepsilon > 0$. Then for any θ such that $|\theta - \theta_\star| < \varepsilon$, we have

$$U_n(\theta) = U_n(\theta_\star) + \left(\frac{1}{n}\sum_{k=1}^{n}\nabla_\theta H(\theta_\star, Z_k) + \delta_n(\theta)\right)(\theta - \theta_\star)\sqrt{n}$$

and

$$|\delta_n(\theta)| \leq \zeta_n(\varepsilon)\ ,$$

where

$$\zeta_n(\varepsilon) = \sup_{|\theta'-\theta_\star|<\varepsilon}\left|\frac{1}{n}\sum_{k=1}^{n}\nabla_\theta H(\theta_\star, Z_k) - \nabla_\theta H(\theta', Z_k)\right|\ . \tag{16.27}$$

and lemma 2.1 implies that

$$\lim_{\varepsilon\longrightarrow 0}\sup_n \zeta_n(\varepsilon) = 0, \qquad a.s. \tag{16.28}$$

Since $U_n(\widehat{\theta}_n) = 0$, we finally get

$$0 = U_n(\theta_\star) + \left(\frac{1}{n}\sum_{k=1}^{n} \nabla_\theta H(\theta_\star, Z_k) + \delta_n(\widehat{\theta}_n)\right)(\widehat{\theta}_n - \theta_\star)\sqrt{n}.$$

Note that $\delta_n(\widehat{\theta}_n)$ tends to zero in probability since for any $\alpha, \varepsilon > 0$:

$$\begin{aligned} P(\left|\delta_n(\widehat{\theta}_n)\right| > \alpha) &\leq P(\zeta_n(\varepsilon) > \alpha, \left|\widehat{\theta}_n - \theta_\star\right| < \varepsilon) + P(\left|\widehat{\theta}_n - \theta_\star\right| \geq \varepsilon) \\ &\leq P(\zeta_n(\varepsilon) > \alpha) + P(\left|\widehat{\theta}_n - \theta_\star\right| \geq \varepsilon) \end{aligned}$$

and the convergence of $\widehat{\theta}_n$ to $\theta_\star$ in probability implies

$$\limsup_n P(\left|\delta_n(\widehat{\theta}_n)\right| > \alpha) \leq P(\zeta_n(\varepsilon) > \alpha)$$

which is arbitrarily small, for ε small enough (cf equation (16.28)).

Finally, $\widehat{\theta}_n - \theta_\star$ rewrites

$$\sqrt{n}\,(\widehat{\theta}_n - \theta_\star) = -\left(\frac{1}{n}\sum_{k=1}^{n} \nabla_\theta H(\theta_\star, Z_k) + \delta_n(\widehat{\theta}_n)\right)^{-1} U_n(\theta_\star).$$

This proves the result since $U_n(\theta_\star)$ converges to $\mathcal{N}(0, \Sigma(\theta_\star))$ (theorem 2.3 in the simple case where $P_\theta = P$ and $\widetilde{\theta} = 0$) and matrix $\widehat{h}'_n(\theta_\star)$ defined by

$$\widehat{h}'_n(\theta_\star) = \frac{1}{n}\sum_{k=1}^{n} \nabla_\theta H(\theta_\star, Z_k) + \delta_n(\widehat{\theta}_n)$$

converges in probability to the invertible matrix:

$$E[\nabla_\theta H(\theta_\star, Z_k)] = h'(\theta_\star).$$

This proves the theorem. ∎

References

1. I.V. Basawa (1985). Neyman-Le Cam tests based on estimating functions. In *Proc. Berkeley Conf. in Honor of Neyman and Kiefer*, vol.2, Le Cam and Olshen (Eds), pp.811-825.
2. M. Basseville (1997). On-board component fault detection and isolation using the statistical local approach. *Research Report* IRISA no 1122/ INRIA no 3252. ftp://ftp.irisa.fr/techreports/1997/PI-1122.ps.gz. *Automatica*, under revision.
3. M. Basseville and I. Nikiforov (1993). *Detection of Abrupt Changes – Theory and Applications*. Prentice Hall, N.J.

4. M. Basseville, M. Abdelghani and A. Benveniste (1997). Subspace-based fault detection and isolation methods - Application to vibration monitoring. *Research Report* IRISA no 1143/ INRIA no 3299. ftp://ftp.irisa.fr/techreports/1997/PI-1143.ps.gz.
5. A. Benveniste, M. Basseville and G. Moustakides (1987). The asymptotic local approach to change detection and model validation. *IEEE Trans. Automatic Control*, vol.AC-32, no 7, pp.583-592.
6. A. Benveniste, M. Mtivier and P. Priouret (1990). *Adaptive Algorithms and Stochastic Approximations*. Springer, NY.
7. B. Delyon, A. Juditsky and A. Benveniste (1997). On the relationship between identification and local tests. *Research Report* IRISA no 1104. ftp://ftp.irisa.fr/techreports/1997/PI-1104.ps.gz.
8. A.R.Gallant, *Nonlinear Statistical Model*, Wiley, 1987.
9. C.C. Heyde (1997). *Quasi-Likelihood and Its Application*. Springer Series in Statistics.
10. P. J. Huber, The behavior of maximum likelihood estimates under nonstandard conditions, *Proc. Fifth Berk. Symp. Math. Statist. and Prob.* , 221-233, Univ. California Press, Berkeley, 1967.
11. P. J. Huber, *Robust statistics*, Wiley, 1981.
12. A. F. Kushnir, A. I. Pinskii Asymptotically Optimal Tests for Testing Hypotheses for Independent Samples of Observations, *Theory of Probability and its Applications, vol XVI, No 2, 1971.*
13. L. Ljung (1987). *System Identification - Theory for the User*. Prentice Hall, NJ.
14. D. L. McLeish, On the Invariance Principle for Non-stationary Mixingales, *Ann. Prob.* 1977, vol 5, No 4, 616-621.
15. G. Moustakides and A. Benveniste (1986). Detecting changes in the AR part of a non-stationary ARMA process. *Stochastics*, vol.16, pp.137-155.
16. Q. Zhang, M. Basseville and A. Benveniste (1994). Early warning of slight changes in systems and plants with application to condition based maintenance. *Automatica*, vol.30, no 1, pp.95-114.

Applications of Digital and Adaptive Control Theories to Mechatronic Systems

M. Tomizuka

Department of Mechanical Engineering, University of California, Berkeley, CA 94720-1740, U.S.A., e-mail: tomizuka@me.berkeley.edu

Summary. This essay touches the past, present and future while reviewing my experience in the area of mechanical system controls over the last two decades. Applications of adaptive control theory to mechanical systems are emphasized throughout this article; Ioan Landau introduced me to the world of adaptive control twenty years ago, which is an important "past" event in my professional career. The word "Mechatronics" is gaining increasing popularity among systems and control engineers, in particular in the mechanical engineering community. While my professional career has been in the field of dynamic systems and control, many of my past and present works are relevant to mechatronics. As long as the common understanding of "mechatronics" has evolved to intersect with the analysis and design of mechanical control systems, it provides an excellent framework to discuss the "present" and "future" direction of control of mechanical systems.

1. Introduction

A Yaskawa Electric engineer introduced the word "mechatronics" in 1969, and Yaskawa obtained trademark rights in 1972 [15]. As the word became popular, Yaskawa released all rights pertaining to the trademark in 1982 and it is now widely used. In the 1970's, mechatronics meant the combination of mechanics and electronics. This is a rather narrow definition, and those working in the area of electro-mechanical systems often questioned "what is new about mechatronics?" The definition of mechatronics has evolved since then. The Institute of Electrical and Electronics Engineers (IEEE) and the

American Society of Mechanical Engineers (ASME) introduced a new periodical, the IEEE/ASME Transactions on Mechatronics in 1996 [9]. Mechatronics is defined in this journal as "the synergistic integration of mechanical engineering with electronics and intelligent computer control in the design and manufacture of industrial products and processes," which is very close to the definition by the International Federation for the Theory of Machines and Mechanism (IFTMM) [3]. This is a good definition although it may still be argued. For example, if we strictly apply "in the design and manufacture of industrial products and processes," most of the university research will immediately be excluded from mechatronics. In my view, the key words in this definition are "synergistic integration." "Intelligent computer control" should be understood in a broader context of "information" technology, which includes software engineering for real time control, signal processing, fault detection and management and system optimization. Having said this, control theory plays a significant role in modern mechatronic systems. A recent paper by Isermann presents a review of mechatronics from the viewpoint of control engineers [11]. In this paper, I will present applications of digital control theory and adaptive control theory as well as other control theories to mechanical systems, which, in my opinion, contribute either directly or indirectly to the design of mechatronic systems. Application examples are drawn from research conducted by my research group at the University of California at Berkeley.

2. Adaptive Control of Robots

A robot can be defined as a multidegree of freedom open loop chain of mechanical linkages and joints. These mechanisms driven by actuators, normally one at each joint, are capable of moving an object in space to accomplish a variety of tasks. The mechanical part is accompanied by a robot controller, which provides kinematic transformations, servo control algorithms and other functions. The robot system can be made progressively more intelligent by adding sensors such as force sensors and vision cameras and by making the software of the robot controller more sophisticated. Thus, the robot system is an excellent mechatronics example. The dynamics of a robot are described by a set of coupled nonlinear differential equations, which has made the robot an ideal test bed for a variety of control algorithms based on the adaptive, nonlinear and robust control theory. The research on adaptive control of robots started in the late 1970's, my work with Horowitz [8], which was published in 1980, was perhaps the first paper, that applied the "modern adaptive control theory" to robot manipulators. I should note that our association with Ioan Landau made it possible for us to make this timely contribution. The adaptive control algorithm in [8] was not complete and it required several assumptions; in particular, because of the way the robot equation was written, we had to assume that dynamic parameters were slowly varying. Craig,

Hsu and Sastry [4], Horowitz and Sadegh [18], Slotine and Li [19] among others later devised more complete versions and removed the assumptions. Many versions of robot adaptive control have been reported during the past fifteen years. Yao and Tomizuka [31] made a synergistic integration of adaptive control and nonlinear robust control, and verified its advantages on a two-link robot. In the early days, adaptive control of robot manipulators was for adaptive servo control of a rigid single manipulator. Since then, adaptive control theory has been applied to many kinds of robot control problems such as force and compliance control, coordination of multiple robots and flexible arms. One of the most recent complete studies in this field was conducted by Yao, and his dissertation lists more than 35 representative papers on this subject [30]. Industrial robots have not yet utilized these advances. However, as the requirements for robots become more versatile including operations in uncertain environments, the role of adaptive control theory along with other methodologies for intelligent control may only increase in the field of robotics. At the same time, the mechatronics nature will be more pronounced in the field of robotics. Space exploration, remote surgery, undersea exploration and maintenance of power plants are but a few examples, which require intelligent robots.

3. Digital and Adaptive Control of Machining Processes

3.1 Control of Cutting Force

Digital computers opened a new era in the history of automation. Computers made it possible to make the old idea of a punched card control system, first introduced in 1725 by Bouchom for a loom, practical as numerical control (NC) of machine tools, which was developed at MIT in the early 1950's [7]. NC had become quite sophisticated even before the word "mechatronics" was invented. Computer controlled machine tools indeed represent another mechatronics example. As the performance specifications for machine tools both in terms of accuracy/precision and machining speed become more stringent and the necessity for flexible operations has increased, the mechatronics nature has become stronger. At the same time, there started a number of attempts to introduce modern control algorithms to the machining arena. It is interesting to note that the terminology "adaptive control" was introduced by manufacturing engineers to imply several different approaches to machining including feedforward control and additional sensor feedback loops added to the basic servo control loop [29, 28]. Control of cutting force is important for operating machine tools at high material removal rates while ensuring the high quality of machined surfaces with no occurrences of tool breakage and failure. The controlling input for cutting force control is the feed-rate of the cutting tool. The input/output dynamics between feed-rate and cutting force depend on tool and cutting conditions as well as material properties. Furthermore, the output force depends on the feedrate applied over the preceding complete revolution of the tool (in milling) or the workpiece (in turning). These aspects of dynamics make discrete time adaptive control theory an attractive tool for cutting force control. The explicit discrete time model reference adaptive control algorithm by Landau and Lozano [16] was found to be suited for force control in milling by Tomizuka, et al. [26], who also considered adaptive force control in turning. An interesting aspect of the milling and turning studies reported in [26] is that one control approach is not necessarily the best for all problems. In fact, an innovative model for turning combined with the indirect adaptive control with a single parameter identification algorithm was found to work quite well for turning. The work in [26] was followed by other researchers, among which the most innovative mechatronics idea to date, in my opinion, is the one by Altintis, who integrated a priori knowledge available form CAD (Computer Aided Design) database with adaptive control [20].

3.2 Noncircular Machining on Lathe

A lathe is a machine tool for machining cylindrical workpieces. It is traditionally used to cut the workpiece in circular cross sectional shape: i.e. the tool position is fixed in the direction normal to the machined surface. The cross sectional shape can be noncircular if the tool is moved in the direction normal to the surface. Cam mechanisms are a traditional mechanical solution for moving the tool to generate desired shapes. This solution is not flexible since different cams must be prepared if the desired shape changes from one workpiece to another. The cam solution does not work unless the desired shape is fixed from the starting point to the ending point in the axial direction of one workpiece. Furthermore, it is not possible to apply closed loop control to this mechanism. Higuchi proposed a mechatronics solution to this problem by adding a servo-control system to move the tool in the direction normal to the surface of workpiece [6]. From the viewpoint of servo-control, noncircular machining is a tracking problem, and the tracking performance of the servo system directly influences the dimensional accuracy of finished workpieces. The digital tracking control algorithm based on Zero Phase Error Tracking Control (ZPETC) and its adaptive version were found to be effective to ensure high tracking performance for arbitrary desired shapes, which may even vary as the tool moves from the start to end one workpiece [23, 27]. When the desired shape remains fixed from the starting and ending point, the desired position of the tool is repetitive or periodic as it is traversed. For such cases, discrete time repetitive control theory offers a simple effective solution for noncircular machining [25]. Repetitive control is based on the internal model principle, and it is useful in general when the disturbance and/or reference input is periodic. This situation is quite common in mechanical systems which involve rotation motions. The mechatronics nature of noncircular machining is evident: the software cam, which improves the performance and flexibility of noncircular machining, has replaced the mechanical cam. Representative applications of noncircular machining may be found in automotive industry. Tsao and his colleague at the University of Illinois conduct a project, "Noncircular Turning Process for Camshaft Machining," in cooperation with the Saginow Machine Systems, Inc.

4. Advanced Vehicle Control Systems (AVCS)

Automotive vehicles represent another area where active controls are utilized in many ways and mechatronics play a significant role [24]. Adaptive control has been studied for various subsystems of the vehicle including intelligent cruise control [10], anti-lock braking and traction control [21], engine control [2] and active suspension [1]. The California PATH (Partners for Advanced Transit and Highways) Program, in which automated highway systems (AHSs) have been an emphasis area, has motivated my recent work on AVCS. In the AHS context, I have been studying with my students automated steering for lateral control of both the passenger vehicle and heavy duty vehicle [5, 13]. While the steering angle is the control input to the vehicle lateral motion, it is an indirect means to control the tire forces, which are the true actuation forces of the vehicle motion. The tire force depends on the road and tire conditions, which are time varying and uncertain. Furthermore, the vehicle dynamics for lateral control depend on other variables such as the vehicle speed. These aspects make the vehicle lateral control problem challenging. Other researchers in the PATH program investigate the vehicle longitudinal control problem [5, 13]. In AHSs, it is critical that the vehicle may follow a car in front of it with a close spacing. In the longitudinal control problem, the engine throttle position and brake actuation signal are the control inputs, but the tire forces again cause the vehicle motion. The vehicle longitudinal control problem for AHSs is another challenging problem. PATH was a major player in the AHS demonstration conducted in San Diego last year by the National Automated Highway Systems Consortium (NAHSC). The PATH's demonstration was platooning (convoying) of eight passenger vehicles fully automated both in the longitudinal and lateral direction. It clearly demonstrated how the advances in the control theory might be integrated with computers and other modern technologies such as communication. Independent of the fundamental question on whether AHSs will be fully implemented in future, technological achievements demonstrated by NAHSC will significantly influence the AVCS for next generation automobiles.

5. Computer Data Storage Devices

Accurate positioning is a requirement in modern precision machines such as computer data storage devices and semi-conductor assembly and inspection systems. This is analogous to machining systems as discussed earlier in this paper. In hard disk drives (HDDs), data is recorded on a data track on a rotating disk surface. The memory capacity of commercial HDDs has roughly doubled every year in recent years. In terms of the storage density, 1 Gigabit per square inches is standard in these days, and drives with higher densities are available. The density is the product of the number of data tracks per inch (TPI) in the radial direction and the linear recording density in terms of the number of bits per inch (BPI). The recording head is integrated into the slider, which is mounted at the end of the suspension. The voice coil motor (VCM) swings the carriage and suspension unit and positions the recording head on the target track. As TPI increases, specification for positioning accuracy of the recording head becomes more and more stringent. Mechatronics technologies, which include servo controls, have and will play an indispensable-supporting role contributing to this dramatic increase [22].

In HDDs, attention must be given to the pivot friction of VCM. Other disturbance sources in HDDs include the tension of the flexible printed circuit, which is attached to the carriage, and the spindle bearing. Other disturbance sources may be due to any errors introduced at the time when the encoded track position reference signal was written on the disk surface and at the time when the HDD unit was assembled. These errors show up as periodic disturbances when the recording head is tracking any particular track. A unique aspect in the servo system of HDD is that the position reference signal is not continuous on the disk surface but is only intermittent at equally spaced angles, i.e. sectors. This makes the servo system inherently a discrete time system. The number of sectors and the spindle frequency determine the sampling frequency for digital control. For example, 80 sectors and a spindle speed of 7,800 rpm will result in approximately 10 kHz sampling rate. The number of sectors should be minimized so that a larger area can be used for recording data. It is a challenge to ensure acceptable positioning performance with low sampling rates. Kempf, et al. compared four different repetitive control algorithms for rejecting these periodic disturbances [14]. The four algorithms included one based on internal model principle [27] and others, which utilized the adaptive control theory. For non-periodic disturbances such as pivot friction force, the disturbance observer [17] the adaptive robust control [23] have been shown to be effective. Ishikawa and Tomizuka [12] recently proposed an innovative friction compensation method, which is based on an accelerometer mounted on the carriage. This sensor-based approach is an interesting departure from the traditional design, which has relied only on the position error signal obtained at each sector. Since the sampling rate for the accelerometer loop is not constrained by the number of sectors, a dramatic performance improvement can be expected. MEMS

(Micro-Electro-Mechanical System) technology has made it possible to manufacture small and inexpensive accelerometers. MEMS is a powerful enabling technology for the next generation HDDs, the target density for which exceeds 20 Gigabit per square inches. For example, Horowitz and his colleague at the University of California at Berkeley conduct a research project on MEMS actuators for HDDs in cooperation with IBM.

6. Conclusions

Mechanical systems are becoming increasingly reliant on sophisticated methods of sensing, signal processing, and control to improve performance. Robots, machine tools, vehicles and computer data storage devices are but a few examples. At the same time, rapid developments in microelectronics and computing have made it possible to use a broad spectrum of signal processing and control methods in real time in a very cost-effective fashion. It is likely that many future advances in mechanical systems will depend upon further application of sophisticated control systems and mechatronics thinking.

My research program at the University of California at Berkeley addresses a broad range of topics in the general area of control systems. Theoretical research is pursued in control methodologies such as adaptive and learning control, digital control, robust control and nonlinear control and is blended with application problems. The Berkeley environment makes it possible to let my students prepare for such advanced research and development work. They can select and take courses in all kinds of control theories, real-time software, and application of micro-controllers for signal processing and control. They can also take application domain course such as manufacturing and robotics. Similar environments are now found at many other places. Control engineers are required to be interdisciplinary more than ever.

References

1. Alleyne, A. and Hedrick, J. K., "Adaptive Control for Active Suspension," DSC-Vol. 5.2, Advanced Automotive Technology, ASME 1991.
2. Choi, S-B. and Hedrick, J. K., "An Observer Based Controller Design Method for Automotive Fuel Injection System," Proceedings of the 1993 American Control Conference, pp. 2567-2571, June 1993.
3. Comerford, R., "Mech. What?" IEEE Spectrum, pp. 46-49, August 1994.
4. Craig, J. J., Hsu, P. and Sastry, S., "Adaptive Control of Mechanical Manipulators," International Journal of Robotics Research, Vol. 6, No. 2, 16-28, 1987.
5. Hedrick, J. K., Tomizuka, M. and Varaiya, P., "Control Issues in Automated Highway Systems," IEEE Control Systems, pp. 21-32, December 1994.
6. Higuchi, T. et al., "Primary Study on Applications of Electro-hydraulic Servo Mechanism to Noncircular Cutting by a Lathe," SEISAN KENKYU, Vol. 36,

No. 2, Institute of Industrial Science, University of Tokyo, February 1984 (in Japanese).
7. Hitomi, K., "Manufacturing Systems Engineering," Taylor & Francis Ltd., London, 1979.
8. Horowitz, R. and Tomizuka, M., "An Adaptive Control Scheme for Mechanical Manipulators - Compensation of Nonlinearity and Decoupling Control," ASME Publication, 80-WA/DSC-6 (also in the ASME Journal of Dynamic Systems, Measurement and Control, Vol. 108, No. 2, 127-135, June 1986).
9. IEEE/ASME Transactions on Mechatronics, Vol. 1, No. 1, March 1996.
10. Ioannou, P. and Xu, Z., "Throttle and Brake Control Systems for Automatic Vehicle Following," IVHS Journal, Vol. 1, No. 4, pp. 345-377/
11. Isermann, R., "MECHATRONIC SYSTEMS - A challenge for control engineering," Proceedings of the 1997 American Control Conference, Albuquerque, New Mexico, June 1997.
12. Ishikawa, J. and Tomizuka, M., "Pivot Friction Compensation Using an Accelerometer and a Disturbance Observer for Hard Disk Drives," presented at the 1997 ASME International Mechanical Engineering Congress and Exposition, Dallas, TX, November 1997.
13. Kanellakopoulos, I. and Tomizuka, M., "Commercial Trucks and Buses in Automated Highway Systems," Chapter 11 of Automated Highway Systems, Edited by P. Ioannou, Plenum Press, New York, pp. 213-246, 1997
14. Kempf, C., Messner, W., Tomizuka, M. and Horowitz, R., "Comparison of Four Discrete-Time Repetitive Control Algorithms," IEEE Control Systems, pp. 48-54, 1993.
15. Kyura, N. and Oho, H., "Mechatronics – An Industrial Perspective," IEEE/ASME Transactions on Mechatronics, Vol. 1, No. 1, pp. 10-15, March 1996.
16. Landau, I. D. and Lozano, R., "Unification of Explicit Discrete Time Model Reference Adaptive Controller Design," Automatica, Vol. 17, No. 4, pp. 593-611, July 1981.
17. Ohnishi, K., "A New Servo Method in Mechatronics," Trans. Of Japanese Society of Electrical Engineering, Vol. 177-D, pp. 83-86, 1987.
18. Sadegh, N. and Horowitz, R., "Stability and Robustness Analysis of a Class of Adaptive Controllers for Robotic Manipulators," International Journal of Robotics Research, Vol. 9, No. 3, pp. 74-92, June 1990.
19. Slotine, J-J. E. and Li, W-P., "Adaptive Manipulator Control: A Case Study," IEEE Transactions on Automatic Control, Vol. 33, No. 11, 1988.
20. Spence, A., Altintas, Y., "CAD Assisted Adaptive Control for Milling", ASME Journal of Dynamic Systems, Measurement and Control, Vol. 113, pp. 444-450, September 1991.
21. Tan, H-S. and Tomizuka, M., "An Adaptive Sliding Mode Vehicle Traction Controller Design," Proceedings of the 1990 American Control Conference, pp. 1856-1861, June 1990.
22. Tomizuka, M., "Mechatronics for Computer Data Storage Devices," Proceedings of the 1997 IEEE International Conference on Intelligent Engineering Systems, Budapest, Hungary, pp. 25-30, September 1997.
23. M. Tomizuka, M. S. Chen, S. Renn and T. C. Tsao, "Tool Positioning for Noncircular Cutting with Lathe," Proceedings of the 1986 American Control Conference, June 1986 (also in the ASME Journal of Dynamic Systems, Meas. and Control, Vol. 109, No. 2, June 1987).
24. M. Tomizuka and J. K. Hedrick, "Advanced Control Methods for Automotive Applications," Vehicle System Dynamics, Vol. 24, pp. 449-468, July 1995.

25. M. Tomizuka, T-C. Tsao and K-K. Chew, "Discrete-Time Domain Analysis and Synthesis of Repetitive Controllers," Proceedings of the 1988 American Control Conference, pp. 860-866, June 1988. (also in the ASME Journal of Dynamic Systems, Meas. and Control, Vol.111, No. 3, pp. 353-358, Sept. 1989).
26. Tomizuka, M., Zhang, S-J., Oh, J-H. and Chen, M-S., "Modeling of Metal Cutting Processes for Digital Control," 1985 Manufacturing Engineering Transactions, SME, pp. 575-580, May 1985.
27. T-C. Tsao and M. Tomizuka, "Robust Adaptive and Repetitive Digital Tracking Control and Application to a Hydraulic Servo for Noncircular Machining," ASME Journal of Dynamic Systems, Meas. and Control, Vol. 116, No. 1, pp. 24-32, March 1994.
28. Ulsoy, A. G. and Koren, Y., "Control of Machining Processes," ASME Journal of Dynamic Systems, Measurement and Control, 50 the Anniversary Issue, Vol. 115, pp. 301-308, June 1993.
29. Watanabe, T. and Iwai, S., "A Control System to Improve the Accuracy of Finished Surfaces in Milling," ASME Journal of Dynamic Systems, Measurement and Control, Vol. 105, No. 3, pp. 192-199, September 1983.
30. Yao, B., "Adaptive Robust Control of Nonlinear Systems with Application to Control of Mechanical Systems," Ph. D. Dissertation, University of California at Berkeley, 1996.
31. B. Yao and M. Tomizuka, "Smooth Robust Adaptive Sliding Mode Control of Manipulators with Guaranteed Transient Performance," Proceedings of the 1994 American Control Conference, pp. 1176-1180, Baltimore, June 1994 (also in the ASME Journal of Dynamic Systems, Measurement and Control, Vol. 118, No. 3, pp. 764-775, December 1996).

Advanced Estimation and Optimization for Air Traffic Surveillance

H. Wang, T. Kirubarajan, and Y. Bar-Shalom

Department of Electrical & Systems Engineering, University of Connecticut, Box U-157, Storrs, CT 06269-2157, e-mail: ybs@ee.uconn.edu

Summary. [1] In this chapter we present the development and implementation of a multisensor-multitarget tracking algorithm for large scale air traffic surveillance based on IMM state estimation combined with a 2-dimensional assignment for data association. The algorithm can be used to track a large number of targets from measurements obtained with a large number of radars. The use of the algorithm is illustrated on measurements obtained from 5 FAA radars, which are asynchronous, heterogeneous and geographically distributed over a large area. Both secondary radar data (beacon returns from cooperative targets) as well as primary radar data (skin returns from non-cooperative targets) are used. The target IDs from the beacon returns are not used in the data association. The surveillance region includes about 800 targets that exhibit different types of motion. The performance of an IMM estimator with linear motion models is compared with that of the Kalman filter. A number of performance measures that can be used on real data without knowledge of the ground truth are presented for this purpose. It is shown that the IMM estimator performs better than the Kalman filter. The advantage of fusing multisensor data is quantified. It is also shown that the computational requirements in the multisensor case are *lower* than in single sensor case. Finally, an IMM estimator with a nonlinear motion model (coordinated turn) is shown to further improve the performance during the manoeuvering periods over the IMM with linear models.

1. Introduction

There are a number of issues related to tracking in multisensor ATC (air traffic control) systems. First, with the Kalman filter being the most widely

[1] Research supported under ONR/BMDO Grant N00014-91-J-1950, AFOSR Grant F49620-94-1-0150 and ONR Grant N00014-97-1-0502.

used state estimator, it is of interest whether other estimators have a better performance than the Kalman filter. A related issue is the handling of the problem that there is a large number of targets in an ATC region and these targets undergo different types of motion. Last, but not least, is the algorithm used to carry out data association and whether the same association algorithm is suitable for both sparse and dense ATC environments. A comparison of several estimators (alpha-beta, adaptive Kalman and IMM — interacting multiple model estimator) was carried out in [8] on simulated data (with association not considered). The superiority of the IMM emerged clearly and the severe limitations of the alpha-beta filter were pointed out.

Previous work dealing with the use of IMM to ATC was reported in [6] and [9]. The recent work [12] presented the algorithm pertaining to the software MATSurv 2 (Multisensor Air Traffic Surveillance) that uses the IMM (interacting multiple model) state estimator [1] and carries out association of ATC radar data (both primary and secondary[2]) via a 2-D assignment. The assignment was implemented using the generalized Auction algorithm [10] and associates the measurements from the most recently arrived scan to the existing tracks, i.e., it handles 2 "lists" (the dimension of the assignment). The algorithm did not use the target ID information (squawk number or IFF — identification friend or foe — from the "cooperative" targets). Other algorithms for 2-D assignment can be found in [3] and [4], and a comparison of the performance of the various algorithms can be found in [7] and [11]. The IMM algorithm [1] was running two Kalman filter modules: one based on a nearly constant velocity motion model and the other on a "manoeuvering model" and was evaluating in real time weights for these models. Both models were of the "white noise acceleration" type [1] with low process noise level for the first and higher level for the second, i.e., linear. Thus the IMM acted as a self-adjusting variable bandwidth filter because the first filter is a low bandwidth filter while the second is a high bandwidth one.

The conclusion from the above work (which dealt with data from Dansville and Remsen FAA radars in upstate New York, a relatively sparse air traffic space) were as follows. Even though ATC is a relatively benign environment, the IMM estimator outperformed the (single model based) Kalman filter alternative in two aspects:

1. It yielded smaller RMS prediction errors (innovations — the only quantity available for performance assessment on real data).
2. It associated more measurements to the existing tracks.

For the cooperative targets with transponder both algorithms had a perfect score of association when compared to the "ground truth" in the form of the squawk numbers (IFF, not used by the algorithm). The only exception was

[2] Primary radar returns consist of reflection from the target skin (skin returns) and they provide range and azimuth information only. Secondary radar returns are from the target transponder (or beacon, if available and functioning, i.e., the target is "cooperative") and they also provide target altitude.

one case where the transponders of two neighboring targets interfered with each other and swapped IDs; the algorithm not using the IDs associated correctly the measurements with obviously smaller errors than what would have been obtained by relying on the IDs. The extra measurements incorporated by the IMM vs. KF version were from non-cooperative targets, which yielded only skin returns (without ID and altitude; the altitude of such targets was taken as a fixed one, FL50=5000ft). Total reliance on IDs is obviously a risky affair because transponder can malfunction or can be turned off or not available (e.g., the non-cooperative target that landed on the White House lawn).

Four questions of interest were not answered by the previous investigation:

1. Is the 2-D assignment sufficient in a denser ATC space?
2. Is the IMM estimator better and more reliable than the KF in a denser space?
3. What are the benefit from multisensor fusion and its cost?
4. Can one improve further the performance by using a nonlinear (coordinated turn) motion model?

The above questions motivated the present study which deals with FAA data from 5 radars (courtesy of USAF Rome Laboratory) from Southern Maine to Maryland along the coast and New York, Pennsylvania inland, which is a much more crowded region. The results presented in this chapter were obtained with the software MATSurv 4, which has numerous modifications and extensions beyond version 2.

One important problem in the case of multiple radars that send full scans of data to the "central processing location" (a centralized processing architecture was used) is the presence of "out-of-sequence measurements" [2]. While this was already observed in [12], it was a significant problem in the present database where it occurred quite often. An algorithm dealing with this situation was introduced. The conclusion of the current work is that positive answers were obtained to the first two and last questions above. The answer to the third question is discussed below. The chapter is organized as follows. Section 2. discusses the procedure used to incorporate out-of-sequence measurements into the state updating. Section 3. describes the target motion models used in the IMM estimators, namely, IMM–L (with linear models) and IMM–CT (with a coordinated turn model). The sensor models are also discussed in this section. The cost used in the 2-D assignment algorithm was, as in [12], the likelihood function of the IMM estimators. Section 4. presents the procedure of fusion of primary and secondary radar data. Section 5. presents the results obtained with MATSurv 4 on the 5 radar database and some of the unusual situations encountered with real data. The advantage of fusing multisensor data is quantified. It is also shown that the computational requirements in the multisensor case are *lower* than in single sensor case. Finally, conclusions are given in Section 6..

2. State Update with Out-of-Sequence Measurements

When the measurements from sensors arrive to the central processor in batches (e.g., corresponding to scans or frames — a common situation) it is quite likely to have an earlier measurement from a target arrive after a later one. This is the out-of-sequence measurements situation. In dealing with out-of-sequence detections, it is critical to update the most recent state with the most recently arrived (but from an earlier time) measurement rather than update the earlier state. For example, assume that scan j of sensor 1 covers the time interval $[T_{1,j-1}, T_{1,j}]$ and scan l of sensor 2 corresponds to the interval $[T_{2,l-1}, T_{2,l}]$, where

$$T_{1,j-1} < T_{2,l-1} < T_{1,j} < T_{2,l}, \tag{18.1}$$

i.e., they have a certain overlap.

The data from each sensor are sent to the central processor at the *end of each scan*, thus the measurements from scan l of sensor 2 will arrive after those of scan j of sensor 1. Now, if sensor 1 obtains a measurement from a target at t_1 near the end of its scan j and sensor 2 obtains a measurement from the same target at t_2 near the beginning of its scan l, i.e.,

$$T_{1,j-1} < T_{2,l-1} < t_2 < t_1 < T_{1,j} < T_{2,l}, \tag{18.2}$$

then the (earlier) measurement at t_2 will arrive to the information processor after the one at t_1, i.e., out of sequence.

Thus one has to carry out what can be called a "negative-time measurement update" because one has a negative sampling interval

$$\delta_k = t_{k+1} - t_k < 0 \tag{18.3}$$

In the standard case, when $\delta_k > 0$, the sufficient statistic for prediction of the measurement is the latest state estimate $\hat{x}(k|k)$. In the present case, the (backward) "prediction" to t_{k+1} would require smoothing or "retrodiction" [1], which in turn, requires the availability of past measurements. In practice this is quite complicated since it also involves arbitrary times and accounting for the process noise in the state equation.

A simple technique that can be used is the following. The state is "predicted" (actually, retrodicted) from t_k to t_{k+1} ignoring the process noise. For a linear system with transition matrix $F(t_{k+1}, t_k)$, with current state estimate $\hat{x}(k|k)$ and covariance $P(k|k)$, the (suboptimal) prediction is

$$\hat{x}(k+1|k) = F(t_{k+1}, t_k)\hat{x}(k|k) \tag{18.4}$$

$$P(k+1|k) = F(t_{k+1}, t_k)P(k|k)F(t_{k+1}, t_k)' \tag{18.5}$$

The goal is to update the state at time t_k using the measurement $z(k+1)$ at (the earlier) time t_{k+1}. This is done using the above "predicted" state as follows. The innovation at t_{k+1} and its covariance are, with the standard notations,

$$\nu(k+1) = z(k+1) - H(k+1)\hat{x}(k+1|k) \tag{18.6}$$

$$S(k+1) = H(k+1)P(k+1|k)H(k+1)' + R(k+1) \tag{18.7}$$

The filter gain will be

$$W(k+1) = P(k+1|k)H(k+1)S(k+1)^{-1} \tag{18.8}$$

and the new updated state at t_k (*the most recent time*, which now incorporates the measurement from the *earlier* time t_{k+1}) and its covariance are

$$\hat{x}(k|k+1) = \hat{x}(k|k) + W(k+1)\nu(k+1) \tag{18.9}$$

$$P(k|k+1) = P(k|k) - W(k+1)S(k+1)W(k+1)' \tag{18.10}$$

3. Motion Models and IMM Parameters

3.1 IMM Estimator with Linear Motion Models

The motion models used in the IMM estimator with linear models (IMM–L) in the horizontal plane and the vertical axis are presented in this section. The models in the horizontal and vertical are treated separately, i.e., the filtering is decoupled, for the purpose of simplicity and, more importantly, due to the fact that the target motions in horizontal plane and vertical plane are comparatively independent. The motion model j in the horizontal plane is specified by the state space matrices $F_{h,j}(\cdot)$, $G_{h,j}(\cdot)$ and $Q_{h,j}(\cdot)$, i.e., according the this model, the state $x_h(t)$ of the target (in the horizontal plane), evolves from time t_l to t_k as:

$$x_h(t_k) = F_{h,j}(\delta)x_h(t_l) + G_{h,j}(\delta)v_{h,j} \tag{18.11}$$

where $\delta = t_k - t_l$ is the sampling interval. The process noise vector $v_{h,j}$ is modeled as a white, zero-mean Gaussian random variable with covariance matrix $Q_{h,j}$. For the horizontal motion models used in this application, the process noise $v_{h,j}$ consists of acceleration components along the local east, north directions (see Figure 3.1). Similarly, the state of the target in the vertical axis $x_v(t)$ evolves (in model j) as:

$$x_v(t_k) = F_{v,j}(\delta)x_v(t_l) + G_{v,j}(\delta)v_{v,j} \tag{18.12}$$

and the process noise vector along the vertical axis $v_{v,j}$ is modeled as a white, zero-mean Gaussian random variable with covariance $Q_{v,j} = \sigma^2_{v,j}$.

In general, civilian aircraft motion can be broadly divided into a uniform motion phase and a manoeuver phase. Hence, two motion models, one with high process noise level to model the manoeuver phase and the other with almost zero process noise to model the uniform motion phase are considered to describe the aircraft motion. In addition, civilian aircraft rarely undergo sustained acceleration over long periods of time; hence, second order motion models along all three directions should suffice. Based on the above considerations, the IMM estimator design shown in Table 18.1 was implemented. The IMM estimators, both in the horizontal and vertical, use second order motion models.

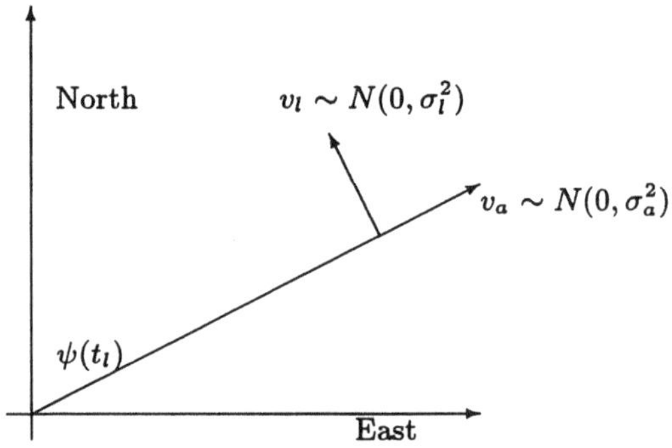

Fig. 18.1. Modeling lateral and axial accelerations in horizontal plane

The mechanism for injecting the process into the two axes in the horizontal (i.e., East and North) is as follows. The lateral and axial accelerations of the target are modeled as white processes, with the variance σ_a^2 in the axial component maintained at a small value, both in the high noise and low noise models. On the other hand, in the lateral direction the high process noise model uses a $2g$ standard deviation. Hence, the process noise covariance for model j in the horizontal would be

$$Q_{h,j} = \begin{bmatrix} -\cos\psi & \sin\psi \\ \sin\psi & \cos\psi \end{bmatrix} \begin{bmatrix} \sigma_{l,j}^2 & 0 \\ 0 & \sigma_{a,j}^2 \end{bmatrix} \begin{bmatrix} -\cos\psi & \sin\psi \\ \sin\psi & \cos\psi \end{bmatrix} \tag{18.13}$$

where $\psi = \psi(t_l)$ is the estimated course (obtained from the velocity components). We have found that by injecting the process noise in this manner we could obtain considerable improvement in the ground speed estimates without increasing errors during manoeuvers. The process noise variance in the vertical is simply given by $Q_{v,j} = \sigma_{v,j}^2$, and the values used for σ_v are shown in Table 18.1.

In Table 18.1, g=9.81 m/s^2 is the acceleration due to gravity. The matrices $F(n_1, ..., n_d, \delta)$ and $G(n_1, ..., n_d, \delta)$ are the state matrices corresponding to a direct discrete kinematic motion model [1], with blocks of order $n_1, ..., n_d$ along the d dimensions, respectively. For the IMM design the following model transition probabilities are used in the horizontal and vertical:

Table 18.1. IMM–L Model parameters.
M=1: uniform motion model, M=2: manoeuvering model

M	State Space Matrices		Process Noise		
	Horizontal	Vertical	Horizontal		Vertical
	$F_h\&G_h$	$F_v\&G_v$	σ_l	σ_a	σ_v
1	$F(2,2,\delta)\&G(2,2,\delta)$	$F(2,\delta)\&G(2,\delta)$	$0.01g$	$0.01g$	$0.01g$
2	$F(2,2,\delta)\&G(2,2,\delta)$	$F(2,\delta)\&G(2,\delta)$	$2g$	$0.4g$	$0.1g$

$$\pi_h = \begin{bmatrix} 0.95 & 0.05 \\ 0.10 & 0.90 \end{bmatrix} \qquad \pi_v = \begin{bmatrix} 0.95 & 0.05 \\ 0.05 & 0.95 \end{bmatrix} \tag{18.14}$$

We will compare the performance of the IMM estimator with the (single model) KF described in Table 18.2.

Table 18.2. KF Design Parameters

State Space Matrices		Process Noise		
Horizontal	Vertical	Horizontal		Vertical
$F_h\&G_h$	$F_v\&G_v$	σ_l	σ_a	σ_v
$F(2,2,\delta)\&G(2,2,\delta)$	$F(2,\delta)\&G(2,\delta)$	$2g$	$0.4g$	$0.1g$

3.2 IMM Estimator with Coordinated Turn Model

Further exploration on IMM estimator is the introduction of nonlinear coordinated turn model (IMM–CT) in the horizontal plane. Since the manoeuvers of civilian aircrafts in the horizontal plane are mainly coordinated turns, we can use a linear model with low process noise (same as in Section 3.1) to model the uniform motion phase and a nonlinear coordinated turn model to model the manoeuvering phase. The vertical model selection remains the same as that in Section 3.1.

The state of the CT model consists of the position and velocity elements in the two horizontal axes and the turning rate ω. For the non-manoeuvering model the last component is identically zero. The state propagation equation is

$$x_h(t_k) = f_{h,j}(\delta, x_h(t_l)) + G_{h,j}(\delta)v_{h,j} \tag{18.15}$$

See equation (18.18) in the Appendix for the second (CT) model. Three approaches for IMM–CT were implemented: 1st order EKF (extended Kalman Filter), 2nd order EKF and Kastella's method [8]. A detailed discussion of these three approaches is given in the Appendix. The IMM-CT design parameters are shown in Table 18.3, where F_{CT} and G_{CT} are the (linearized) transition matrix and noise gain for the coordinated turn model, respectively (the argument δ is omitted here). The model transition probabilities of IMM–CT are the same as those of IMM–L. The initial estimate of the turn rate is zero.

Table 18.3. IMM–CT Model Parameters
(M=1: uniform motion model, M=2: coordinated turn model)

M	State Space Matrices		Process Noise			
	Horizontal	Vertical	Horizontal			Vert.
	$F_h \& G_h$	$F_v \& G_v$	σ_l	σ_a	σ_ω	σ_v
1	$F(2,2)\&G(2,2)$	$F(2)\&G(2)$	$0.01g$	$0.01g$	—	$0.01g$
2	$F_{CT}(\cdot)\&G_{CT}(\cdot)$	$F(2)\&G(2)$	$0.2g$	$0.1g$	$0.2°/s^2$	$0.1g$

3.3 Sensor Parameters

The data are from 5 FAA radars covering most of NE of USA, which is a dense air traffic region. More than 800 targets were detected in this surveillance region. The validation gate threshold [2] for the horizontal plane was $\gamma = 4^2$ while for altitude it was $\gamma = 3.5^2$.

The sensor parameters, i.e., the pulsewidth (pw), horizontal beamwidth (hbw) and error statistics are given in Table 18.4. The calculation of σ_r and σ_θ (i.e., errors of range and azimuth) is based on the assumption that measurement errors are uniformly distributed within a range resolution cell and horizontal beam, respectively. Each radar surveillance region overlaps with at least two other radars' regions.

Table 18.4. Sensor Models

sensor ID	sensor location	pw (μs)	hbw (deg)	σ_r (km)	σ_θ (mrad)
D	Dansville, NY	1.8	1.2	0.08	6.05
R	Remsen, NY	6.0	1.3	0.26	6.55
5	N. Truro, MA	6.0	1.3	0.26	6.55
6	Riverhead, NY	2.0	1.1	0.09	5.54
7	Trevose, PA	6.0	1.3	0.26	6.55

The measurements are obtained in the radar coordinates and they are transformed to ECR (earth centered rotating, or ECEF – earth fixed). The state update is carried out in a local Cartesian system and the results are transformed back into ECR for display in geographic coordinates.

4. Fusion of Primary and Secondary Radar Data

The procedure of incorporation of skin returns (i.e., fusion of primary and secondary radar data) is as follows:

1. Prior to the assignment, a skin return is augmented with a nominal altitude in order to be transformed into the local Cartesian system in which the assignment is carried out (in the horizontal plane only).

2. Once the assignment is done, if a skin return is assigned to an existing track which has an altitude estimate from previous beacon measurements, the original altitude (the nominal one) will be replaced by this altitude estimate, and then the recomputed transformed measurement is used in the update of the track state. This reduces the errors due to the nominal altitude being different from the actual one. If the track to which the measurement was assigned has no altitude estimate (i.e., it consists of only skin returns) then no recomputation is done and the track state update is carried out with the original transformation.

3. If a skin return is not assigned to any existing track, it starts a new track with the altitude being the nominal value.

Since there are two types of measurements (skin and beacon), the 2-D assignment between the list of the existing tracks and of the measurements from the most recently arrived scan is implemented as follows. The existing tracks are separated into two lists, T_S and T_B, where T_S consists of tracks composed of skin returns only and T_B is the rest. The most recently arrived scan returns are separated into two lists, R_S (skin returns) and R_B (beacon returns). Then the 2-D assignment is carried out in the following steps:

1. Between T_B (beacon tracks) and R_B (beacon returns). The part of T_B left unassigned is denoted as T_B^1. The part of R_B left unassigned is denoted as R_B^1.

2. Between T_B^1 (beacon tracks which have no current beacon measurements) and R_S. The part of R_S left unassigned is denoted as R_S^1.

3. Between T_S and R_S^1. The part of R_S^1 left unassigned is denoted as R_S^2.

4. New tracks are initialized from R_B^1 and R_S^2.

5. Results

5.1 Comparison of IMM–L estimator with KF

In this subsection the performance of the IMM estimator with 2 models and the (single model based) KF are compared on the 5 radar data that span a time interval of 7 minutes. The measurements plotted on a U.S. map are shown in Figure 18.2. The actual prediction errors and number of associations obtained using these two algorithms are tabulated. These results not only demonstrate the error reduction obtained with the multiple model estimator, but also indicate the magnitude of the actual errors in a typical ATC scenario.

The RMS prediction errors in the horizontal plane and vertical (altitude) are given in Tables 18.5 and 18.6, respectively. Both beacon and skin returns are included in comparing the performance of these two algorithms in the horizontal plane, while in the altitude comparisons only beacon returns are included.

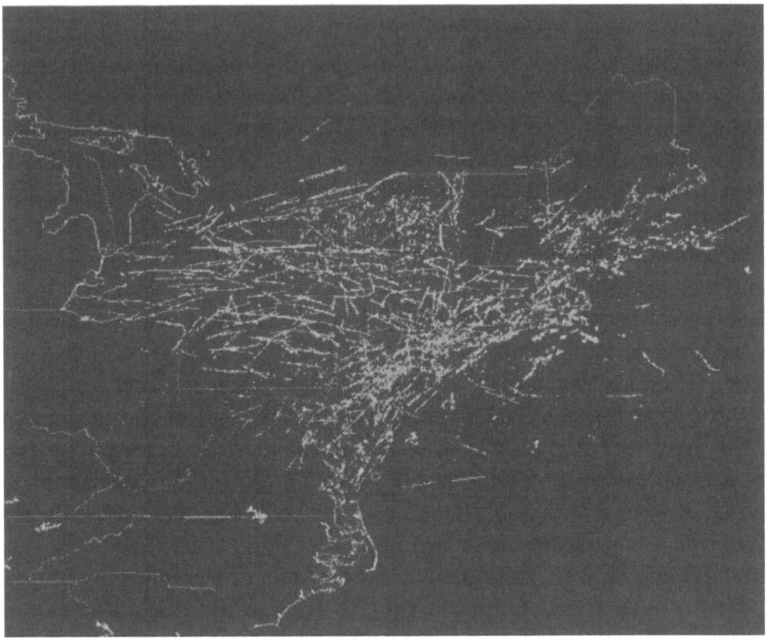

Fig. 18.2. Measurements over a period of 7 minutes from 5 sensors

In Tables 18.5 and 18.6 the number of data points in each sampling interval bin are listed for both schemes. Since the number of tracks formed (and the tracks themselves) are different for different estimators, the number of data points in each bin are not the same for the two estimators. The rows marked with the symbol "$\star$" do not contain a sufficient number of data points, hence, one cannot make any statistically significant inference about the errors in these bins. Some uncommon situations of the measurements will be discussed later in this section. The bins with sufficient data points indicate approximately 20%–50% RMS prediction error reduction for the IMM design over the KF. Also, the IMM estimator associates more measurements than the KF.

5.2 Examples of Uncommon Situations

The following uncommon situations are encountered when doing associating and filtering:

Table 18.5. RMS Prediction errors in the horizontal plane per sampling interval bin in nautical miles (nmi)

Sampling Bins (sec)	Detections per Bin IMM	KF	RMS error (nmi) IMM	KF	% Reduction
[0,5)	3720	3692	0.36	0.38	5.6
[5,10)	2205	2201	0.38	0.42	9.5
[10,15)	4237	4004	0.42	0.52	19.2
[15,20)	222	213	0.43	0.61	28.6
[20,25)	477	417	0.55	1.29	57.4
[25,30)	33	27	0.87	2.70	⋆
[30,35)	22	14	1.45	1.13	⋆
[35,40)	125	77	0.79	1.47	46.3
[40,45)	9	2	0.83	3.80	⋆
[45,50)	57	11	0.83	3.80	⋆
[50,55)	6	1	0.65	6.23	⋆
[55,60)	11	1	5.67	0.25	⋆
[60,65)	13	1	0.87	0.87	⋆
[65,70)	1	0	1.53	—	—
[70,75)	13	0	1.52	—	—
[75,∞)	12	0	1.52	—	—

1. Bad altitude measurements.
   a) Several altitude measurements are "zero". This is the most common case concerning bad altitude measurements. The method of dealing with this situation in MATSurv 3 is to treat all measurements with zero altitude as skin returns and then apply the algorithm of incorporating skin returns as discussed before. This method has a better performance than simply not associating such bad measurements, especially for those targets that have comparatively fewer measurements. However, if the first beacon measurement of a certain target has zero altitude (there are a few such cases), the algorithm will not associate subsequent measurements to this track because beacon returns are not associated to skin tracks.
   b) Measurements that have non-zero but clearly bad altitude (The assessment that it is "bad" is done using the ID). The algorithm (which does not use the ID) will initialize a new track from such a measurement, which will die out quickly (a track is terminated if no returns are assigned to this track for more than 90 seconds.).
2. Bad measurements in the horizontal plane. A couple of measurements of certain targets are far away (more than 100 nautical miles) from the main track. This can be a different target with the same ID or a spurious detection. In any case the assignment algorithm ignores them.
3. Some targets changed their ID midflight, e.g., a target with IFF 1355 changed its IFF into 1452 after about 60 seconds.
4. There are a few IFFs that are shared by different targets which are far away from each other in that there might be more targets in the air

Table 18.6. RMS Prediction errors in the altitude per sampling interval bin in feet (ft)

Sampling Bins (sec)	Detections per Bin IMM	KF	RMS error in ft IMM	KF	% Reduction
[0,5)	3592	3586	34	34	0.0
[5,10)	2128	2117	60	75	20.0
[10,15)	3238	3186	90	95	5.3
[15,20)	201	193	106	115	7.8
[20,25)	210	210	120	133	9.8
[25,30)	6	6	138	123	⋆
[30,35)	12	12	147	516	⋆
[35,40)	27	25	152	223	⋆
[40,45)	2	2	191	212	⋆
[45,50)	13	11	376	420	⋆
[50,55)	0	1	—	2213	—
[55,60)	3	1	234	0	⋆
[60,65)	0	1	—	1	—
[65,70)	0	0	—	—	—
[70,75)	1	0	300	—	—
[75,∞)	0	0	—	—	—

⋆ indicates that the results are not statistically significant due to the small number of data points.

than the IFFs available. This situation, together with the previous one, reinforces the motivation of our approach of not using IDs when doing association and filtering.

5. Two measurements of a certain target are detected in a single scan. This situation takes place when the target was flying in the opposite direction of the scanning direction of radar and the target was detected once at the beginning and once at the end of the scan. The effect of this situation is as follows:
   a) In most cases the extra detections did not cause any problems in associations or filtering. They simply initialize new tracks which are terminated because of lack of subsequent associations.
   b) Sometimes such extra detections "steal" one or two subsequent detections. This situation occurs when the extra detection happens to be in a proper location and the measurement from the next scan will be assigned to this newly initialized track instead of being associated to the main track of this target.
   c) Very rarely tracks are broken at the point of the extra measurements, i.e., the extra measurement initializes a new track and all the subsequent detections from this target are associated to this newly initialized track. There were two such occurrences observed in our database.

In real radar data there might be uncommon situations happening. The MATSurv 4 algorithm seems to be able to deal with such situations with acceptable robustness.

5.3 Advantages of Fusing Multisensor Data

This subsection illustrates the advantage of multisensor tracking over single sensor tracking in the sense of RMS prediction error as well as discusses the computational complexity of the multisensor algorithm.

In Figure 5.3, the comparison of the RMS position prediction errors obtained using multisensor data and single sensor data is presented. It can be seen clearly that the RMS errors in multisensor tracking decrease as the number of sensors increases. The reason is straightforward: there are a few measurements detected in the multisensor case that fill in the "gap" of two consecutive measurements detected by a single sensor and thus the prediction time interval is shortened and the information based on which the prediction is made is more recent. Note that the improvement in the RMS prediction error is becoming smaller as the number of sensors increase due to the fact that as the error of state update decreases along with the increasing of the number of sensors, the measurement noise (which is approximately the same for single sensor tracking or multisensor tracking) accounts for more of the RMS prediction error.

The second advantage of fusing multisensor data is that multisensor tracking has (surprisingly) *less* computation per scan than single sensor tracking does. Since the calculation of velocity gate (coarse gating) is computationally negligible, the computational complexity is primarily determined by the time spent fine-gating (validation [2]), which is many orders of magnitude costlier than velocity-gating. Table 18.7 presents the comparison of computation for fine-gating (validation) between multisensor tracking and single sensor tracking. As discussed in [11], the validation takes up 95% of the total computations.

A measurement will be processed with fine-gating if it is in the velocity gate of a certain track, i.e., comparing the value of $\| z(k) - \hat{x}_p(k-1|k-1) \|$ with $v_{max}T_k$, where $\hat{x}_p(k-1|k-1)$ is the position estimate at time $k-1$. Taking the measurement error and state estimation error into consideration, the modified velocity gate size of a certain track in horizontal and vertical (in the k-th scan) are as follows:

$$V_h = \left[v_{x_{\max}}T_k + 2\sqrt{R_{xx}(k) + P_{xx}(k-1|k-1)}\right] \cdot \left[v_{y_{\max}}T_k + 2\sqrt{R_{yy}(k) + P_{yy}(k-1|k-1)}\right] \quad (18.16)$$

$$V_v = v_{z_{\max}}T_k + 2\sqrt{R_{zz}(k) + P_{zz}(k-1|k-1)} \quad (18.17)$$

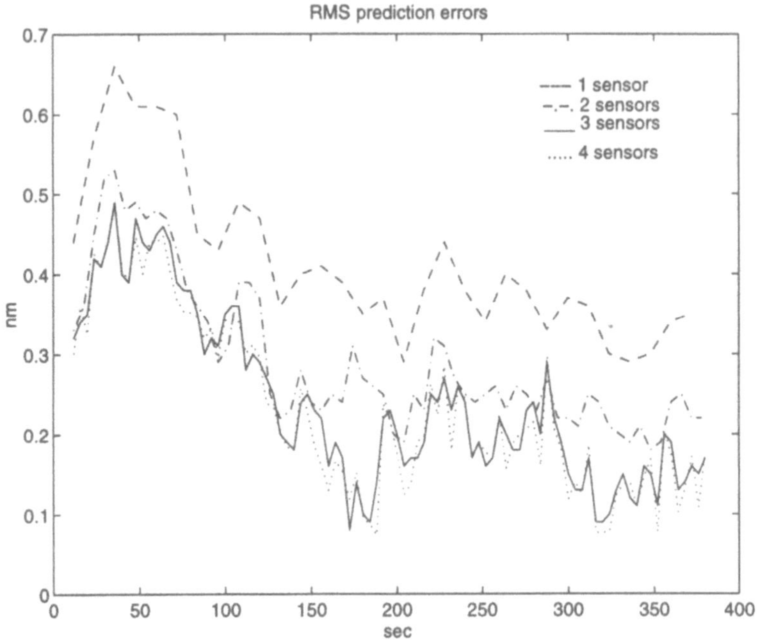

Fig. 18.3. Comparison of RMS errors in horizontal plane using different number of sensors

where $R(k)$ and $P(k-1|k-1)$ are the measurement noise and state estimation covariance (with subscripts indicating the axes) respectively, which account for the uncertainty in gating due to measurement and estimation errors.

In single sensor tracking, the time interval T and state estimation error P are larger than in multisensor tracking, and consequently, the velocity gate size is larger than in the multisensor case and there will be more candidate measurements to be passed into fine-gating, especially in dense air traffic regions. This accounts for the reason of the reduction in computation using multisensor tracking. Thus, the computations for a multisensor tracking are less than the total computations for single sensor tracking.

5.4 Comparison of IMM–L and IMM–CT estimators

In this subsection the performances of IMM–L (with linear models) and IMM–CT (with coordinated turn model) are compared on the 5 radar database over a time interval of 15 minutes. All the three approaches of IMM–CT (1st order EKF, 2nd order EKF and Kastella's method) were found to have minor differences in the sense of RMS position prediction errors and number of associations, although the 1st order EKF is much simpler than the other two. Table 5.4 shows the comparison of IMM–L and IMM–CT in

Table 18.7. Comparison of computational complexity in multisensor and single sensor tracking for the measurements from sensor R. R_i represents the i-th scan from sensor R.

Scan Number	number of fine-gating computation		time of fine-gating computation (sec)	
	5 sensors	R	5 sensors	R
R_2	1779	1819	5.60	5.95
R_3	2254	2679	7.10	8.44
R_4	1442	2003	4.54	6.30
R_5	1601	1988	5.05	6.26
R_6	4088	4992	12.88	15.42
R_7	2859	3441	9.01	10.84
R_8	1799	2045	5.67	6.46
R_9	1788	2168	5.66	6.83
R_{10}	3742	4821	11.79	15.18
R_{11}	1672	1826	5.27	5.75
R_{12}	2975	3774	9.37	11.89
R_{13}	1730	2116	6.06	6.67
R_{14}	1883	2364	5.93	8.27

the horizontal plane based on the measurements of the manoeuvering periods. The periods were selected as those for which the CT model probability exceeded 0.5.

The bins with sufficient data points indicate approximately 10%–50% RMS prediction error reduction in the critical manoeuvering periods for the IMM–CT over the IMM–L. Also, IMM–CT associates (slightly) more manoeuvering measurements during these periods than IMM–L. During the non-manoeuvering periods, these two designs have identical performance.

5.5 Computation time

Processing time for the estimation/association algorithm disscussed above was evaluated on a PC with Pentium-2 processor running at 300 MHz. For the data spanning a time interval of 6 minutes (consisting of 130 radar scans), the system running time was 8 minutes and 30 seconds. Since this included time spent on file reading/writing and data swapping between disk and memory, real-time processing is clearly feasible.

6. Conclusions and Summary

A multisensor-multitarget tracking algorithm based on an IMM estimator and the "generalized Auction" 2-D assignment for large scale ATC system has been presented. The algorithm is illustrated on measurements obtained from 5 FAA radars covering a large dense ATC area. The following conclusions can be drawn:

Table 18.8. RMS Prediction errors in the horizontal plane during manoeuvering periods per sampling interval bin in nautical miles (nmi)

Sampling Bins(sec)	Detections per Bin		RMS error in nmi		% Reduction
	IMM–CT	IMM–L	IMM–CT	IMM–L	
[0,5)	420	412	0.42	0.46	8.7
[5,10)	353	350	0.48	0.61	21.0
[10,15)	689	661	0.60	0.84	48.0
[15,20)	83	79	0.55	1.06	48.0
[20,25)	116	104	0.89	1.77	50.0
[25,30)	10	8	0.95	1.94	⋆
[30,35)	9	9	1.42	2.06	⋆
[35,40)	35	36	1.46	2.58	40.0
[40,45)	4	2	1.65	1.88	⋆
[45,50)	14	13	1.80	2.25	⋆
[50,55)	0	0	—	—	—
[55,60)	9	6	2.04	3.35	⋆
[60,65)	3	3	1.70	2.18	⋆
[65,70)	0	0	—	—	—
[70,75)	0	0	—	—	—
[75,∞)	1	0	1.81	—	—

1. 2-D assignment proves to be sufficient even for a dense ATC space.
2. The IMM estimator has a significantly better performance than the single model based Kalman Filter in a dense ATC space in the sense of RMS prediction error.
3. The IMM estimator associates more measurements to existing tracks than the KF.
4. There is a significant performance improvement in multisensor tracks vs. single sensor tracks and the computational complexity for multisensor tracking is less than the total in single sensor tracking.
5. Further performance improvement can be achieved by using a coordinated turn (CT) model.
6. Different designs of IMM–CT (1st order EKF, 2nd order EKF and Kastella's method) yield same performance and 1st order EKF is recommended in an ATC system due to its simplicity.

It should be noted that a (spaghetti logic based) "switching gain" KF is unsuitable when assignment has to be carried out. In contrast, the IMM is a self-adjusting variable bandwidth estimator, which has a well-defined likelihood function that is directly usable as a cost function for a large scale assignment algorithm. An algorithm dealing with out-of-sequence measurements was also applied in the current work due to the fact that such measurements occur quite often in a large scale multisensor-multitarget ATC system.

Acknowledgement. We are grateful to Rick Gassner from Rome Laboratory for providing the real radar data.

Appendix A. EKF for Coordinated Turn Model

For the coordinated turn model, the state vector is $x = [\xi, \eta, \dot{\xi}, \dot{\eta}, \omega]'$ and the dynamic equation is given by

$$
\begin{aligned}
x(k+1) &= \begin{bmatrix} 1 & 0 & \frac{sin(\omega(k)\delta)}{\omega(k)} & -\frac{1-cos(\omega(k)\delta)}{\omega(k)} & 0 \\ 0 & 1 & \frac{1-cos(\omega(k)\delta)}{\omega(k)} & \frac{sin(\omega(k)\delta)}{\omega(k)} & 0 \\ 0 & 0 & cos(\omega(k)\delta) & -sin(\omega(k)\delta) & 0 \\ 0 & 0 & sin(\omega(k)\delta) & cos(\omega(k)\delta) & 0 \\ 0 & 0 & 0 & 0 & 1 \end{bmatrix} x(k) + \\
&+ \begin{bmatrix} \delta^2/2 & 0 & 0 \\ 0 & \delta^2/2 & 0 \\ \delta & 0 & 0 \\ 0 & \delta & 0 \\ 0 & 0 & \delta \end{bmatrix} v(k) \qquad (18.18)
\end{aligned}
$$

To obtain the predicted state $\hat{x}(k+1|k)$, the nonlinear function in Eq (18.18) is expanded in Taylor series around the latest estimate $\hat{x}(k|k)$ with terms up to first or second order to yield the 1st order EKF or 2nd order EKF, respectively. The vector Taylor series expansion of Eq (18.18) up to second order terms is:

$$
\begin{aligned}
x(k+1) &= f(k, \hat{x}(k|k)) + f_x(k)[x(k) - \hat{x}(k|k)] + \\
&\frac{1}{2}\sum_{i=1}^{n_x} e_i[x(k) - \hat{x}(k|k)]' f^i_{hxx}(k)[x(k) - \hat{x}(k|k)] + HOT + G_{CT}(k)v(k)
\end{aligned} \qquad (18.19)
$$

where e_i is the i-th n_x-dimensional Cartesian basis vector and HOT represents the higher order terms,

$$f_x(k) = [\nabla_X f(k,x)']' \mid_{x=\hat{x}(k|k)} \equiv F_{CT}(\delta) \qquad (18.20)$$

is the Jacobian of the vector f evaluated at the latest estimate of the state and

$$f^i_x(k) = [\nabla_x \nabla'_x f^i(k,x)']' \mid_{x=\hat{x}(k|k)} \qquad (18.21)$$

is the Hessian of the i-th element of f.

Based on the above expansion, the state prediction and state prediction covariance of 1st order EKF are:

$$\hat{x}(k+1|k) = f[k, \hat{x}(k|k)] \tag{18.22}$$

$$P(k+1|k) = f_x(k)P(t_k)f_x(k)' + Q(k) \tag{18.23}$$

where Q is given by (13) (with parameters from Table 3.2) augmented with $q_{33} = \sigma_\omega^2$, also from Table 3.2. and for 2nd order EKF, these two equations become:

$$\hat{x}(k+1|k) = f[k, \hat{x}(k|k)] + \frac{1}{2}\sum_{i=1}^{n_x} e_i tr[f_{hxx}^i(k)P(k|k)] \tag{18.24}$$

$$P(k+1|k) = f_x(k)P(k|k)f_x(k)' + \frac{1}{2}\sum_{i=1}^{n_x}\sum_{j=1}^{n_x} e_i e_j{}' tr[f_{xx}^i(k)P(k|k)f_{xx}^j(k)P(k|k)] + Q(k) \tag{18.25}$$

Appendix B. Kastella's Method for Coordinated Turn

In Kastella's method [8], the dynamic equation of the coordinated turn model is:

$$x(k+1) = \Theta(\omega(k+1))x(k) + \begin{bmatrix} \delta^2/2 & 0 & 0 \\ 0 & \delta^2/2 & 0 \\ \delta & 0 & 0 \\ 0 & \delta & 0 \\ 0 & 0 & \delta \end{bmatrix} v(k) \tag{18.26}$$

where $\Theta(\omega(k+1))$ is defined as

$$\Theta(\omega(k+1)) = \begin{bmatrix} 1 & 0 & \frac{sin(\omega(k+1)\delta)}{\omega(k+1)} & -\frac{1-cos(\omega(k+1)\delta)}{\omega(k+1)} & 0 \\ 1 & 0 & \frac{1-cos(\omega(k+1)\delta)}{\omega(k+1)} & \frac{sin(\omega(k+1)\delta)}{\omega(k+1)} & 0 \\ 0 & 0 & cos(\omega(k+1)\delta) & -sin(\omega(k+1)\delta) & 0 \\ 0 & 0 & sin(\omega(k+1)\delta) & cos(\omega(k+1)\delta) & 0 \\ 0 & 0 & 0 & 0 & 1 \end{bmatrix} x(k) \tag{18.27}$$

This function is different from the previous one in that it uses $\omega(k+1)$ instead of $\omega(k)$ for the nonlinear dynamic equation. The linearization process of the dynamic equation is accomplished as follows. In Eq.(18.27), $\Theta(\omega(k+1))$ is expanded in a Taylor series using the previous estimated $\hat{x}(k|k)$ yielding

$$\Theta(\omega(k+1)) \approx \Theta(\hat{\omega}(k|k) + (\omega(k+1) - \hat{\omega}(k|k))\Theta_\omega(\hat{\omega}(k|k) \tag{18.28}$$

where Θ_ω denotes the derivative with respect to ω. The predicted state $\hat{X}(k+1|k)$ at time $k+1$ is obtained by inserting (18.28) into (18.27) and taking its expectation, resulting in

$$\hat{x}(k+1|k) \approx [\Theta(\hat{\omega}(k|k) + (\hat{\omega}(k+1|k) - \hat{\omega}(k|k))\Theta_\omega(\hat{\omega}(k|k)]\hat{x}(k|k) \tag{18.29}$$

which is linear in $\hat{x}(k+1|k)$. To solve this for $\hat{x}(k+1|k)$, define the vector $Y = \Theta_\omega(\hat{\omega}(k|k))\hat{x}(k|k)$ and the matrix

$$M = \begin{bmatrix} I_{4\times4} & -Y \\ O_{1\times4} & 1 \end{bmatrix} \tag{18.30}$$

which leads to the linearized state prediction equation

$$\hat{x}(k+1|k) = F_{KT}(k)\hat{X}(k|k) \tag{18.31}$$

where

$$F_{KT}(k) = M^{-1}[\Theta(\hat{\omega}(k|k)) - \hat{\omega}(k|k)\Theta_\omega(\hat{\omega}(k|k))] \tag{18.32}$$

and the state prediction covariance is

$$P(k+1|k) = F_{KT}(k)P(k|k)F_{KT}(k)' + M^{-1}Q(k)(M^{-1})' \tag{18.33}$$

References

1. Y. Bar-Shalom and X. R. Li, *Estimation and Tracking: Principles, Techniques and Software*, Dedham, MA: Artech House, 1993 (Reprinted by YBS Publishing, 1998).
2. Y. Bar-Shalom and X. R. Li, *Multitarget-Multisensor Tracking: Principles and Techniques*, Storrs, CT: YBS Publishing, 1995.
3. D. P. Bertsekas, *Linear Network Optimization: Algorithms and Codes*, MIT Press, Cambridge, MA, 1991.
4. S. S. Blackman, *Multiple-Target Tracking with Radar Applications*, Artech House, Dedham, MA, 1986.
5. S. S. Blackman, M.T. Busch and R. F. Popoli, "IMM/MHT Tracking and Data Association for Benchmark Problem", *Proc. ACC*, pp. 2606-2610, Seattle, WA, June 1995.
6. H. A. P. Blom, R. A. Hogendoorn and B. A. van Doorn, "Design of a Multisensor Tracking System for Advanced Air Traffic Control", Chapter 2 in *Multitarget-Multisensor Tracking: Applications and Advances*, Vol. II, Y. Bar-Shalom (editor), Artech House, Norwood, MA, 1992.
7. O. E. Drummond, D.A. Castanon, and M.S. Bellovin, "Comparison of 2-D assignment algorithm for sparse, rectangular, floating point matrices", *Journal of the SDI Panels on Tracking*, Issue No. 4/1990, pp. 4-81 to 4-91, Institute for Defense Analysis, Alexandria, VA, December 1990.
8. K. Kastella and M. Biscuso, "Tracking Algorithms for Air Traffic Control Applications", *Air Traffic Control Quarterly*, 3(1): 19-43, January 1996.
9. X. R. Li and Y. Bar-Shalom, "Design of and Interacting Multiple Model Algorithm for Air Traffic Control Tracking", *IEEE Trans. Control Systems Technology*, 1(3): 186-194, September 1993.
10. K. R. Pattipati, S. Deb, Y. Bar-Shalom and R. Washburn, "A new relaxation algorithm and passive sensor data association", *IEEE Trans. Automatic Control*, 37(2): 197-213, February 1992.

11. R. L. Popp, K. R. Pattipati, Y. Bar-Shalom, "Dynamically Adaptable m-best 2D Assignment and Multi-level Parallelization", submitted to *IEEE Trans. AES*, 1997.
12. M. Yeddanapudi, Y. Bar-Shalom and K. R. Pattipati, "IMM Estimation for Multitarget-Multisensor Air Traffic Surveillance", *Proc. IEEE*, 85(1): 80-94, January 1997.

Why Wide Webs Wrinkle

C.D. Mote, Jr. and E.M. Mockensturm

Department of Mechanical Engineering, University of California, Berkeley, CA 94720, USA,
email: {cdmote,eric}@mote.me.berkeley.edu

1. Introduction

Our webs are wide, very thin, two-dimensional, continua arising commonly as a material in transport in machine systems. Though they are thin, webs often require modeling as plates or shells, albeit very thin and flexible ones, because even small bending stiffness distinguishes their behaviors from membranes in critical ways. Technological applications generating interest in web mechanics include the manufacture and handling of paper where wide sheets are transported by rollers at the highest possible speed, thereby maximizing productivity. Material handling problems of a similar type also arise with tissues, films, magnetic media, textiles, fabrics, polymer sheets, and the like. In a technical world pressing for ever higher speeds of stable transport, the web instabilities of wrinkling and flutter commonly bound the productivity of a technological process. These instabilities can result in permanent damage or rupture of the material and disruption of the manufacturing process. Consequently, this paper addresses some findings on webs that relate to their modeling, the prediction and causes of wrinkling, and the effect of roller misalignment on web behavior during transport. To allow us to maintain this focus, the rich collection of works on membrane structures, including modeling of inflated membranes, space structures and fiber reinforced membranes will not be part of our discussion. For attention to interesting membrane developments readers might examine Steigmann and Pipkin [14], Jenkins and Leonard [4], Li and Steigmann [5], and Haseganu and Steigmann [3]. The most important topic of web flutter caused by air flows over the web surfaces will not be addressed, though recent experimental findings can be found in Chang and Moretti [1] and Nguyen [12]. Theories modeling the flow and web in these coupled problems tend to underestimate the flutter speeds. This problem area is one of great importance.

2. Modeling of Webs

Webs in transport are normally rectangular in form, have two free and two supported edges, and travel across unsupported spans; see Fig. 19.1. Membrane

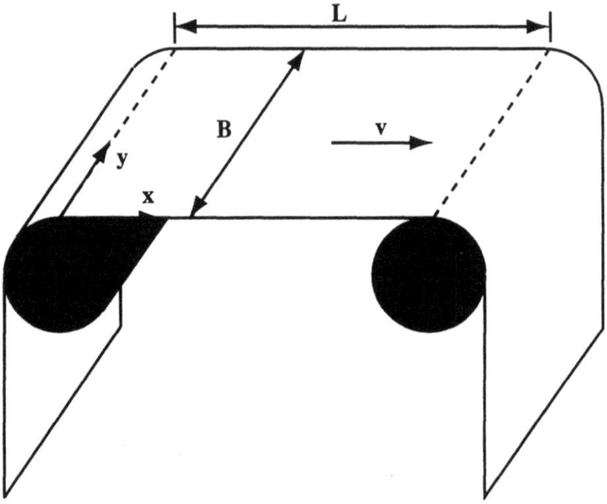

Fig. 19.1. Axially moving web

stresses due to transverse loading (aerodynamic, weight, large deflection, et alia) can often not be neglected. Coupling between the stresses in the web and the deflection is often significant, and is difficult to predict in closed form because of the nonlinear relationships. The equilibrium of these webs cannot be modeled using membrane, or linear plate, theories.

Viewed longitudinally the web is somewhat 'string-like' in that it exists under tension at the ends. The modeling of the supported ends, where shear, bending and displacement constraints exist, challenge the membrane behavior there. Viewed across the width, the web is not obviously 'string-like' for no boundary tractions are present and the free edges also create modeling complications. It would seem that at a minimum, modeling of the web must include fourth-order effects at the edges if not throughout the domain.

Translating continua have been studied for many years. In the great majority of this work, the continua have been modeled as a one-dimensional structure, i.e., a string or beam. While ever more sophisticated one-dimensional models have been developed, Ulsoy and Mote [15] appear to be the first to have modeled a translating continua as a two-dimensional structure. This allowed for width-dependent effects to be considered. In this case, the model was a simple, tensioned, axially-moving Kirchhoff plate,

$$D\nabla^4 w + 2\rho c w_{,xt} + \rho w_{,tt} - [(N_x - \rho c^2) w_{,xx} + 2N_{xy} w_{,xy} + N_y w_{,yy}] = 0. \quad (19.1)$$

D is the flexural stiffness, ∇^4 is the biharmonic operator, c is the translation speed, ρ is the lineal density, $N_x(y)$ is the pretensioning in the web, and $w(x, y)$ is the transverse displacement.

Ulsoy and Mote used this model to study how non-uniform stress profiles across the width of the web affect the first transverse and torsional natural frequencies of the plate. Linear, parabolic and cubic stress distributions were considered. Because it was assumed that all other components of the membrane stress tensor vanished, it can be shown that only the linear profile satisfies the compatibility equations governing the pretensioning.

In 1996, Lin and Mote again used a tensioned, Kirchhoff plate to model a web. Instead of assuming membrane stress distributions throughout the web, they used a stress function to determine the membrane stresses caused by nearly arbitrary boundary tractions. Thus, in this formulation, according to Lin and Mote, the equations governing the response of the plate are

$$D\nabla^4 w - (N_x w_{,xx} + 2N_{xy} w_{,xy} + N_y w_{,yy}) = 0 \tag{19.2}$$

where the membrane stresses are found from the stress function, F, such that

$$N_x = F_{,yy}, \quad N_{xy} = -F_{,xy}, \quad N_y = F_{,xx}. \tag{19.3}$$

F is determined from the applied boundary tractions and is biharmonic, $\nabla^4 F = 0$, throughout the domain of the web. Thus, in this formulation, the membrane stresses satisfy both the equilibrium and compatibility equations. This model was then used to determine if nonlinearly distributed edge loading could cause the web to wrinkle.

Under transverse loading, the deflections of a thin web are often large compared to the thickness. Thus, in order to study the response of an axially-moving web under transverse loading, Lin and Mote [7] used a modified von Karman theory:

$$\begin{aligned} D\nabla^4 w - L[w, F] + 2\rho c w_{,xt} + \rho w_{,tt} + \rho c^2 w_{,xx} &= p \\ \nabla^4 F &= -6(1 - \nu^2) D L[w, w] \end{aligned} \tag{19.4}$$

where F is again an Airy stress function, p is the transverse loading, ν is Poisson's ratio, and the operator L is

$$L[\gamma, \delta] = \gamma_{,xx}\delta_{,yy} - 2\gamma_{,xy}\delta_{,xy} + \gamma_{,yy}\delta_{,xx}. \tag{19.5}$$

Singular perturbation and boundary layer techniques can lead to useful analytical expressions. In this analysis the edges are modeled as 4th order regions (inner solution) and the central web is a 2nd order membrane (outer solution). The equilibrium solutions satisfy nondimensional equations

$$\begin{aligned} \varepsilon\nabla^4 W - L[W, \hat{F}] + C^2 W_{,XX} &= P \\ \nabla^4 \hat{F} &= -6(1 - \nu^2)\varepsilon L[W, W] \end{aligned} \tag{19.6}$$

where

$$W = \frac{w}{h}, \quad X = \frac{x}{L}, \quad Y = \frac{y}{L},$$

$$\hat{F} = \frac{F}{TL^2}, \quad P = \frac{pL^2}{Th}, \quad \varepsilon = \frac{D}{TL^2}, \quad C = \sqrt{\frac{\rho}{T}}c \tag{19.7}$$

and T is the constant longitudinal tension applied at the boundaries, L is the length of the web between the rollers, and h is the thickness.

As $\varepsilon \to 0$ the equations become linear and decoupled. The widths of the boundary layers are $\varepsilon^{1/2}$ at the supported ends and $\varepsilon^{1/4}$ at the free edges. The Airy stress function, $\hat{F}$, has no boundary layer in this model. A uniform approximation to W has been developed for $\varepsilon < 10^{-2}$ and is found in Lin and Mote [7]. As a matter of reference, the order of ε found in paper sheets is typically 10^{-3} to 10^{-5} and approximately 1 in wide band saw plates used to cut timber. Depending on the web tension, the bending stiffness and the curvature, the equilibrium predicted by this model can be quite different from the one predicted by membrane or linear plate theories. Decreasing tension or increasing stiffness enlarges ε and, thus, the boundary layer widths. Confidence in the use of membrane theories can be bolstered with parameters ζ_1 and ζ_2 found in Lin and Mote [7] that relate the magnitudes of first perturbation term to the membrane amplitude prediction of the generating solution.

In work recently completed, Mockensturm and Mote [10, 11] developed two models to describe small deformations superposed on large ones. Both models are based on work by Naghdi [13] and use the Cosserat formulation. The first, termed the general, single-director model, includes the effects of both transverse shear and normal strains. The second, termed the restricted, single-director model, assumes that the material cannot undergo transverse stretching. The discussion here will focus on the general model. Details about the restricted model can be found in the references.

In developing these models three configurations of the body are considered; see Fig. 19.2. In the *reference* configuration, the web undergoes a rigid body transla-

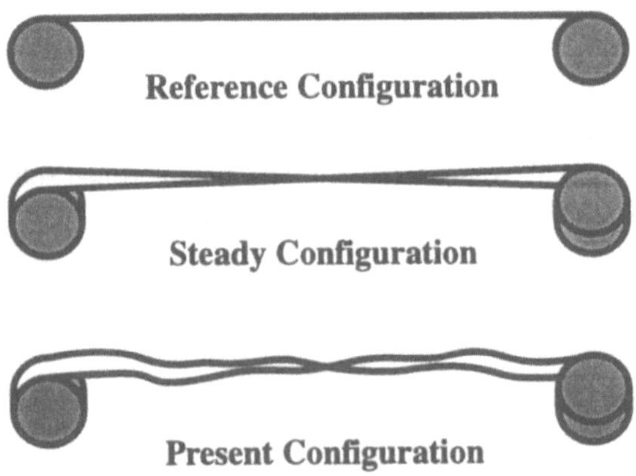

Fig. 19.2. The reference, steady, and present configurations of a twisted web.

tion with constant planar velocity, $\mathbf{v}$. The intermediate, non-trivial, *steady* configuration is determined next. Motions from the reference to the steady configuration are assumed to be large. Superposed on the steady configuration are infinitesimal displacements that describe the *present* configuration. The dynamic response and stability of the web in its steady configuration are predicted in the present configuration. In the following, Latin indices take the values 1, 2, and 3, and Greek indices take the values 1 and 2. Summation over all values occurs when indices, Latin or Greek, are repeated.

The motion taking the reference configuration to the steady configuration is assumed to have the form

$$\mathbf{x}^*(\psi^i(t)) = \bar{\mathbf{x}}(\psi^\alpha(t)) + \psi^3\mathbf{D}(\psi^\alpha(t)) \tag{19.8}$$

where $\psi^i(t)$ are convected (but not material), Cartesian coordinates, $\dot{\psi}^i = v^i$, $\bar{\mathbf{x}}(\psi^\alpha(t))$ describes the mid-surface of the web in its steady configuration and the director field, $\mathbf{D}(\psi^\alpha(t))$, describes the motion of the web through it thickness.

Let $N^{\beta 3}$ and $M^{\beta 3}$ be the components of the first and second resultant stress tensors in the steady configuration, respectively. These are obtained by integrating the second Piola–Kirchhoff stress tensor, with weights 1 and ψ^3, through the thickness of the web. The fully nonlinear equations governing the steady configuration are then

$$\begin{aligned}
N^{\beta\Gamma}_{,\Gamma} + N^{\alpha\Gamma}\Gamma^{\beta}_{\alpha\Gamma} + N^{3\Gamma}_{,\Gamma}D^{\beta} + N^{3\Gamma}D^{\beta}{}_{|\Gamma} + M^{\alpha\Gamma}_{,\Gamma}D^{\beta}{}_{|\alpha} + M^{\alpha\Gamma}D^{\beta}{}_{|\alpha\Gamma} &= h\rho_0\Gamma^{\beta}_{11}v^1v^1, \\
N^{\alpha\Gamma}B_{\alpha\Gamma} + N^{3\Gamma}_{,\Gamma}D^3 + N^{3\Gamma}D^3{}_{|\Gamma} + M^{\alpha\Gamma}_{,\Gamma}D^3{}_{|\alpha} + M^{\alpha\Gamma}D^3{}_{|\alpha\Gamma} &= h\rho_0B_{11}v^1v^1, \\
M^{\beta\Gamma}_{,\Gamma} + M^{\alpha\Gamma}\Gamma^{\beta}_{\alpha\Gamma} + M^{3\Gamma}_{,\Gamma}D^{\beta} - (N^{\beta 3} + N^{33}D^{\beta}) &= \frac{\rho_0h^3}{12}D^{\beta}_{|11}v^1v^1, \\
M^{\alpha\Gamma}B_{\alpha\Gamma} + M^{3\Gamma}_{,\Gamma}D^3 - N^{33}D^3 &= \frac{\rho_0h^3}{12}D^3_{|11}v^1v^1
\end{aligned} \tag{19.9}$$

where $\Gamma^{\beta}_{\alpha\Gamma}$ and $B_{\alpha\Gamma}$ are the Christoffel symbols and second fundamental form of the mid-surface in the steady configuration, $(\bullet)_{,\Gamma}$ denotes partial differentiation with respect to ψ^Γ, $(\bullet)_{|\Gamma}$ denotes covariant differentiation with respect to ψ^Γ in the 3-space induced by the steady configuration, h is the initial thickness of the web, ρ_0 is the volumetric density of the web in the reference configuration, and the vectors $\mathbf{v}$ and $\mathbf{D}$ have been resolved into their components along the contravariant basis on the steady configuration, $\mathbf{A}^i$. When a constitutive equation is specified these become six equations for the components of $\bar{\mathbf{x}}(\psi^\alpha(t))$ and $\mathbf{D}(\psi^\alpha(t))$.

Once the equations of steady motion have been solved, attention turns to the present configuration which describes infinitesimal deformations from the steady configuration. Because we have not assumed that the steady configuration is flat, the equations governing the present configuration are those of a linear, prestressed shell. If the infinitesimal deformation taking the steady to the present configuration is given by

$$\mathbf{u}^*(\psi^i, t) = \bar{\mathbf{u}}(\psi^\alpha, t) + \psi^3 \delta(\psi^\alpha, t), \tag{19.10}$$

the linearized equations of motion, in vector form, for the present configuration are

$$\begin{aligned}
&[N^{\alpha\Gamma}\mathbf{u}_{,\alpha} + N^{3\Gamma}\delta + N_R^{\alpha\Gamma}\mathbf{A}_\alpha + N_R^{3\Gamma}\mathbf{D} + M^{\alpha\Gamma}\delta_{,\alpha} + M_R^{\alpha\Gamma}\mathbf{D}_{3,\alpha}]_{,\Gamma} \\
&\quad = h\rho_0[\mathbf{u}_{,tt} + 2v^1\mathbf{u}_{,1t} + v^1v^1\mathbf{u}_{,11}], \\
&[M^{\alpha\Gamma}\mathbf{u}_{,\alpha} + M^{3\Gamma}\delta + M_R^{\alpha\Gamma}\mathbf{A}_\alpha + M_R^{3\Gamma}\mathbf{D}]_{,\Gamma} \\
&- [N^{\alpha 3}\mathbf{u}_{,\alpha} + N^{33}\delta + N_R^{\alpha 3}\mathbf{A}_\alpha + N_R^{33}\mathbf{D} + M^{\alpha 3}\delta_{,\alpha} + M_R^{\alpha 3}\mathbf{D}_{,\alpha}] \\
&= \frac{1}{12}h^3\rho_0[\delta_{,tt} + 2v^1\delta_{,1t} + v^1v^1\delta_{,11}]
\end{aligned} \tag{19.11}$$

where $N_R^{3\Gamma}$ and $M_R^{3\Gamma}$ represent the additional stresses resulting from the infinitesimal deformation $\mathbf{u}^*(\psi^i, t)$ and $\mathbf{A}_\alpha$ are the covariant base vectors on the mid-surface of the steady configuration. Since the constitutive equations for $\hat{N}^{3\Gamma}$ and $\hat{M}^{3\Gamma}$ are linear, and $N^{\alpha\Gamma}$, $M^{\alpha\Gamma}$, $\mathbf{A}_\alpha$, and $\mathbf{D}$ are known from the solution of the steady problem, these are six linear equations governing $\bar{\mathbf{u}}(\psi^\alpha, t)$ and $\delta(\psi^\alpha, t)$.

This model thus allows for consideration of equilibria that are non-planar and result from possibly large deformations, $\mathbf{x}^*(\psi^i)$, away from the stress-free reference configuration. The equations describing the present configuration are those of a pre-stressed, linear, axially-moving shell. In this sense, this model is the two-dimensional extension of translating cable and rod models.

While the modeling of the translating web itself has progressed significantly in the past 20 years, the interaction of the web with its surroundings has not been given a great deal of consideration. Of particular interest are fluid–structure coupling that causes flutter and web–roller contact.

Flutter is one of the most obvious physical phenomena observed in experiments on thin translating webs. Because the longitudinal edges are free and the web offers little resistance to bending, a significant amount of flutter occurs as the translation speed is increased. While some experimental studies on edge flutter have been conducted [1, 12], rigorous modeling of the fluid–structure interaction has not been attempted. Edge flutter is a serious problem during manufacturing and handling of webs such as paper and fabric. Thus, this area seems like a natural one for future work.

The interaction of the web and the rollers that guide it is another open area in the modeling of translating webs. Slipping that occurs between the web and the rollers is also a common problem. While a number of empirical methods have been developed to account for this effect, no mathematical model has come to our attention. The problems lie in understanding the friction between the web and the rollers and how axial transport plays a role. Because new material is continually coming in contact with the rollers, the usual Coulomb friction models would seem to be inadequate. Web–roller contact is a modeling issue that deserves much greater consideration in the future.

3. Prediction of Wrinkling

Four equilibrium states are possible for a web. When both in-plane principal stresses are tensile, the web is taut [14]. When both in-plane principal stresses are compressive, the web is slack. When one principal stress is tensile and the other is compressive with magnitude less than a critical value, the web is smooth. And when one principal stress is tensile and the other is compressive with magnitude greater than or equal to the critical value, the web is wrinkled. The critical compressive stress at impending wrinkling depends on the flexural stiffness of the web and vanishes in the case of a membrane. An infinite number of wrinkle lines are predicted in that case. Finite stiffness leads to a finite number of wrinkle waves, and the number of waves increases as the aspect ratio of the web increases, the tension increases and stiffness decreases [8].

Transport of webs by rollers or similar devices can easily create in-plane shear by the small misalignments normally occurring in a process, and consequently a principal compressive stress will be created in the web. Large deflections also create principal compression. The prediction of wrinkling and the shape of the wrinkled region for thin, flat rectangular webs reduces to an eigenvalue problem. For a uniform principal tension, the equilibrium is described by

$$\varepsilon(\hat{w}_{,\hat{x}\hat{x}\hat{x}\hat{x}} + 2\hat{w}_{,\hat{x}\hat{x}\hat{y}\hat{y}} + \hat{w}_{,\hat{y}\hat{y}\hat{y}\hat{y}}) - \hat{w}_{,\hat{x}\hat{x}} + b\hat{w}_{,\hat{y}\hat{y}} = 0 \tag{19.12}$$

with

$$\hat{w} = \frac{w}{h}, \quad \hat{x} = \frac{x}{L}, \quad \hat{y} = \frac{y}{L},$$
$$\varepsilon = \frac{D}{TL^2}, \quad \eta = \frac{B}{L} \tag{19.13}$$

where B is the width of the web, the uniform tension is T and compression is bT. The eigenvalues of this problem are

$$b_{mn} = \frac{m^2 + \varepsilon[m^2 + (n/\eta)^2]^2\pi^2}{(n/\eta)^2} \tag{19.14}$$

whose minimum depends on aspect ratio η (= width/length) and mode number (m,n). The minimum eigenvalue b (for integer n) denotes the first impending wrinkling solution to the equilibrium problem.

While the number of wrinkle lines increases with aspect ratio, the critical compression b_{cr} is nearly independent of aspect ratio when $\eta > 0.3$. For $\varepsilon < 10^{-4}$, the solution b_{cr} is essentially independent of the boundary conditions as well facilitating greatly development of approximate solutions to the eigenvalue problem for various cases.

When the in-plane principal tension is oriented off the longitudinal axis by angle θ, shear stress is generated in the web and the equilibrium problem becomes

$$D(w_{,xxxx} + 2w_{,xxyy} + w_{,yyyy}) - (T_x w_{,xx} + 2T_{xy} w_{,xy} + T_y w_{,yy}) = 0 \qquad (19.15)$$

where

$$T_x = \frac{T - bT}{2} + \frac{T + bT}{2}\cos(2\theta)$$
$$T_y = \frac{T - bT}{2} - \frac{T + bT}{2}\cos(2\theta)$$
$$T_{xy} = \frac{T + bT}{2}\sin(2\theta)\,. \qquad (19.16)$$

Appending simple support boundary conditions and solution by the Galerkin method leads to solutions like that shown in Fig. 19.3 for $\theta = 45$ degrees. The

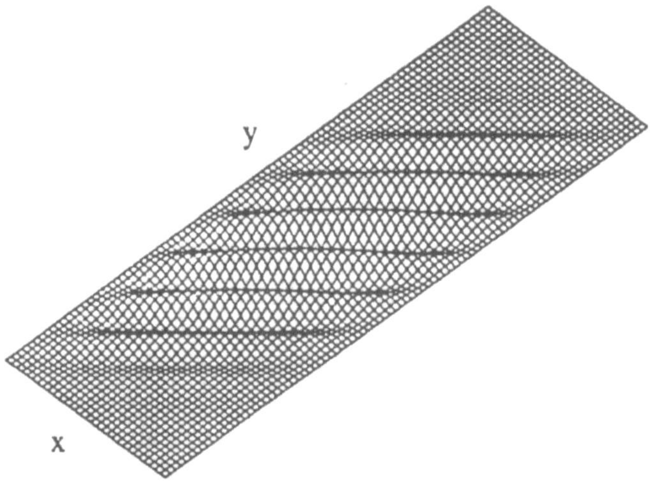

Fig. 19.3. The fundamental wrinkling mode corresponding to $b_{cr} = 0.014$ of a rectangular web with $\varepsilon = 10^{-5}$, $\eta = 3$, $\theta = 45°$, and all edges simply supported.

wrinkle lines lie parallel to the directions of principal compression, perpendicular to the directions of principal tension, and exist only over part of the domain in a parallelogram. The parallelogram is formed by the minimum principal stress directions (extending from the web corners) and the longitudinal edges.

As an alternative to seeking the equilibrium at impending wrinkling through solution of an eigenvalue problem, the criterion of wrinkling can be described by conditions on the stretch, H ($0<H<1$), in the second in-plane direction where stretch $\alpha > 1$ is the first in-plane stretch. Steigmann and Pipkin [14] developed wrinkling criteria for a membrane using principal stretches along with a relaxed energy density in the wrinkled region.

4. Wrinkling under Boundary Traction

Though counter to the intuition of many people, application of normal tensile forces on the edges of a web can wrinkle it. Consider a rectangular web that is pulled in normal tension at opposite ends or pulled in normal tension along all four edges in such a manner that it remains in equilibrium. We conclude that a normal, tensile, end loading that is either uniformly distributed or linearly distributed along the edges cannot wrinkle the web at any loading magnitude. This results because the Airy stress function that satisfies these boundary conditions is simply

$$F = B_1 y^2 + B_2 y^3 \tag{19.17}$$

where B_1 and B_2 are constants. Thus, compression and shear will not appear in the web because $F_{,xx} = 0$ and $F_{,yx} = 0$. However, when the normal tensile edge loading is nonlinearly distributed on the ends, or is even uniformly distributed over a part of opposing boundaries, the biharmonic Airy stress function cannot be a function of y alone. Thus, in these cases, either lateral normal stresses or shear stresses must exist in the web and may cause wrinkling.

If the Airy stress function in the rectangular web is represented in series form, then the coefficients of the series, say λ_m, can be represented in terms of the coefficients of the series representations of the boundary tractions [2]. The stress function can be separated into its symmetric and antisymmetric representations and any stress field in the web, representable in a series form, can be described pending sufficient computational capacity and patience. It remains to examine eigensolutions of the equilibrium problem, given by

$$\varepsilon\nabla^4 w - [(1 + F_{,yy})w_{,xx} - 2F_{,xy}w_{,xy} + F_{,xx}w_{,yy}] = 0 \tag{19.18}$$

under various normal and shear boundary tractions. When the boundary tractions are nonlinearly distributed, compression can be produced in the web and wrinkling can be predicted as above. For instance, if the longitudinal boundary traction is $T[1 + \lambda\sin(2y)]$ on a web of length 1 and width .5, the normal stress across the width N_{yy} is compressive away from the boundaries when $\lambda = 1$, as shown in Figs. 19.4(a) and 19.4(b). Compression sufficient to wrinkle the web is predicted by the eigensolution, which for this traction and $\varepsilon = 10^{-6}$, gives two solutions: $\lambda_P = 0.157$ and $\lambda_N = -0.264$. Accordingly, the web is either taut or smooth for $\lambda_P > \lambda > \lambda_N$, but wrinkles for all λ outside this range. Because $N_{yy} = \lambda N^*_{yy}$ (N_{yy} with $\lambda = 1$) in Figs. 19.4(a) and 19.4(b), the wrinkle waves at λ_P are stretched longitudinally in the middle of the web, perpendicular to the principal width compression, and those at λ_N are stretched longitudinally but at the traction boundaries where the principal compression occurs under this loading, as shown in Fig. 19.5. The antisymmetric shear stresses do not affect the directions of the wrinkle lines.

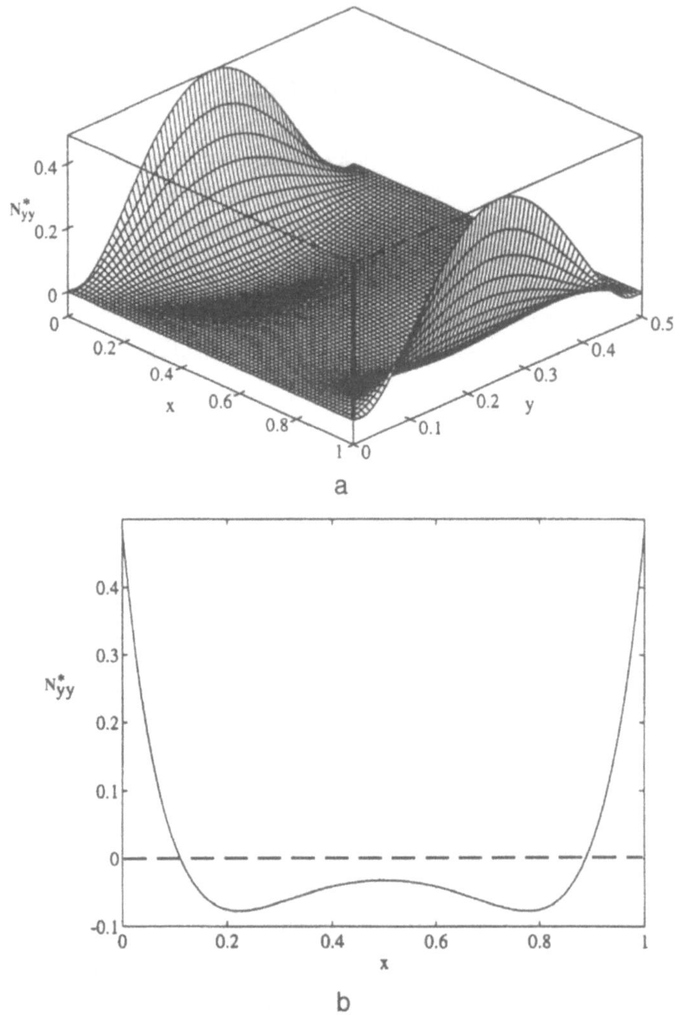

Fig. 19.4. The stress state in the web under edge loading $T[1 + \sin(2y)]$ along x=0 and x=1; (**a**) the normal stresses in the width direction ($N^*_{yy} = F_{,xx}$), and (**b**) the stresses N^*_{yy} along y=0.25.

A boundary traction that is not distributed along the entire boundary can be represented by a Fourier series and an eigenvalue analysis similar to that above can be undertaken. For our web of dimension 1 by 0.5 with the traction applied only, and uniformly, over the mid 10% of two opposing ends and vanishing elsewhere, the eigenvalues α, scaling the applied traction at wrinkling, become $\alpha_P = 127613$ and $\alpha_N = -4941$. Wrinkling is cause by width compression at α_P and by longitudinal compression at α_N. Accordingly, either a taut or a smooth web is predicted

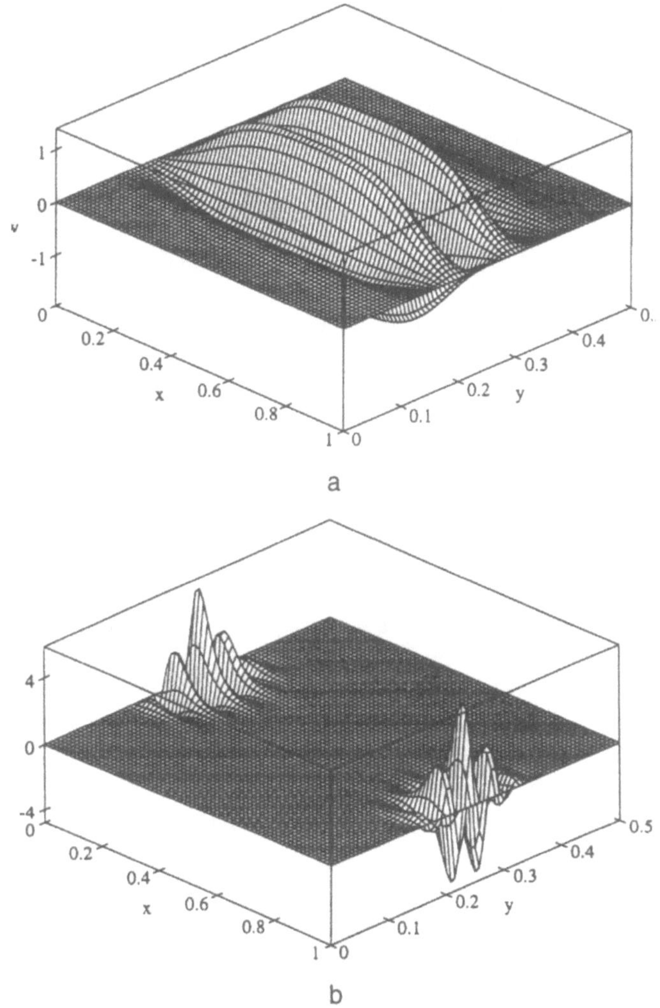

Fig. 19.5. The wrinkling shapes of the web under the edge loading $T[1 + \lambda \sin(2y)]$ with $\varepsilon = 10^{-6}$ and (**a**) $\lambda_P = 0.157$ and (**b**) $\lambda_N = -0.264$.

when $\alpha_P > \alpha > \alpha_N$.

While all nonlinearly distributed edge loadings may not cause wrinkling, they are a major source of wrinkling. Thus, removing shear tractions may not be sufficient to prevent a web from wrinkling.

5. Roller Misalignment

In the studies by Ulsoy and Mote [15] and Lin and Mote [9], no physical source of the nonuniform boundary tractions were given. One possible source is roller misalignment. If the rollers are not aligned perfectly, the tractions they impart to the web will not be constant along the length of the roller. Also, if tilting of either roller twists the web out of plane, the steady configuration will not be flat as is typically assumed.

In research to appear, Mockensturm and Mote [10, 11] investigated the effects of (possibly large) out-of-plane roller misalignment on the stress distribution, natural frequencies and stability of translating webs. Both the general and restricted, fully-nonlinear, single–director models discussed earlier were first analyzed to determine the steady configuration. The material was modeled using the St. Venant–Kirchhoff strain energy function

$$\frac{\hat{\phi}(\bar{\mathbf{C}}, \Lambda)}{\lambda} = \frac{h^3}{24}\left\{\frac{2\sigma}{1+2\sigma}\left[\frac{1}{4}(\Lambda\cdot\mathbf{I})^2\right] + 2\sigma\left[\frac{1}{4}(\Lambda\cdot\Lambda)\right]\right\} \qquad (19.19)$$

$$+\frac{h}{2}\left\{\frac{\bar{\sigma}}{4}(\bar{\mathbf{C}}\cdot\mathbf{I}-3)^2 + 2\sigma\left[\frac{1}{4}(\bar{\mathbf{C}}-\mathbf{I})\cdot(\bar{\mathbf{C}}-\mathbf{I})\right]\right\}.$$

where $\bar{\mathbf{C}}$ is the mid-surface stretch tensor, Λ is the bending strain tensor, σ is the ratio of the Lamé constants, μ/λ, and $\bar{\sigma} = 1$ for the general, director theory and $(2\sigma)/(1+2\sigma)$ for the restricted theory.

It is assumed that out-of-plane roller misalignment causes the web to deform into a right helicoid,

$$\bar{\mathbf{x}}(\psi^i) = \lambda_1\psi^1\hat{\mathbf{E}}_1 + f(\psi^2)\cos(a\lambda_1\psi^1)\hat{\mathbf{E}}_2 + f(\psi^2)\sin(a\lambda_1\psi^1)\hat{\mathbf{E}}_3 \qquad (19.20)$$

where 'a' measures the amount of out-of-plane roller misalignment, λ_1 is the initial longitudinal stretch in the web, and $f(\psi^2)$ governs the lateral normal stretch in the web. Because of symmetries the director field must possess, the equations of steady motion from the general theory reduce to two nonlinear, ordinary differential equations governing the lateral and transverse stretch in the web. If the restricted theory is used, the equations of steady motion reduce to a single nonlinear, ordinary differential equation. In either case, the problem to be solved is now a two-point boundary value problem whose solution was obtained using perturbation techniques.

While the general, single-director theory provides more precise stress distributions, especially near the free edges of the web, the restricted theory is adequate for very thin webs. The equations of the restricted theory are also much simpler than those of the general theory. We will, thus, address results obtained from the restricted theory.

Variation of the mid-span value of the nondimensional, lateral, normal, resultant stress, $\tilde{N}^{22}(0) = N^{22}(0)/(\lambda h)$, with the angle of twist of the rollers, a, is shown in Fig. 19.6. It is negative for all non-zero twist. Thus, a critical value of twist is

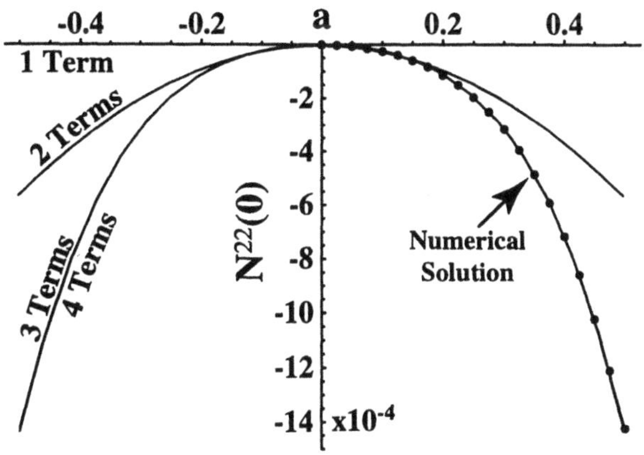

Fig. 19.6. Prediction, by the restricted, single-director theory, of nondimensional $\tilde{N}^{22}$ at the mid-span of the web under roller misalignment, a. Solid dots represent values obtained numerically.

expected where the smooth, right-helicoidal solution is no longer stable. At this critical value, the web wrinkles with wrinkle lines extending in the direction of axial motion [8]. Also illustrated in this figure is a measure of convergence of the perturbation series. The one-term solution describes the tensioned, flat web solution where $\tilde{N}^{22}$ vanishes throughout the web. For a < 0.1, a two-term expansion predicts accurately the lateral, normal stress resultant, and a three term expansion, permits prediction of $\tilde{N}^{22}$ for a < 0.5.

While the effect of translation speed on $\tilde{N}^{11}$ is small, it affects $\tilde{N}^{22}$ significantly as shown in Fig. 19.7. As the translation speed increases, $\tilde{N}^{22}$ increases and becomes positive throughout the web. This effect is attributed to the acceleration of the particles that are away from the span centerline. A particle on the free edge of the web traverses a helicoidal path and accelerates towards the span centerline. Thus, like centrifugal tensioning in spinning disks, the axial motion of a twisted web has a tensioning effect.

As the aspect ratio, $\eta = B/L$, and the pre-stretch, λ_1, increase, the magnitude of the compressive $\tilde{N}^{22}$ increases. Thus, one would expect the likelihood of wrinkling under roller misalignment to increase as the span becomes shorter and the web more highly tensioned.

The consequences of out-of-plane roller misalignment on the natural frequen-

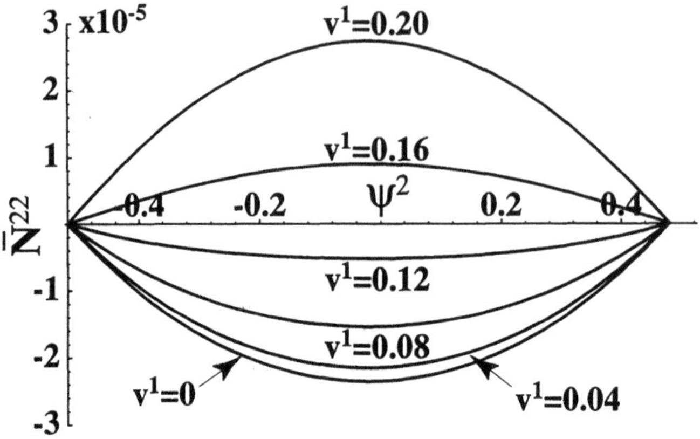

Fig. 19.7. Prediction, by the restricted, single-director theory, of nondimensional $\tilde{N}^{22}$ across the width of the twisted web for nondimensional translation speed, v^1.

cies and stability of the web are two-fold. First, the membrane stresses induced can have a significant effect on the both the natural frequencies and stability (with regard to wrinkling and critical speed) of the web. If the roller misalignment is great enough, the web will wrinkle. The roller misalignment also effects the response of the web by creating a non-planar configuration.

The free vibration and wrinkling of a non-translating web were studied by using the Galerkin method to discretize (19.11) with $v^1 = 0$. Because the compressive stress is the greatest when $v^1 = 0$ (see Fig. 19.7), bounds on the unwrinkled region were determined. For the case when $\lambda_1 = 1$, $\tilde{\varepsilon} = h/L = 1/5000$, $\eta = 1$, and n, the longitudinal mode number, equals 1, the natural frequencies increase substantially with twists less than 0.12 as seen in Fig. 19.8. This occurs because $\tilde{N}^{11}$ increases with twist angle. For $\tilde{a} > 0.12$, the compressive lateral normal stress becomes sufficiently large that the two lowest natural frequencies (indistinguishable from one another in the figure) reach their maximums and begin to fall. At $\tilde{a} = 0.170$, these two frequencies vanish and the web wrinkles into a linear combination of the even and odd modes shown for $0 < \psi^2 < 1/2$ in the figure.

The value of twist at which a natural frequency vanishes, $\tilde{a}_{cr}$, is a function of the system parameters. The surface of critical twist angles depends on dimensionless thickness, $\tilde{\varepsilon}$, and initial tension, $\hat{T} = \hat{T}(\lambda_1)$ as shown in Fig. 19.9. For webs with $\varepsilon > 10^{-2.5}$ the critical twist angle is greater than 0.5 radians for the range of initial tensions studied. The sensitivity of the critical twist angle to the initial tension is greater for thinner webs. For $\tilde{\varepsilon} = 10^{-4}$ the critical twist angle is 0.129 radians at $\hat{T} = 0$, rising to 0.148 at $\hat{T} = 205$. For $\tilde{\varepsilon} = 10^{-2.6}$, the critical twist angle falls from its maximum of 0.490 at $\hat{T} = 0$ to 0.480 at $\hat{T} = 1500$. The initial tension that maximizes the critical twist angle at each $\tilde{\varepsilon}$ is denoted by the curve on the surface. Although $\tilde{a}_{cr}$ does not vary greatly with $\hat{T}$ for thicker webs, the optimal initial tension, $\hat{T}^*$, initially increases with increasing ε, reaching a maximum at

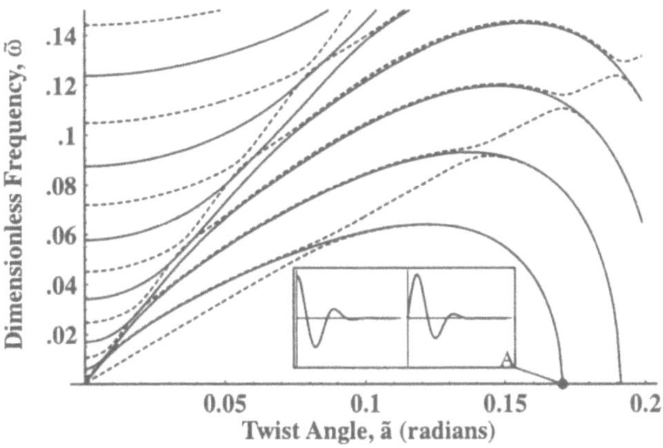

Fig. 19.8. The natural frequencies of the first, transverse, longitudinal mode (n = 1) as a function of twist angle, ã, with $\lambda_1 = 1$, $\tilde{\varepsilon} = 1/5000$, and $\eta = 1$. Solid lines denote odd modes; dashed lines denote even modes.

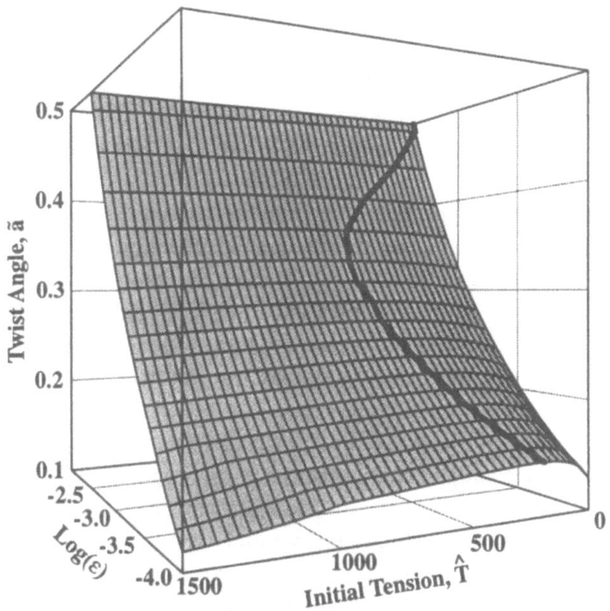

Fig. 19.9. The critical twist angle, $\tilde{a}_{cr}$, with $\eta = 1$. The maximum $\tilde{a}_{cr}$ for specified $\tilde{\varepsilon}$ denoted by the curve on the surface.

$\tilde{\varepsilon} \approx 10^{-3}$. At a dimensionless thickness of $10^{-2.6}$, $\hat{T}^* = 0$.

Because the natural frequencies of two modes vanish together, the wrinkling mode will be a linear combination of them. The half width of the modes are shown inset in Fig. 19.8; the modes are either even or odd functions of ψ^2. In this case the relative amplitude of the wrinkles is the largest near the center of the web because the compressive stress is the greatest there. As $\tilde{\varepsilon}$ increases from 1/5000 to 1/500 the number of wrinkle lines decreases. For the thicker web, the wrinkle extend across the width of the web, and are not concentrated near the center. The number of wrinkle lines is constant as $\tilde{\varepsilon}$ is decreased from 1/5000 to 1/10000 but the spatial frequency of the modes as well as the relative amplitudes of the wrinkles–the amplitudes of the secondary wrinkles to those closest to the mid-width–continues to increase as the wrinkled region concentrates near the mid-width.

For low initial tensions, $0 < \hat{T} < 100$, the number of wrinkle lines is constant but the relative amplitudes and spatial frequency of the wrinkles increases with increasing tension because the wrinkled region concentrates near the mid-width. Increasing $\hat{T}$ to 1000 causes the number of wrinkle lines, as well as the relative amplitudes and spatial frequency of the wrinkles, to increase. In this range, the portion of the web over which wrinkles occur is nearly constant, increasing slightly with increasing $\hat{T}$. For $\hat{T} = 2000$, the web wrinkles in a mode with wrinkles near the free edges. The initial tension level where the transition from center to edge wrinkles occurs depends on the thickness. For $\varepsilon = 1/5000$, the transition occurs around $\hat{T} \approx 1370$. The shift in wrinkling mode can be seen as a kink in the surface of critical twists in Fig. 19.9.

Increasing the aspect ratio causes the relative amplitudes and number of wrinkle lines to increase. For $\eta < 1/2$ the wrinkles extend over the entire width of the web. As η is increased above 1/2, the relative surface of the web that is wrinkled decreases.

A preliminary study of the effect of in-plane roller misalignment has also been undertaken. With this type of roller misalignment, modeling of roller–web contact is very significant. If the web is free to slide along the length of the roller without friction, and if it cannot compress the web, then a global force balance shows that the web slides down the roller until all traction vanishes. However, if friction between the roller and the web prevents sliding, the web will remain tensioned.

If one assumes that there is no friction between the web and the rollers and that the rollers can compress the web, the problem can be solved using a method similar to that for out-of-plane roller misalignment. In this case, the web is assumed to take the shape of a circular sector, and the fully nonlinear equations of motion reduce to a single, nonlinear ordinary differential equation. In this case, the web again slides down the rollers until the compressive traction balances the tensile traction. As the compression along the shortened free edge of the web is very large, even for small roller misalignment, wrinkling is likely.

In processes where webs are driven at constant speed by rollers, periodic boundary loading will occur and consequently the opportunity for parametric instability is ever present [6]. Following modal analysis and modal decomposition and deduction of the transition boundaries through the method of multiple scales, the web is stable under boundary shear excitation when ω is away from the web eigenvalues $\lambda_{ab} \pm \lambda_{cd}$ and when ω is close to $\lambda_{ab} - \lambda_{cd}$. However, when ω is close to $\lambda_{ab} + \lambda_{cd}$, where (a–c) and (b–d) are odd, then a combination resonance of the sum-type can be excited. This frequency happens to be within the range of reasonable operations. For a typical, thin paper web with length 1.2m, width .6m, tension 55 N/m, the excitation range for parametric instability is 47 to 50 Hz when the amplitude of periodic shear equals 5 N/m. As the aspect ratio increases the sum-type combination resonance frequency decreases.

References

1. Y.B. Chang and P.M. Moretti. An Experimental Study of Edge Flutter in Webs. *ASME AMD-Vol 149, Web Handling*, ASME 1992:67-78, 1992.
2. D.J. Gorman and R.K. Singhal. A Superposition-Rayleigh-Ritz Method for Free Vibration Analysis of Non-Uniformly Tensioned Membranes. In *J Sound Vib*, 62.3, pages 489-501, 1993.
3. E.M. Haseganu and D.J. Steigmann. Analysis of partly wrinkled membranes by the method of dynamic relaxation. In *Comp Mech*, 14, pages 596-614, 1994.
4. C.H. Jenkins and J.W. Leonard. Nonlinear dynamic response of membranes: State of the art. In *Appl Mech Rev*, 44.7, pages 319-328, 1991.
5. X. Li and D.J. Steigmann. Finite Plane Twist of an Annular Membrane. In *Q J Mech Appl Math*, 46.4, pages 601-625, 1993.
6. C.C. Lin. Stress Distribution, Wrinkling and Parametric Instability of Thin, Rectangular Webs. PhD Dissertation, University of California, Berkeley, CA, 1995.
7. C.C. Lin and C.D. Mote, Jr. Equilibrium Displacement and Stress Distribution in a Two-Dimensional, Axially Moving Web Under Transverse Loading. In *ASME J Appl Mech*, 62.3, pages 772-779, 1995.
8. C.C. Lin and C.D. Mote Jr. The Wrinkling of Thin, Flat, Rectangular Webs. In *ASME J Appl Mech,* 63.3, pages 774-779, 1996.
9. C.C. Lin and C.D. Mote, Jr. The Wrinkling of Rectangular Webs Under Nonlinearly Distributed Edge Loading. In *ASME J Appl Mech*, 63.3, pages 655-659, 1996.
10. E.M. Mockensturm and C.D. Mote, Jr. Steady Motions of Translating, Twisted Webs. To appear in *Int J N-L Mech*, 1998.
11. E.M. Mockensturm and C.D. Mote, Jr. Free Vibrations and Wrinkling of Twisted Webs. To be submitted, 1998.
12. D.D. Nguyen On-line Sensing of Sheet Flutter using Infrared Light. In *ASME AMD-Vol 149, Web Handling*, ASME 1992, pages 79-86, 1992.
13. P.M. Naghdi. The Theory of Shells and Plates. C. Truesdell (ed) S. Flügge's Handbuch der Physik VIa/2. Springer-Verlag, Berlin, pages 641-703, 1972.
14. D.J. Steigmann and A.C. Pipkin. Finite Deformations of Wrinkled Membranes. In *Q J Mech Appl Math*, 42.3, pages 427-440, 1989.
15. A.G. Ulsoy and C.D. Mote, Jr. Vibrations of Wide Band Saw Blades. In *ASME J Engr Ind*, 104.1, pages 71-78, 1982.

Classical/Advanced controls for hydraulic power plants

E. Irving[1], *A. Libaux*[2]

[1] Scientific Advisor, Supélec, Plateau du Moulon, 91140 Gif /Yvette, France
[2] Project Engineer, EDF/CNEH, Technolac, 73373 Le Bourget-du-Lac, France

Summary. This paper presents control problem of hydraulic plants. The renovation of the control in hydraulic power plants in France and consultant engineering for new hydraulic plants in foreign countries motivate to examine what is the actual control in these plants and what possibilities of improvement exists with the modern advanced approaches. The need for adequate models is a long standing one. There exists since a long time nonlinear partial derivative models of hydraulic plants which have been carefully experienced. The first approach to have a simplified model was to identify simulated responses of the complicate model. Indeed, this approach was not fully successful. Some dynamical long term effects cannot be identify by this way due to nonlinear effects. Another approach using physical nonlinear differential models was completely successful by representing these long term effects. Linearized state space models were deduced which allow to study carefully classical and modern approaches. It was found that classical approach may be greatly improved by precompensator feedback assuring positivity. LQG control using all measurements available was particularly efficient and gives good hopes that practical experiments will be successful.

1. Introduction

Control problems for hydraulic power plants is a long standing one. In addition with power and rotating speed, the control system of the plant has to master the differents levels of the reservoirs components the hydraulic system. These levels have to be controlled for the following reasons. First, it is necessary that the level be not to high to avoid discharging water which jeopardises hydraulic energy On the other hand, the highest level possible gives the highest rate of energy. Second, a not too low level is necessary to avoid uncovering tunnels which may spoil their stone-work. To realise the coordination of these different specifications, different time scales are used. High frequency control is used for power and speed control. Low frequency control is used for level control. Indeed, it is the power set point which is used as the control variable for the level. What is used generally to control the reservoirs levels is to control the surge shaft level. More precisely, it is a dynamical combination of the plant power and the surge shaft level which is controlled. The effect of this strange controlled variable is to make the level of the surge shaft decreasing for increasing power. This is necessary to avoid increasing levels of the reservoirs produced by a constant level surge shaft and increasing head loss in the tunnels makes discharging the water. It is a PI which is used to control this combination.

A first problem which appears when the surge level is controlled is the fact that, if this level behaves without surges, the levels of the reservoirs take a very long time attaining their permanent operation. To make the reservoir level attaining quicker their permanent operation, it is necessary to produce high overshoots on the surge shaft level which is possible inside given limits. A idea to improve the situation is to include to the PID controller a linear combination of the reservoir and surge shaft levels. The big question concerning this first improvement is to know if the empirical determination of this linear combination coefficients is optimal. A first step of controller design is the determination of a nonlinear very high order partial derivative model representing water pressure surge waves. This model has been updated on the Aston plant in the Pyrénées mountains of South of France. A simplified model of order three has been identified from numerical simulations with this partial derivative model. A QFT approach was used to determine the new controller of the surge shaft level. The different reservoirs levels have been controlled with additional controller using the set-point of the surge shaft as their control variable and adequate antisaturation schemes to realise connection or disconnection of the different controllers in good conditions. This first series of controllers gave disappointing results on simulation on the partial derivative model. Clearly, the identification of an order three model was not adequate. To have a physical idea of the minimum order necessary to give acceptable results, it was determined a nonlinear differential model of order five neglecting water and penstock elasticity and representing only accumulation of water in the two reservoirs and the surge shaft. In addition, were considered the inertias of the water in the different mains of this system. The different inertias of three mains were modelled, which gives an additional order two. A nonlinear state space model of dimension five was elaborated. This gave by linearization a linear state space model. By computing the different transfer functions of this model, it was possible to detect what the identification of the nonlinear partial derivative model had missed. Mainly, what was not considered, was the inter reservoir oscillation mode. This very long period (1 hour) oscillating mode was missed, firstly, because the numerical experiment was necessarily not long enough, due to nonlinear effects, to put in evidence the phenomena. Secondly, this phenomena is quite well damped. The design problem which appears when this low frequency phenomena is neglected is the low frequency phase lag of 180° which exists in reality and is not considered with the identified model of order three. A QFT design ignoring this phase lag cannot be adequate. To proceed with the order five model, a state space LQG was determined. One of the motivations of this methodological choice was the fact, cited above, that a linear combination of output variables improved substantially the performances of the control loop and LQG can be considered as a linear combination with filter of the outputs. The theoretical reason for this improvement can be found in the fact that zeros of the open loop system are changed by additional loops. As is shown in [3] and [4], zeros with high frequency dynamics give good performances. Another motivation was to analyse as the LQG manages with the phase lag resulting of the order five model. Also, what was considered convenient with the state space controller is the fact that the observer of the controlled system is the same for the different state controllers of the two reservoirs and the surge shaft levels. Adequate antisaturation devices were implemented to stabilise the additional integrator of each controller when it is not in operation. To determine a non conservative design for the different weighting coefficients of the quadratic criteria, the following methodology was used. First, a too performant choice was taken for the linearized model. Indeed, the level of the surge shaft presented high overshoot and oscillating behaviour. The simulation on the nonlinear differential model was unstable. Additional weightings on the surge level variations were used which stabilised the behaviour of the closed loop model with the nonlinear differential model. What was remarkable is the fact that the nonlinear partial derivative model was also stabilised without use of high frequency weighting, in spite of high frequency modes due to water elasticity. In simulations, the differences of behaviour between the differential and partial derivative models appeared only on the different flow rates and not on the levels which were very similar. These good results on the high complexity model gives good hopes that the experiments on real ground will be successful. The paper will be organised as follows. The second section will be devoted to the description of the plant and the nonlinear differential equations. A third section will describe the identification technics used and the actual control system including the first improvement tentative. A fourth section will give the results of the LQG with one measurement. A fifth section will give the robust and performant LQG results with 3 measurements. The last chapter will be the conclusion.

2. Description of the plant and nonlinear differential state space model

The Aston hydraulic system includes two reservoirs, a surge shaft, two penstocks and a plant with four Pelton turbo-units as described on figure 20.1. First, the two penstocks are globalized in one. Second, the four units are reduced to only one where the control variable is the flow rate into the global penstock. It is the speed and power control which transforms the reference penstock flow rate in the actual control which is the active electric power set point.

First, the reservoirs accumulation laws are given by :

$$\mathrm{d}\, Z_1(t)\, /\, \mathrm{d}t = (1/S_1(Z_1(t))\ (Q_{e1}(t) - Q_1(t) - Q_{dev1}(t)) \qquad (20.1)$$

$$\mathrm{d}\, Z_2(\mathrm{t})\, /\, \mathrm{d}t = (1/S_2(Z_2(t))\ (Q_{e2}(t) - Q(t) - Q_{dev2}(t)) \qquad (20.2)$$

$$\mathrm{d}\, Z_c(t)\, /\, \mathrm{d}t = (1/S_c)\ (-\ Q_c(t)) \qquad (20.3)$$

$S_1(Z_1(t))$ and $S_2(Z_2(t))$ are the reservoirs sections given functions of the levels $Z_1(t)$ and $Z_2(t)$.

The discharging laws are defined by :

if : $Z_1(t) - Z_{dev1} \geq 0$ (20.4), then : $Q_{dev1}(t) = C_{dev1}\,(Z_1(t) - Z_{dev1})^{3/2}$ (20.5)

if : $Z_1(t) - Z_{dev1} < 0$ (20.6), then : $Q_{dev1}(t) = 0$ (20.7)

if : $Z_2(t) - Z_{dev2} \geq 0$ (20.8), then : $Q_{dev2}(\mathrm{t}) = C_{dev2}\,(Z_2(t) - Z_{dev2})^{3/2}$ (20.9)

if : $Z_2(t) - Z_{dev2} < 0$ (20.10), then : $Q_{dev2}(t) = 0$ (20.11)

The levels Z_{dev1} and Z_{dev2} are called the discharging levels of the reservoirs 1 and 2 respectively. It is defined also the uncovering levels Z_{dec1} and Z_{dec2}. The level control has to manage in such a way that the levels remain inside the intervals defined by discharging and uncovering levels. The same kind of levels Z_{devC} and Z_{decC} are defined for the surge shaft.

Pose :

$l_{G1} = S_{G1}\, g\, /\, L_{G1}$ (20.12), $l_{G2} = S_{G2}\, g\, /\, L_{G2}$ (20.13)

$$l_{F1} = S_{F1}\, g \,/\, L_{F1} \qquad (20.14), \qquad l_{F2} = S_{F2}\, g \,/\, L_{F2} \qquad (20.15)$$

$$l_T = S_T\, g \,/\, L_T \qquad (20.16)$$

$$s_{G1} = 1/(2\, g\, S_{G1}^2) \qquad (20.17), \qquad s_{G2} = 1/(2\, g\, S_{G2}^2) \qquad (20.18)$$

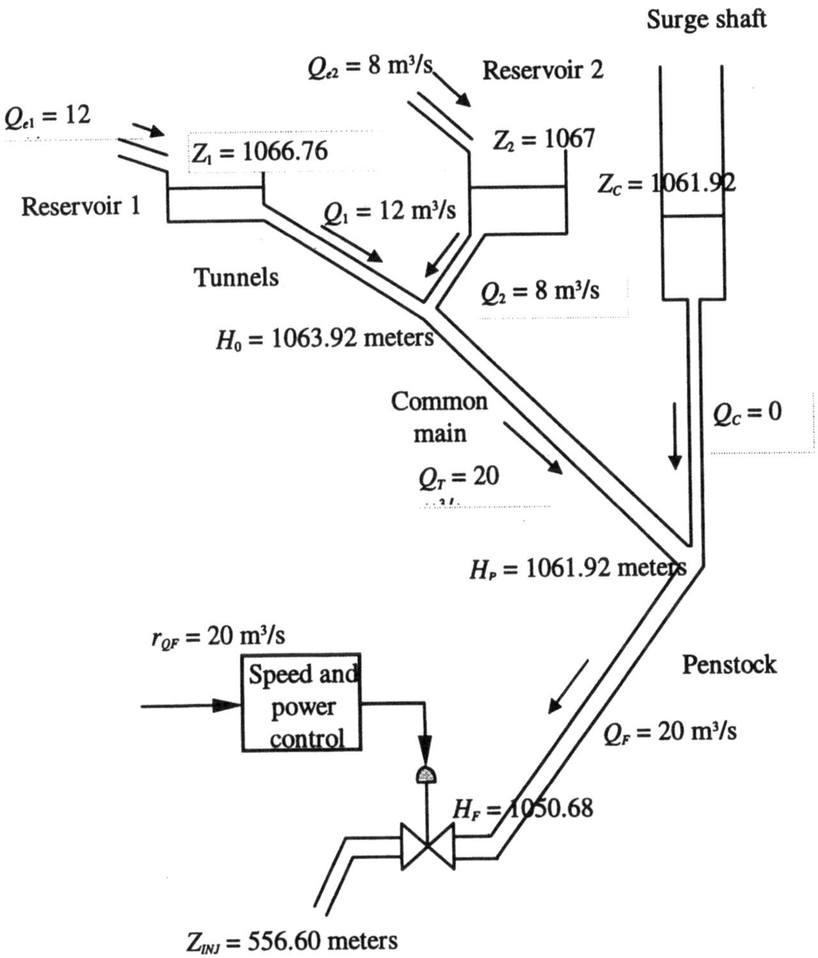

Fig. 20.1: Aston hydraulic system with r_{QF} as control input

$$s_F = 1/(2\, g\, S_F^2) \qquad (20.19)$$

$$s_T = 1/(2\, g\, S_T^2) \qquad (20.20), \qquad s_C = 1/(2\, g\, S_C^2) \qquad (20.21)$$

Inertias of the water in the different tunnels and main, neglecting inertia in the surge shaft, are :

$$dQ_1(t)/dt = l_{G1}\ (Z_1(t) + P_a - H_T(t) - K_{G1}\ Q_1(t)\ |\ Q_1(t)\ |\ ) \quad (20.22)$$

$$dQ_2(t)/dt = l_{G2}\ (Z_2(t) + P_a - H_T(t) - K_{G2}\ Q_2(t)\ |\ Q_2(t)\ |) \quad (20.23)$$

$$dQ_T(t)/dt = l_T\ (H_T(t) - H_P(t) - K_T\ Q_T(t)\ |\ Q_T(t)\ |\ ) \quad (20.24)$$

Total heads at junction points of the hydraulic circuit satisfy :

$$H_T = Z_T + P_{G1}(t) + Q_1(t)\ |\ Q_1(t)\ |\ s_{G1} \quad (20.25)$$

$$= Z_T + P_{G2}(t) + Q_2(t)\ |\ Q_2(t)\ |\ s_{G2} \quad (20.26)$$

$$= Z_T + P_{TH}(t) + Q_T(t)\ |\ Q_T(t)\ |\ s_T \quad (20.27)$$

$$H_P(t) = Z_P + P_{TB}(t) + Q_T(t)\ |\ Q_T(t)\ |\ s_T \quad (20.28)$$

$$= Z_P + P_{CB}(t) + Q_C(t)\ |\ Q_C(t)\ |\ s_C \quad (20.29)$$

$$= Z_P + P_{FH}(t) + Q_F(t)\ |\ Q_F(t)\ |\ s_F \quad (20.30)$$

$$H_F(t) = Z_{INJ} + P_{FB}(t) + Q_F(t)\ |\ Q_F(t)\ |\ s_F \quad (20.31)$$

Head loss in the surge shaft is given by :

$$Z_C(t) + P_a - H_P(t) = K_{Ci}\ (\mathrm{sign}(Q_C(t)))\ Q_C(t)\ |\ QC(t)\ | \quad (20.32)$$

$$K_{Ci}(1) = K_{Ch} \quad (20.33)$$

$$K_{Ci}(-1) = K_{Cb} \quad (20.34)$$

The speed and power system satisfies

$$Q_F(t) = r_{QF}(t) \quad (20.35)$$

Flow rate bilan at the different hydraulic nodes gives :

$$Q_1(t) + Q_2(t) = Q_T(t) \quad (20.36)$$

$$Q_T(t) + Q_C(t) = Q_F(t) \quad (20.37)$$

which implies that :

$$d\,Q_1(t)\,/\,dt \;+\; d\,Q_2(t)/\,dt \;-\; d\,Q_T(t)/d\,t \;=\; 0 \qquad (20.38)$$

This last relation allows elimination of the state equation (20.24) which is replaced by the following algebraic equation :

$$\begin{aligned} & l_{G1}\,(Z_1(t) + P_a - H_T(t) - K_{G1}\,Q_1(t)\,|\,Q_1(t)\,|\,) \\ & + l_{G2}\,(Z_2(t) + P_a - H_T(t) - K_{G2}\,Q_2(t)\,|\,Q_2(t)\,|) \\ & - l_T\,(H_T(t) - H_P(t) - K_T\,Q_T(t)\,|\,Q_T(t)\,|\,) = 0 \end{aligned} \qquad (20.39)$$

Finally, the previous equations are nonlinear state equations in addition with a set of purely algebraic equations of the following form :

$$dx(t)\,/\,dt = f(x(t), y(t), u(t), d(t)) \qquad (20.40)$$

$$0 = g(x(t), y(t), u(t), d(t)) \qquad (20.41)$$

with :

$$x(t) = [Z_1(t)\; Z_2(t)\; Z_C(t)\; Q_1(t)\; Q_2(t)]^{T} \qquad (20.42)$$

$$y(t) = [Q_C(t)\; Q_T(t)\; H_T(t)\; H_P(t)]^{T} \qquad (20.43)$$

$$u(t) = r_{QF}(t) \qquad (20.44)$$

$$d(t) = [Q_{e1}(t)\; Q_{e2}(t)]^{T} \qquad (20.45)$$

$$f_1(t) = (1/S_1(Z_1(t))\;(Qe_1(t) - Q_1(t)) \qquad (20.46)$$

$$f_2(t) = (1/S_2(Z_2(t))\;(Qe_2(t) - Q_2(t)) \qquad (20.47)$$

$$f_3(t) = (1/S_C)\;(-\,Q_C(t)) \qquad (20.48)$$

$$f_4(t) = l_{G1}\,(Z_1(t) + P_a - H_T(t) - K_{G1}\,Q_1(t)\,|\,Q_1(t)\,|\,) \qquad (20.49)$$

$$f_5(t) = l_{G2}\,(Z_2(\mathrm{t}) + P_a - H_T(t) - K_{G2}\,Q_2(t)\,|\,Q_2(t)\,|) \qquad (20.50)$$

$$g_1(t) = Q_1(t) + Q_2(t) - Q_T(t) \qquad (20.51)$$

$$g_2(t) = Q_T(t) + Q_C(t) - r_{QF}(t) \qquad (20.52)$$

$$g_3(t) = Z_C + P_a - H_P(t) - K_{ci}\,(\mathrm{sign}(Q_C(t)))\,Q_C(t)\,|\,Q_C(t)\,| \qquad (20.53)$$

$$g_4(t) = l_{G1}\,(Z_1(t) + P_a - H_T(t) - K_{G1}\,Q_1(t)\,|\,Q_1(t)\,|\,)$$

$$+ \; l_{G2} \; (Z_2(t) + P_a - H_T(t) - K_{G2} \, Q_2(t) \mid Q_2(t) \mid) \\ - \; l_T \; (H_T(t) - H_P(t) - K_T \, Q_T(t) \mid Q_T(t) \mid) \qquad (20.54)$$

To solve the system (20.40) and (20.41), it is first necessary to solve (20.41) in $y(t)$ with $x(t)$, $u(t)$ and $d(t)$ given, then :

$$y(t) = h(x(t), u(t), d(t)) \qquad (20.55)$$

We deduce :

$$y_2(t) = Q_T(t) = h_2(x(t), u(t), d(t)) = Q_1(\mathrm{t}) + Q_1(\mathrm{t}) \qquad (20.56)$$

$$y_1(t) = Q_C(t) = h_1(x(t), u(t), d(t)) = r_{QF}(t) - y_2(t) \qquad (20.57)$$

$$y_4(\mathrm{t}) = H_P(t) = h_4(x(t), u(t), d(t))$$

$$= Z_C(t) + P_a - K_{ci} \;\; y_1(t) \mid y_1(t) \mid \qquad (20.58)$$

$$y_3(t) = H_T(t) = h_3(x(t), u(t), d(t))$$

$$= (1/(l_{G1} + l_{G2} + l_T)) \, (\, l_{G1}(\, Z_1(t) + P_a - K_{G1} \, Q_1(t) \mid Q_1(t) \mid) \\ + \; l_{G2} \, (\, Z_2(t) + P_a - K_{G2} \, Q_2(t) \mid Q_2(t) \mid) \\ + \; lT \, (\; y_4(t) + K_T \; y_2(t) \mid y_2(t) \mid) \,) \qquad (20.59)$$

with :

$$0 = g(x(t), h(x(t), u(t), d(t)), u(t), d(t)) \qquad (20.60)$$

which allows to solve explicitly the state equation :

$$\mathrm{d}x(t) \, / \, \mathrm{d}t = f(x(t), h(x(t), u(t), d(t)), u(t), d(t)) \qquad (20.61)$$

Using the Euler scheme :

$$x(t + T_e) = x(t) + T_e \;\; f(x(t), h(x(t), u(t), d(t)), u(t), d(t)) \qquad (20.62)$$

T_e being the sampling interval.

The permanent operation relations

We need to linearize the different nonlinear differential equations at some equilibrium feasible working points. So, before linearizing, determine the way to obtain feasible equilibrium working points. Note by index $_0$ variables in their permanent operations.

The total flow rate is noted Q_{FC0}. In permanent operations, the sum of the flow rates of the water inputs in the reservoirs is equal to Q_{FC0} :

$$Q_{e10} + Q_{e20} = Q_{FC0} \tag{20.63}$$

The ratio R_{DP0} is given :

$$R_{DP0} = Q_{e10} / Q_{e20} \tag{20.64}$$

then :

$$Q_{e10} = (R_{DP0}/(1 + R_{DP0}))\, Q_{FC0} \tag{20.65}$$

$$Q_{e20} = (1/(1 + R_{DP0}))\, Q_{FC0} \tag{20.66}$$

Due to permanent operations, we deduce :

$$Q_{10} = Q_{e10} \qquad (20.67), \qquad Q_{20} = Q_{e20} \qquad (20.68)$$

Then Q_{T0} flow rate in the common main is deduced :

$$Q_{T0} = Q_{FC0} \tag{20.69}$$

and :

$$Q_{C0} = 0 \tag{20.70}$$

Flow rates being determined in all the tunnels and main, it remains to choose the different levels. For the reservoir 1 and 2 the highest level Z_{MAX} is chosen.

Then, the reservoir where this level is realised is determined in the following. The level difference D_{Z0} is defined by :

$$D_{Z0} = Z_{20} - Z_{10} \tag{20.71}$$

From (20.12) and (20.13), by imposing :

$$\mathrm{d}\, Q_1(t) / \mathrm{d}t = 0 \qquad (20.72), \qquad \mathrm{d}\, Q_2(t) / \mathrm{d}t = 0 \qquad (20.73)$$

it is obtained :

$$H_{T0} = Z_{10} + P_a - K_{G1} Q_{10} \mid Q_{10} \mid$$

$$= Z_{20} + P_a - K_{G2} Q_{20} \mid Q_{20} \mid \quad (20.74)$$

then :

$$D_{Z0} = K_{G1} Q_{10} \mid Q_{10} \mid - K_{G2} Q_{20} \mid Q_{20} \mid \quad (20.75)$$

As Q_{10} and Q_{20} are determined, difference D_{Z0} with its sign is obtained by the following relations :

If :

$$D_{Z0} \geq 0 \quad (20.76)$$

then :

$$Z_{20} = Z_{MAX} \quad (20.77), \quad Z_{10} = Z_{20} - D_{Z0} = Z_{MIN} \quad (20.78)$$

if :

$$D_{Z0} < 0 \quad (20.79)$$

then :

$$Z_{10} = Z_{MAX} \quad (20.80), \quad Z_{20} = Z_{10} + D_{Z0} = Z_{MIN} \quad (20.81)$$

$$H_{T0} = Z_{10} + P_a - K_{G1} Q_{10} \mid Q_{10} \mid \quad (20.82)$$

$$H_{P0} = H_{T0} - K_T Q_{FC0} \mid Q_{FC0} \mid \quad (20.83)$$

$$H_{F0} = H_{P0} - K_F Q_{F0} \mid Q_{F0} \mid \quad (20.84)$$

Using injector law :

$$pol(X_{INJ0}) = Q_{F0} /(sign\ (H_{F0} - Z_{INJ}) \mid H_{F0} - Z_{INJ} \mid^{1/2}) \quad (20.85)$$

The Newton algorithm allows to invert the nonlinear function *pol* :

$$X_{INJ0} = pol(Q_{F0} /(sign\ (H_{F0} - Z_{INJ}) \mid H_{F0} - Z_{INJ} \mid^{1/2}))^{-1} \quad (20.86)$$

The surge shaft level is obtained from (20.25) by imposing :

$$Q_{C0} = 0 \quad (20.87)$$

$$Z_{C0} = H_{P0} - P_a \qquad (20.88)$$

The choice of adequate feasible levels

For any given total flow rate Q_{FC0}, we have to determine the feasible variation interval of R_{DP0}. The feasible levels satisfy :

if :

- $D_{Z0} \geq 0$ (20.89)

then :

$$Z_{MAX} \leq Z_{dev2} \quad (20.90), \qquad Z_{MIN} \geq Z_{dec1} \quad (20.91)$$

if :

$$D_{Z0} < 0 \qquad (20.92)$$

then :

$$Z_{MAX} \leq Z_{dev1} \quad (20.93), \qquad Z_{MIN} \geq Z_{dec2} \quad (20.94)$$

The previous inequalities are not satisfy for any R_{DP0} function of Q_{FC0}. Typically, it can be shown that :

for : $Q_{FC0} = 5$ m3/s $\quad \min\{ R_{DP0} \} = 0, \quad \max\{ R_{DP0} \} = \infty$ (20.95)

for : $Q_{FC0} = 10$ m3/s $\quad \min\{ R_{DP0} \} = 0.4, \quad \max\{ R_{DP0} \} = \infty$ (20.96)

for : $Q_{FC0} = 15$ m3/s, $\quad \min\{ R_{DP0} \} = 0.8, \quad \max\{ R_{DP0} \} = 11$ (20.97)

for : $Q_{FC0} = 20$ m3/s $\quad \min\{ R_{DP0} \} = 1.1, \quad \max\{ R_{DP0} \} = 3.3$ (20.98)

We deduce the nominal points for R_{DP0} :

$$R_{DP0n}(5) = 1 \quad (\text{II}99), \qquad R_{DP0n}(10) = 1.333 \quad (\text{II}100)$$

$$R_{DP0n}(15) = 1.666 \quad (\text{II}101), \qquad R_{DP0n}(20) = 2 \quad (\text{II}102)$$

and the relation :

$$R_{DP0n} = 1 + (Q_{FC0} - 5) / 15 \tag{20.103}$$

Nonlinear differential equations linearization

We pose :

$$x_0 = [Z_{10}\ Z_{20}\ Z_{C0}\ Q_{10}\ Q_{20}]^T \tag{20.104}$$

$$y_0 = [Q_{C0}\ Q_{T0}\ H_{T0}\ H_{P0}]^T \tag{20.105}$$

$$u_0 = r_{QFC0} \tag{20.106}$$

$$d_0 = [Q_{e10}\ Q_{e20}]^T \tag{20.107}$$

From the relations of the previous section defining permanent operation, it is possible to linearize the nonlinear state space equations (20.47) and the algebraic equation (20.48) :

$$\mathrm{d}\,\Delta x(t) / \mathrm{d}t = A\ \Delta x(t) + B\ \Delta u(t) + C\ \Delta d(t) + D\ \Delta y(t) \tag{20.108}$$

$$0 = E\ \Delta x(t) + F\ \Delta u(t) + G\ \Delta d(t) + H\ \Delta y(t) \tag{20.109}$$

with :

$$\Delta x(t) = x(t) - x_0 \quad (20.110), \qquad \Delta u(t) = u(t) - u_0 \quad (20.111)$$

$$\Delta d(t) = d(t) - d_0 \quad (20.112), \qquad \Delta y(t) = y(t) - y_0 \quad (20.113)$$

$$0 = f(x_0, y_0, u_0, d_0) \quad (20.114), \qquad 0 = g(x_0, y_0, u_0, d_0) \quad (20.115)$$

$$A = \mathrm{d}f / \mathrm{d}x\ \big|_{x_0,\, y_0,\, u_0,\, d_0} \quad (20.116), \qquad B = \mathrm{d}f / \mathrm{d}u\ \big|_{x_0,\, y_0,\, u_0,\, d_0} \quad (20.117)$$

$$C = \mathrm{d}f / \mathrm{d}d\ \big|_{x_0,\, y_0,\, u_0,\, d_0} \quad (20.118), \qquad D = \mathrm{d}f / \mathrm{d}y\ \big|_{x_0,\, y_0,\, u_0,\, d_0} \quad (20.119)$$

$$E = \mathrm{d}g / \mathrm{d}x\ \big|_{x_0,\, y_0,\, u_0,\, d_0} \quad (20.120), \qquad F = \mathrm{d}g / \mathrm{d}u\ \big|_{x_0,\, y_0,\, u_0,\, d_0} \quad (20.121)$$

$$G = \mathrm{d}g / \mathrm{d}d\ \big|_{x_0,\, y_0,\, u_0,\, d_0} \quad (20.122), \qquad H = \mathrm{d}g / \mathrm{d}y\ \big|_{x_0,\, y_0,\, u_0,\, d_0} \quad (20.123)$$

If H is invertible, the global relation equivalent to the relations (20.108) and (20.109) is :

$$\mathrm{d}\,\Delta x(t) / \mathrm{d}t = A_{ex}\,\Delta x(t) + B_{ex}\,\Delta u(t) + C_{ex}\,\Delta d(t) \tag{20.124}$$

$$\Delta y(t) = E_{ex}\,\Delta x(t) + F_{ex}\,\Delta u(t) + G_{ex}\,\Delta d(t) \qquad (20.125)$$

with :

$$A_{ex} = A - D\,H^{-1}\,E \quad (20.126), \qquad B_{ex} = B - D\,H^{-1}\,F \quad (20.127)$$

$$C_{ex} = C - D\,H^{-1}\,G \quad (20.128), \qquad E_{ex} = -\,H^{-1}\,E \quad (20.129)$$

$$F_{ex} = -\,H^{-1}\,F \quad (20.130), \qquad G_{ex} = -\,H^{-1}\,G \quad (20.131)$$

Determine the non zeros elements of A, B, C, D, E, F and G :

$$\partial f_1 / \partial x_4 = -1/S_1(Z_{10}) \quad (20.132), \qquad \partial f_1 / \partial d_1 = 1/S_1(Z_{10}) \quad (20.133)$$

$$\partial f_2 / \partial x_5 = -1/S_2(Z_{20}) \quad (20.134), \qquad \partial f_2 / \partial d_2 = 1/S_2(Z_{20}) \quad (20.135)$$

$$\partial f_3 / \partial y_1 = -1/S_C \quad (20.136)$$

$$\partial f_4 / \partial x_1 = l_{G1} \quad (20.137), \qquad \partial f_4 / \partial x_4 = -\,2\,KG_1\,Q_{10}\,l_{G1} \quad (20.138)$$

$$\partial f_4 / \partial y_3 = -\,l_{G1} \quad (20.139)$$

$$\partial f_5 / \partial x_2 = l_{G2} \quad (20.140), \qquad \partial f_5 / \partial x_5 = -\,2\,K_{G2}\,Q_{20}\,l_{G2} \quad (20.141)$$

$$\partial f_5 / \partial y_3 = -\,l_{G2} \quad (20.142)$$

$$\partial g_1 / \partial x_4 = 1 \quad (20.143), \qquad \partial g_1 / \partial x_5 = 1 \quad (20.144)$$

$$\partial g_1 / \partial y_2 = -\,1 \quad (20.145)$$

$$\partial g_2 / \partial y_1 = 1 \quad (20.146), \qquad \partial g_2 / \partial y_2 = 1 \quad (20.147)$$

$$\partial g_2 / \partial u = -\,1 \quad (20.148)$$

$$\partial g_3 / \partial x_3 = 1 \quad (20.149), \qquad \partial g_3 / \partial y_1 = 2\,K_{Cm}\,Q_{C0} \quad (20.150)$$

$$K_{Cm} = (K_{Cb} + K_{Ch})/2 \quad (20.151)$$

$$\partial g_3 / \partial y_4 = -\,1 \quad (20.152)$$

$$\partial g_4 / \partial x_1 = l_{G1} \quad (20.153), \qquad \partial g_4 / \partial x_2 = l_{G2} \quad (20.154)$$

$$\partial g_4 / \partial x_4 = -\,2\,K_{G1}\,Q_{10}\,l_{G1} \quad (20.155), \qquad \partial g_4 / \partial x_5 = -\,2\,K_{G2}\,Q_{20}\,l_{G2} \quad (20.156)$$

$$\partial g_4 / \partial y_2 = 2 K_T Q_{T0} l_T \quad (20.157), \quad \partial g_4 / \partial y_3 = - l_{G1} - l_{G2} - l_T \quad (20.158)$$

$$\partial g_4 / \partial y_4 = l_T \quad (20.159)$$

Numerical results

After differents experimental tests in real ground, the following parameters have been estimated :

$$L_{G1} = 5889.80 \quad (20.160), \quad L_{G2} = 6372.23 \quad (20.161)$$

$$S_{G1} = 8 \quad (20.162), \quad S_{G2} = 7.92 \quad (20.163)$$

$$S_F = 3.974 \quad (20.164), \quad S_T = 8.50 \quad (20.165)$$

$$L_F = 1030.00 \quad (20.166), \quad L_T = 1089.15 \quad (20.167)$$

$$S_C = 78.54 \quad (20.168)$$

$$K_{G1} = 0.0204 \quad (20.169), \quad K_{G2} = 0.0490 \quad (20.170)$$

$$K_F = 0.0281 \quad (20.171), \quad K_T = 0.0050 \quad (20.172)$$

$$K_{Cb} = 0.064 \quad (20.173), \quad K_{Ch} = 0.043 \quad (20.174)$$

$$Z_{dev1} = 1068.15 \quad (20.175), \quad C_{dev1} = 35 \quad (20.176)$$

$$Z_{dev2} = 1068.40 \quad (20.177), \quad C_{dev2} = 40 \quad (20.178)$$

$$Z_{dec1} = 1066 \quad (20.179), \quad Z_{dec2} = 1064.40 \quad (20.180)$$

Laws of sections functions of levels are given on figure 20.2.

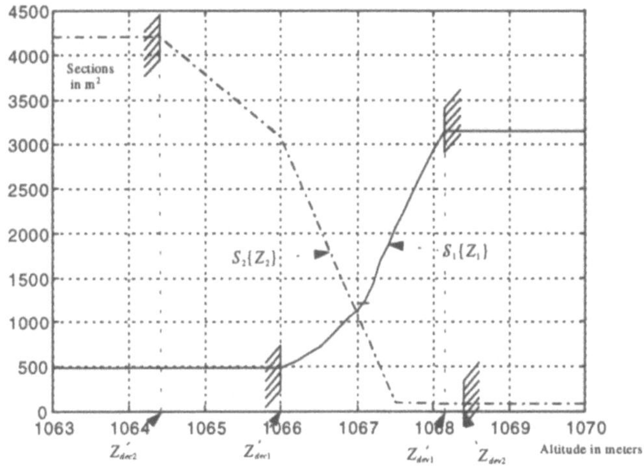

Fig. 20.2 : Reservoirs sections functions of levels

The first numerical experiment with the model and numerical values given previously is an open loop with all the inputs flow varying at different instants, following table 20.1.

The reservoir 1 is initially at the level 1067.75 meters, so 3.37 meters above the level of reservoir 2. Decreasing $Q_{e1}(t)$ input flow rate to 8 m^3/s makes the two branches including tunnels from reservoir 1 and 2 almost at equilibrium, the $Q_{e2}(t)$ being at 5 m^3/s because the level difference between reservoirs 1 and 2 is only 8 cm, see figure 20.4. This is explained by equation (20.75) and numerical values given by (20.169) and (20.170). Then, the level difference increases to finally, decreases when the input flow rate $Q_{e2}(t)$ increase to 10 m^3/s.
To maintain the levels inside their limiting values given by figure 20.2, it is necessary to accept a penstock flow rate $Q_F(t)$ equal to the sum $Q_{e1}(t) + Q_{e2}(t)$:

$$Q_F(t) = Q_{e1}(t) + Q_{e2}(t) \qquad (20.181)$$

Of course, this control is, a priori, not possible, because the flow rates $Q_{e1}(t)$ et $Q_{e2}(t)$ are not measured. Nevertheless, it indicates that if measurement were available, the reservoirs levels be maintained in a very efficient way. This open loop control indicates the surge shaft level $Z_C(t)$ is oscillatory weakly damped and highly overshooting. If overshooting is inherently due to a quick control $Q_F(t)$ which will be necessary to counteract the effect on the reservoirs levels of their inputs flow rates, it is clear, that in addition its basic previous use, the control will have to damp in all measure possible the oscillation of the surge shaft level.

t	Q_{e1}	Q_{e2}	Q_F
0	15	5	20
300	15	5	20
420	8	5	20
5000	8	5	13
5120	15	5	20
6000	15	5	20

Table 20.1 : Variations of reservoirs and penstock inputs flow rates

Another, very important point, if we control the surge shaft level, is to make decrease the level set point when the penstock flow rate is increasing. To make vary the reservoirs levels sufficiently quickly, it is necessary to control the surge shaft level accepting a big overshoot after a set point variation.

The difficult point of the control is the situation when the reservoir 2 input flow rate $Q_{e2}(t)$ is small when $Q_{e2}(t)$ is large. It has been shown previously that it becomes impossible to prevent the both reservoirs levels to discharge water or to uncover the stone work of the tunnels because the term $D_Z(t) = Z_2(t) - Z_1(t)$ becomes too large, see (20.75). It was shown that at $Q_F(t) = 20$ m^3/s, the ratio $r_{DP0}(t)$ cannot be greater than 3.3. At 25 m$_3$/s, this ratio decreases to 2.2. For these situations which can be detected by a large flow rate $Q_F(t)$ and a large absolute value of $r_{DP0}(t)$, the reservoirs levels set points has to be carefully chosen.

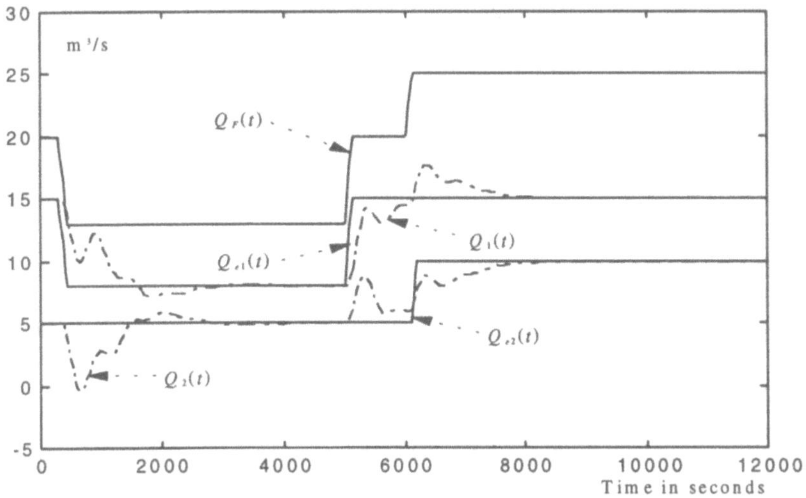

Fig. 20.3 : Reservoirs inputs, outputs and penstock flow rates

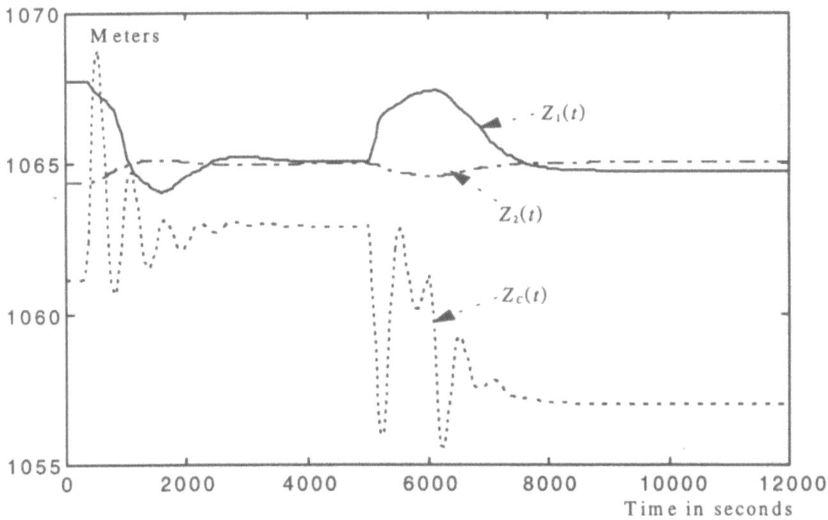

Fig. 20.4 : Reservoirs and surge shaft levels

A safe choice may be to control half the sum of reservoirs levels in the middle of the admissible level zone. However, when $D_Z(t)$ is small, it is necessary to choose levels set points near the limit of discharging water to optimize the energy produced by this water.

With all these physical insights into the system to be controlled, it is time, now, to examine the standard control system which has been implemented in the real plant.

3. The standard control actually implemented

The designers of the standard control system took account of the different delicate points of the system to be controlled to implement something which is acceptable. The first design problem which motivates the utilisation of the model described in the paper is the fact that identification of a simple model from step input on $Q_F(t)$ gives completely wrong results because from responses of figure 20.5, it is not possible of estimating a system of order larger than 3. What can be see on this figure is the fact that the surge shaft level response is clearly the sum of the responses of an integrator and a second order oscillating system. If the response of the surge shaft level $Z_C(t)$ is evidently oscillating at period neat 600 seconds, the fact that the responses of reservoirs levels $Z_1(t)$ and $Z_2(t)$ are oscillating is not evident. What is clear is they are responses of delated integrators. What helps to find an order three structure of system is to notice that the response of the surge level has a non zero derivative at the instant of the step application. Then, the structure directly representing the responses of figure 20.5 is :

$$x_I(t) = 1/(s^3 + a_1 s^2 + a_0 s)\ Q_F(t) \qquad (20.182)$$

$$Z_1(t) = b_{10}\, x_I(t) \qquad (20.183)$$

$$Z_2(t) = b_{20}\, x_I(t) \qquad (20.184)$$

$$Z_C(t) = [\, b_{32} s^2 + b_{31} s + b_{30} \,]\ x_I(t) \qquad (20.185)$$

In the previous equations $x_I(t)$ represents the interne state, see [7].

Now, it is important to ask the question, apparently, the transfer functions of the form :

$$x(t) = C\ (s I_5 + A_{ex})^{-1} B_{ex}\, Q_F(t) \qquad (20.186)$$

with :

$$C = I_5 \qquad (20.187)$$

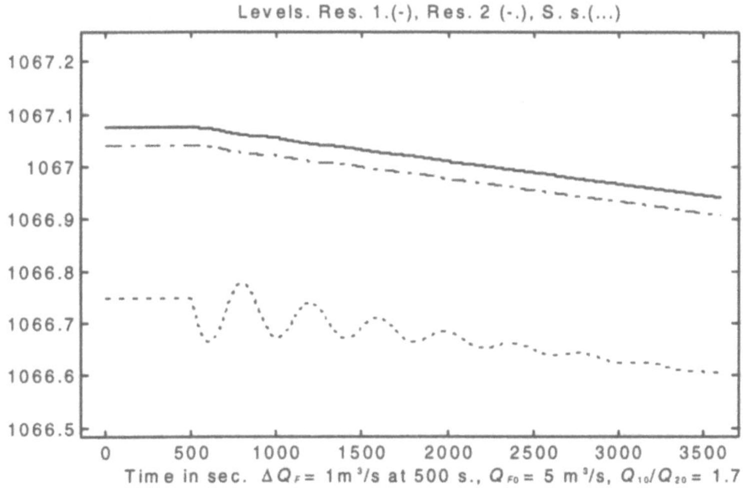

Fig. 20.5 : Levels resulting from a step input on the control

$$Z_1(t) = C_1 \ x(t) \tag{20.188}$$

$$Z_2(t) = C_2 \ x(t) \tag{20.189}$$

$$Z_C(t) = C_3 \ x(t) \tag{20.190}$$

C_i represents the i-th row, $i = 1, \ldots, 5$, of the matrix C the 5 x 5 identity matrix. Due to the fact that A_{ex} is a 5 x 5 matrix, the transfer functions given by (20.186) to (20.190) has a denominator $A(s)$ given by :

$$A(s) = Det\{ \ s \ I_5 \ + \ A_{ex} \ \} \tag{20.191}$$

of degree five. Is the neglected dynamics present when using the transfer functions coming from (20.182) to (20.185) high frequency ? If it is the case, the design will need an adequate roll off. Unhappily, it can be shown that this dynamics is at a lower frequency that the one corresponding to the 600 seconds oscillation, that is to say, this low frequency corresponds to a period of almost one hour. Why, in the identification simulation this oscillation was not seen, is due to the fact it is very well damped. Also, another reason was the duration of the experiment which cannot be larger than one hour due to error risk produced by excitation of the nonlinear effects of the model.

The first step of the classical design is to damp the controlled system by a zero gain filtered derivative feedback compensator (washout filter) of the following form :

$$Q_F(t) = u'(t) - K_d \, T_d \, s \, / \, (1 + T_d \, s) \; Z_C(t) \qquad (20.192)$$

In a second step, $u'(t)$ is given by the P.I. :

$$u'(t) = K \, (1 + 1 \, / \, T_i \, s) \, (r_{ZC}(t) - Z_C(t)) \qquad (20.193)$$

where $r_{ZC}(t)$ is the surge shaft set point level.

The motivation of the filtered derivative in (20.11) is to change the open loop Nyquist plot of figure 6 encircling the critical point in the Nyquist plot of figure 20.7 leaving this point completely outside the curl. Clearly, it is a positivity property which has been brought to the system with the washout filter giving a theoretical infinite gain margin. As is well known, the infinite gain margin is only theoretical due to neglected dynamics. For this reason, the filter gain has been chosen not too large to obtain a reasonable delay margin of 109.5 seconds.

Due to the previous property, the gain of the P.I. can be chosen sufficient to obtain the quite quick closed loop response of figure 20.8.

4. LQG control with one measurement

Because to control the surge shaft level makes the control of the reservoirs levels not precise enough, it has been tried to control directly the reservoirs levels by LQG with an integrator using the linearized state space model given by section 2 and the one measurement of the controlled variable itself. In addition, it has been estimated as an unknown constant the reservoirs input flow rates. The simulation results are given on figure 20.9. What can be seen is this scheme is not robust to nonlinearities as discharging operations, where it can be seen that a disturbance ramp on the reservoir 1 input flow rate at 4000 secondes and set point variation on reservoir 1 level generate oscillations which takes a very long time to be damped.

5. LQG with 3 measurements

To overcome the problems of the previous section, it has been implemented a LQG control with 3 measurements which gives excellent results as can be seen on figure 20.10. The reservoir 1 input flow rate ramp at 4000 seconds is very well damped and the brutal change of the reservoir 1 level set point of 0.5 meter is controlled with only one discharging operation. Comparing with the results of the previous section, there is a huge improvement in the control. The theoretical reason for this improvement when taking more measurements is to be found in the

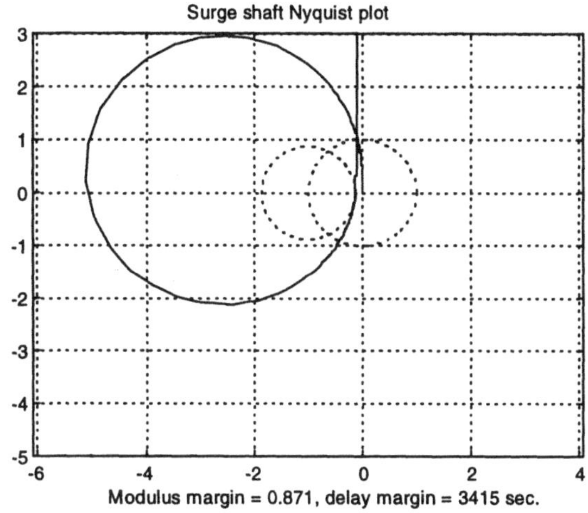

Fig. 20.6 : Controlled system Nyquist plot

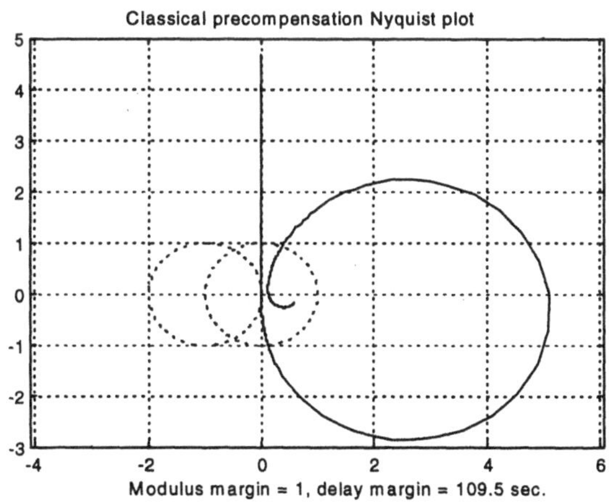

Fig. 20.7 : Compensated controlled system by washout filter

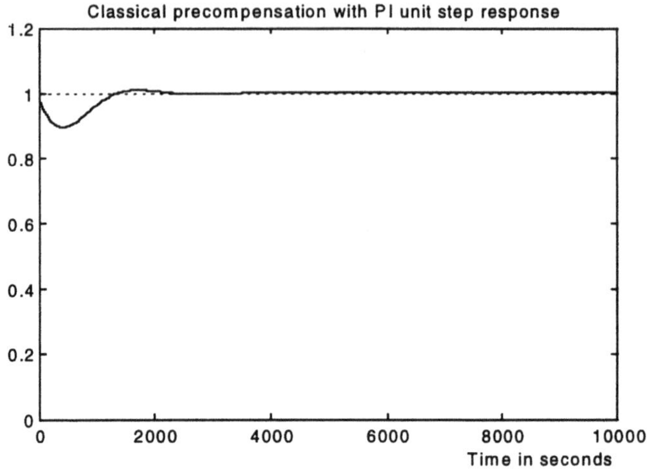

Fig. 20.8 : Classical compensation with PI set point step surge shaft response

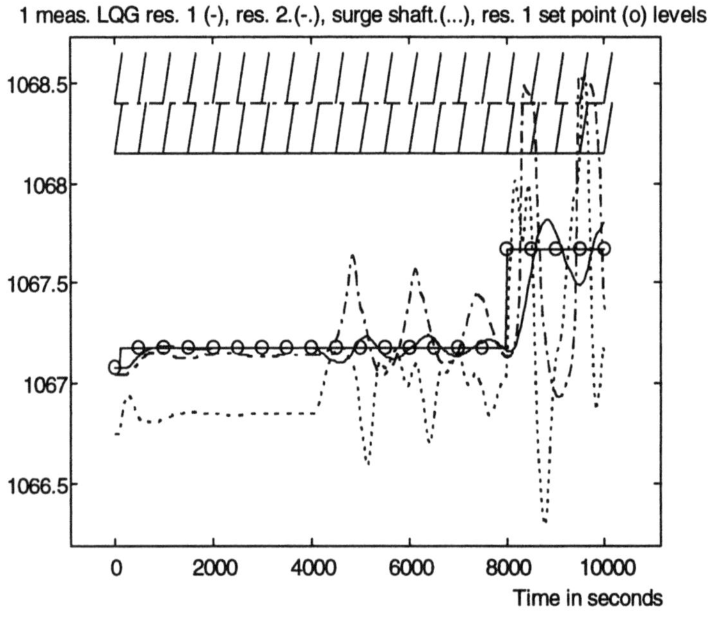

Fig. 20.9 : LQG with one measurement reservoir 1 input flow disturbance at 4000 seconds and level set point variation of 0.5 meter at 8000 seconds.

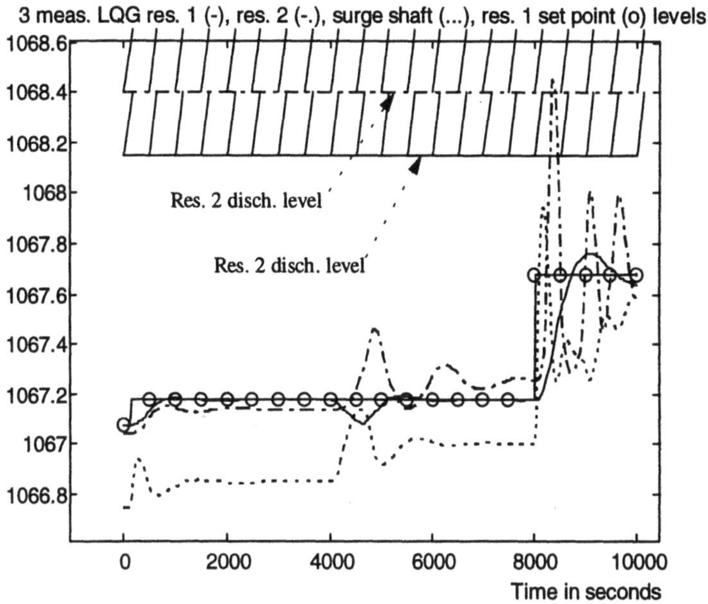

Fig. 20.10 : LQG with 3 measurements reservoir 1 input flow disturbance at 4000 seconds and level set point variation of 0.5 meter at 8000 seconds.

Kalman theorem concerning the LQ control infinite gain margin. Of course, a LQG control with 3 measurements is not exactly a LQ control but it is an intermediate solution which brings some of the performances of the LQ. One the other hand, results as infinite gain margin have to be moderated by the consideration of the delay margin which includes the effects of neglected dynamics. This last point has been particularly usefull for the choice of the LQG weighting coefficients.

6. Conclusions

The lesson of this promizing result is the fact that it is usefull to feedback in the controller the maximum possible disturbed variables. Due to the integrator on the controlled variable, this does not spoil the tracking of the set point. To realize an performant and robust controller, of course, the knowledge of a whole family of linearized models is necessary. The nonlinear model which was presented in section 2 was a basis for it. It has to noted that this nonlinear model was verified on real experiments. The next improvement to these studies on control on hydraulic systems will be the implementation of a nonlinear observer with nonlinear control.

References

1. B. D. 0. Anderson, J. B. Moore (1989). *Optimal Control Linear Quadratic Methods*. Prentice Hall.

2. K. J. Astrom, B. Wittenmark (1990). *Computer Controlled Systems*. Prentice Hall.

3. J . S. Freudenberg, D. P. Looze (1985). Right Half Plane Poles and Zeros and Design Tradeoffs in Feedback Systems. *IEEE Trans. Aut. Control*, vol. AC-30, n° 6, pp. 555-565, June 1985.

4. J. S. Freudenberg, D. P. Looze (1988). *Frequency Domain Properties of Scalar and Multivariable Feedback System*. Springer-Verlag.

5. I. M. Horowitz (1993). *Quantitative Feedback Design Theory (QFT)*, Vol. 1. QFT Publications, 4470 Grinnel Ave., Boulder, Colorado, 80303.

6. E. Irving (1998). *Commande Robuste Adaptative*. Cours Supélec.

7. T. Kailath (1984). *Linear system theory*. Prentice Hall.

8. A. Libaux (1995). *Etude de la conduite d'Aston (1ère partie) : modélisation*. Note EDF/CNEH EH DT DF / 95 097 A.

9. A. Libaux (1996). *Principes de régulation pour une conduite optimale de l'usine d'Aston*. Note EDF/CNEH EH DT ED / 96 051 A.

Competition, Interaction and Control

I.D. Landau

Laboratoire d'Automatique de Grenoble, ENSIEG (CNRS/INPG/UJF), B.P. 46, 38402 Saint Martin D'Hères, France, e-mail: landau@lag.ensieg.inpg.fr

Summary. [1] Automatic Control is a technical discipline which has extremely varied fields of application but is only visible through the field of application where it is used. Control competes with other disciplines as provider of an efficient new technology. Interaction with other technologies is necessary in practical situations.

Competition and interaction occur inside the control discipline as well. Interaction between various control techniques is not only necessary but it is also very beneficial for the development of the discipline itself.

The paper will illustrate these aspects via a few practical examples and research developments.

1. Introduction

Automatic Control is the discipline of analysis, modelling, identification and control of dynamical systems.

Automatic Control is a "horizontal" discipline which has extremely varied fields of application: mechanics, chemistry, power production, aeronautics, automotive, biotechnology ... . Unfortunately, automatic control is not visible by itself, but is only visible through the field of application where it is used.

The growing importance of control "has led to the development of one of the modern world's key technologies" [24]. The development of the control technology aims at developing of a control design methodology and implementation techniques at a competitive cost and applicable to a variety of situations. This development is partly driven by practical needs which may

[1] The original version of this paper was presented as a plenary lecture at the European Control Conference, Bruxelles, July 1-4, 1997.

include important research problems and partly by the advance in the basic research in control.

The wide applicability of control is on the one hand a major opportunity for our discipline and its development. The fact that control is only visible through the application field is on the other hand a serious obstacle for its application and recognition since the control specialists are not the *maîtres d'oeuvre.*

The use of advanced control techniques in industry is based on a competitive analysis with respect to other technologies by evaluating the possible technical, economical and environmental impacts. In many industrial areas, people have understood that improvements no longer result from the traditional technology used. However, control has to compete with other disciplines as provider of an efficient new technology. It is therefore very important that in addition to developing arguments for the use of advanced control techniques, the control discipline interacts with other disciplines in order to provide a combined offer of new technologies for application in a certain field.

Some of the reasons for using advanced control systems are essentially related to:

- assuring a certain performance (which can be expressed in "control" terms)
- improving product quality
- reducing production costs
- decreasing pollution
- increasing process flexibility

One of the basic diagrams used to justify the choice of advanced control technology is illustrated in Fig. 21.1. If the variance of the controlled variable

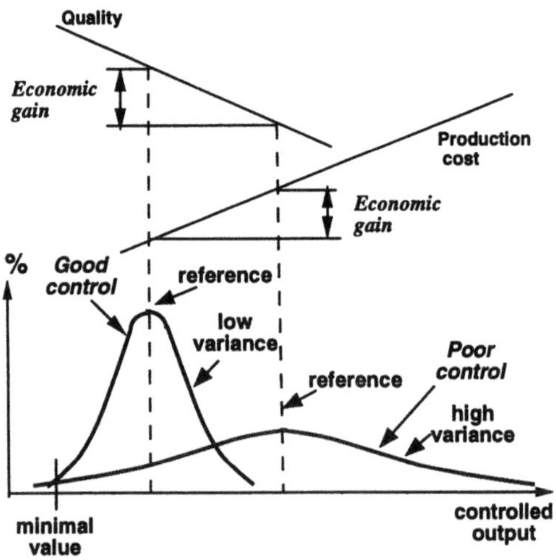

Fig. 21.1. Histograms for good and poor control.

is high, a significant number of measurements are far from the desired value. In a large number of applications, a minimal acceptable value is imposed (e.g.: depth of the coating, humidity of the paper, minimum diameter of a fiber), and the poor quality of the control will require a higher value to be chosen for the reference. As a consequence, more energy or more material will be needed and the direct consequence is the increase in production costs.

If one has a "good" controller which significantly reduces the variance of the controlled variable around the reference value, this will, on the one hand, improve the quality and, on the other hand, allow the reference value to be reduced. This leads in general to energy and material savings which correspond to the reduction in production costs.

Similar conclusions can be drawn when considering the "tracking" aspects which is the other typical task of a control system.

Achieving a better control has normally a quantifiable economic impact taking in account primary and secondary effects of a good control. However it is important to have a short investment return for control improvements which implies, on the one hand, the development of an efficient design methodology and, on the other hand, reduction in implementation costs (including commissioning).

In fact, the problems which are posed in the various potential applications areas are often not posed as control problems (or control issues are just part of the picture). They are posed in general in terms of: quality, production costs, environment impact, customer demand, investment and so on. Evolution of the market in various areas points to the need for better quality, less pollution, higher efficiency and flexibility. On the other hand, industry is forced to operate with low profit margins and short investment return is becoming crucial.

Improvement of a product or process cannot necessarily be translated in an obvious way into a control problem for the specialists of the product or process. Their natural reaction is to try to improve the basic technology of the process or product using the advances in their own field and not by using advanced control. Another important aspect is that it is not obvious to directly apply advanced control. In many cases, additional technological changes have to be considered in order that advanced control becomes a solution (or part of it) for the problem posed. Often, transducers and actuators have to be added and/or redesigned and this is directly related to the technology of the field of application.

In short:

1. Since control engineers are not the *maîtres d'oeuvre*, the control technology competes with the specific technology of the field of application.
2. Control technology cannot be spread without interaction with the technology of the field of application.

We will illustrate these aspects in section 2 for the case of *gas heating systems* for domestic use (this is a wide spread consumer product).

Let us turn now to the control design problem (assuming that the preliminary work for converting the original problem into a control problem has been completed and that the appropriate hardware is available).

Control engineering has been involved in a remarkable research development since the end of the fifties (seminal theoretical development arose around 1960 which was the year of the first IFAC Congress). Of course, in order to produce significant results, the research area should be narrow enough to allow in-depth exploration. A number of subfields of control have been defined and at various periods substantial progress has been achieved. Unfortunately, as a consequence of this research specialization, which had also a strong impact in the academia, each subfield roughly claims that it is the "most important part" in solving a control problem or in other terms "is the key for a successful design". The consequence of this situation is a strong *competition* between the various subfields for obtaining research funds or finding prospective users. In order to globally solve a control problem, one focuses on a particular method which can solve just part of the problem and for the remainder of the problem, one often uses ad-hoc solutions which are not necessarily the best (in the sense that they do not use the major achievements available in the related areas).

Such examples were up to the end of the eighties:

- robust control which ignored system identification,
- system identification which mainly focused on search for the "true system model" rather than the models for control design,
- adaptive control which focused on robust adaptation without taking in account the lack of robustness of the underlying control design.

While in the research community things are changing, the damage caused by this situation is still very perceptible at application level.

Design methodology which is the core of an engineering discipline has an integrating character. Control design problems start from data and end up with data, and the various areas of control have to interact in order to design a controller in the most efficient way. This interaction is also extremely beneficial, on the one hand, for a better understanding of control problems and, on the other hand, for the development of new algorithms and of a more coherent approach to control design.

One such example is the interaction between system identification and control design. *Dual control* in the early sixties and *adaptive control* in the early seventies were attempts to address some of the issues. In the seventies, system identification underwent a tremendous development, while robust control experienced a major development in the eighties. The interaction between system identification and robust control started to attract a consequent attention by the end of the eighties. Important contributions had already been made and even if all the problems had not been solved and a number of questions had not yet been answered, some of the available results have a considerable practical impact. Excellent reviews of the current status of research on the interaction between identification and control are available [6], [4], [7], [30].

One of the key strategies, issued from the work on interaction between identification and control, is the so-called "iterative closed loop identification and control redesign".

Among other consequences of the work on interaction between identification and control is the revival of the interest for plant model identification in closed loop operation. This was motivated, on the one hand, by practical constraints (plants which have an integrator behavior or are unstable in open loop, maintenance of existing controllers without opening the loop, etc) as well as by the fact that plant model identification in closed loop may provide better models for design. New algorithms have been developed and some of them have remarkable properties.

It is interesting to note that the "sensitivity" functions which play a key role in robust control, play also a fundamental role for identification in closed loop. The new results in identification in closed loop also have an impact in adaptive control.

The remainder of the paper is organized as follows. In Section 2, the interaction between control and the specific technology of a field of application will be illustrated for the case of gas heating systems. Then, the problem of controller design will be reviewed in Section 3. Section 4 will illustrate the interaction between identification in closed loop and robust control, and will present some recently developed algorithms for identification in closed loop. The connections between iterative identification in closed loop followed by controller redesign, and adaptive control will be discussed in Section 5. The benefits of identification in closed loop for controller redesign will be illustrated for the case of the control of a flexible trasmission system in Section 6.

2. Control of a Gas Heating System

Gas heating systems are widely used consumer goods. There is a clearly identified market demand for improving their performances and reducing their size. It is therefore interesting to examine why and how control can become a major actor in product improvement. A standard gas heating system used for house heating and production of hot water is depicted in Fig.21.2. For simplicity of presentation, only the problem of providing hot water at a certain temperature is considered. The system is formed by:

1. a gas burner which is fed by gas and air and which has essentially an on/off mode of operation delivering either full power (on mode) or zero power (no flame)
2. a combustion chamber with appropriate evacuation pipes for the smoke

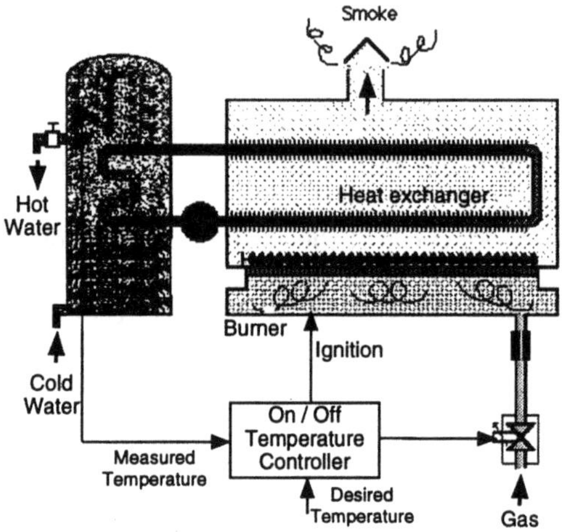

Fig. 21.2. A standard gas heating system for hot water production.

3. a heat exchanger
4. a water balloon
5. a transducer for the water temperature measurement.

The standard controller is, in many gas heating systems, a rudimentary on/off controller and, in general, a large water balloon (large inertia) is used to compensate for the poor quality of the control.

In what follows, we will try to present the approach chosen by Théobald S.A., a manufacturer of gas heating systems, for improving their products. The main reference is [26][2]

The objective is to make a better heating system characterized by:

- A very good control (including transients) of the temperature of the delivered hot water for a large spectrum of flow demands (to avoid the so-called *douche ecossaise*).
- Suppression or significant size reduction of the water balloon.
- Reduction in gas pollution (the burner generates various NOx as a function of the delivered power and CO at each ignition).
- Reduction in noise pollution (noise depends on the power, and additional noise is generated in the ignition phase).
- Achieving the above improvements with little increase in costs.

The first reaction of a control engineer for improving the performances will be to insert a PWM with sufficiently high chopping frequency to command the on/off ratio. This will transform the problem into a standard process

[2] The discussions with C. Pechoux were extremely helpful in preparing this section. Previous contacts with Gaz de France are also acknowledged.

control problem. Unfortunately, even if this seems a feasible solution, it will not allow the objectives to be achieved because:

- A very high on/off frequency has to be considered which is, in general, beyond the limit resulting from the design of the burner.
- The gas pollution will increase due to the increase in on/off frequency.
- The noise pollution will increase.
- In the absence of a control of the air/gas ratio, the dynamic characteristics of the system will be strongly influenced by a number of phenomena such as: outside wind, partial fouling of the smoke pipe, etc...

The natural reaction of a gas burner specialist for improving the heating system is to redesign the burner by reconsidering the combustion problem at the basis (i.e., which implies research in combustion). The idea is to use a different combustion principle with the objective to reduce the gas and noise pollution, as well as the sensitivity with respect to outside wind, partial pipe fouling, etc...

Another approach taken by Théobald S.A. [26] is to use control plus some additional hardware for improving the characteristics of the gas burner which plays the role of the actuator in the heating system and is the source of pollution.

The basic idea is to build a "continuous" type gas burner which can assure a ratio $\frac{\text{max power}}{\text{min power}} \approx 10$ without using an on/off mode of operation. If this can be achieved, one obtains:

- a reduction in noise pollution (for a given power the noise is lower for a continuous type burner - see Fig.21.3),
- a reduction in gas pollution (see Fig.21.3),
- better dynamic characteristics of the system.

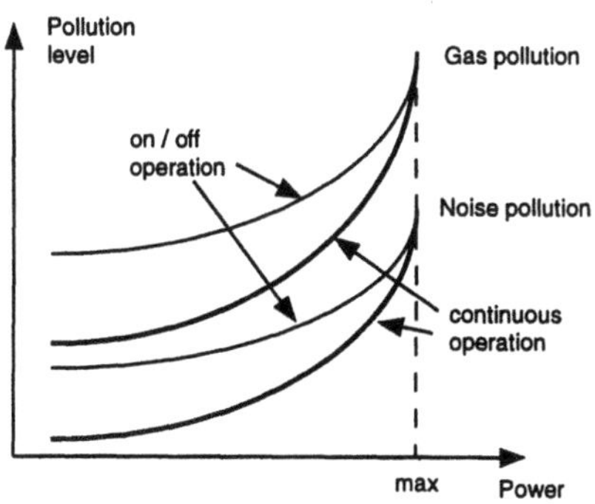

Fig. 21.3. Typical curves of noise and gas pollution for continuous type burners and on/off burners.

The key point for achieving a wide range of delivered power in a continuous type operation of the burner is tight control of the air/gas ratio. This will require the use of secondary actuators to control the air flow (a fan) and the gas flow (a valve) as well as the use of secondary transducers (to control the air and gas flow for a given system is equivalent to the control of a differential pressure). The scheme including the new type of burner with its local control loops is shown in Fig. 21.4. There are two local control loops: a loop on the instantaneous demanded power which acts on the air flow via the fan, and a second loop which controls the air/gas ratio via the differential pressure between the air and the gas and which acts on the gas valve. The particular architecture of these loops takes into account the time scale separation (the gas valve has a time response which is an order of magnitude lower than the time response of the fan). A side effect of the control of the air/gas ratio is

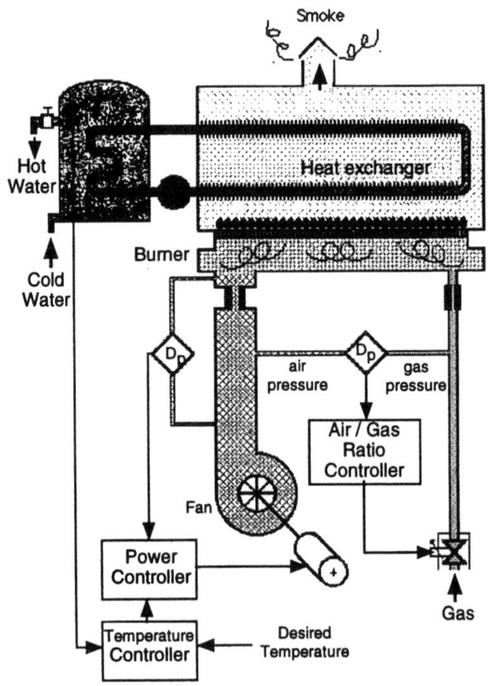

Fig. 21.4. A gas heating system with a continuous type controlled gas burner.

that the parametric disturbances caused to the system depicted in Fig. 21.2 by the external wind and the partial fouling of the smoke pipe (through the variations of the air/gas ratio which was not controlled) are now mainly transformed into disturbances on a controlled variable (the air/gas ratio) which can be removed to a large extent by the control loop.

With this new type of gas burner (which already includes two additional control loops), it becomes possible to make a high performance heating system by solving the control design problems at each level (burner, heating) using the standard route which includes plant model identification followed by controller design based on the identified model.

This example clearly shows:

- Control (more exactly advanced control) competes with other technologies.
- The control designer should interact with the field specialist.

To ignite the interaction, the field specialist has to be aware of the potentiality of control as well as of techniques and tools available for fast and efficient design since these elements are directly related to the expected performances and cost of the project.

3. Controller Design and Validation: From Data to Data

In order to achieve a high performance control system, one needs:

1. the specification of the desired performances
2. the identification (or knowledge) of the dynamic plant model
3. a controller design technique compatible with the desired performances and the plant model
4. a controller implementation technique
5. a procedure for controller validation and on site re-tuning
6. a procedure for controller maintenance

Step 1 concerns not only typical time domain or frequency domain specifications, but also specifications on the various sensitivity functions which allow the robustness of the resulting control system to be assessed with respect to model uncertainities.

Step 2 requires in general identification from data of a dynamic model (or several models) for the plant to be controlled.

Step 3 raises the question of the compatibility of the design method used with the type of model which results from system identification and with the desired performances. Since robustness with respect to plant model uncertainities is one of the key issues in practice, it is reasonable to examine the compatibility between robust control design techniques and the plant models obtained by system identification. Typically, robust control design techniques [5] require a nominal model and uncertainity bounds while system identification provide a plant model which best fit a set of input-output data. Various attempts have been made in order to try to identify a model but also the uncertainities bounds. See [30]. The problem is without doubt difficult and constitutes a current area of research. One way to get around this incompatibility is to consider that what we need is:

- a plant model identified from a set of data which is as close as possible to the data which will be generated by the controller in closed loop,
- an identified (nominal) model which should allow the desired performances to be achieved (or approached) on the true plant,
- to design a controller which in addition to achieving the nominal performances possesses good robustness margins, i.e., the sensitivity functions are constrained within templates which in turn guarantee robust stability (at least) with respect to certain classes of model uncertainities.

This approach can be extended to the case of several models (resulting for example in different operational conditions).

More detailed discussions on the various points raised above can be found for example in [20], [6], [30], [16], [17].

Step 4, controller implementation is beyond the scope of this paper, and will not be discussed.

Step 5 concerns the validation of the designed controller, i.e., comparison of the achieved performances on the real system with the desired performances achieved on the "design" system. (See Fig. 21.5) This may involve time domain comparisons, frequency domain comparisons, pole closeness examination, statistical tests. In fact, since the controller is the same on the real system and on the design system, *controller validation* checks to what extent the model used for the design allows achievement of the desired nominal performances and the constraints on the sensitivity functions. It can also be considered as a validation procedure for the plant model used for the design.

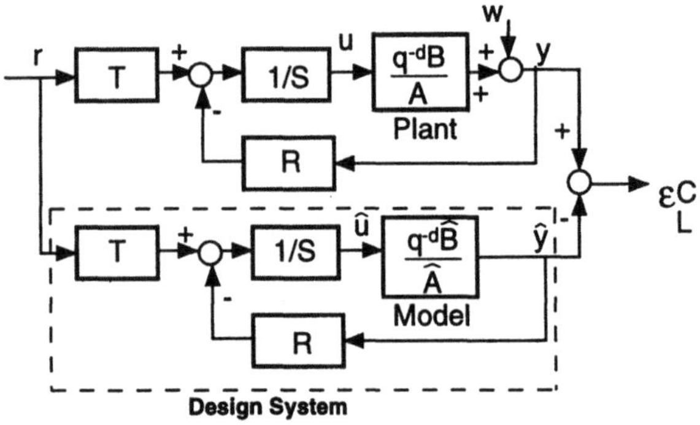

Fig. 21.5. The real system and the design system.

If the real achieved performances do not sufficiently match the desired ones, what can be done in order to obtain a better model for design ? One answer is: make an identification of the plant model in closed loop using the previously designed controller (provided that it stabilizes the real system) and then redesign the controller. One of the reasons for doing this is the fact that the new data which will be used for identification will be closer to those which will be generated by the final controller than the one used in a standard open loop identification. There are also other reasons which will be discussed later.

One can also ask (for the purpose of model validation) what will be the sense of a "good" model in the new identification setting. A good model should allow to give an accurate image of the current closed loop behavior, i.e., replacing the model used for the design in Fig. 21.5 by the new identified model, the discrepancy between the two systems should be reduced (for example, the new model should minimize the error between the output of the two systems called *closed loop output error*).

The use of plant model identification in closed loop in order to improve the design model raises a number of questions:

3.1) Why can plant model identification in closed loop give a better model for design than open loop identification?

3.2) Under what conditions can we obtain the same results using an open loop identification?

3.3) What is the effect of noise on the quality of the plant model identified in closed loop?

3.4) Is there any guarantee that the new plant model identified in closed loop with a given controller will be a better *design model*?

Answers to these questions will be given in Sections 4 and 5.

Controller maintenance (step 6) is nothing more than a new validation of the controller after a certain time of operation and re-tuning (if necessary) of the controller based on a new identified model in closed loop operation.

An important practical remark is that validation of the controller and identification of a new plant model in closed loop can be performed with the same set of data (i.e., same experiment).

If now, one considers repeating a number of times the plant model identification in closed loop and re-design of the controller based on the new identified model, one may expect to come closer and closer to the best design model which will allow to obtain the desired performances and to meet the constraints on the sensitivity functions. This is the fundamental idea of the "iterative identification in closed loop and controller redesign" which was prompted at the beginning of the nineties [2], [32], [27] and had a significant impact in practice. Recent results on this subject are reported in [30], [33].

However, as most of the practical results have shown, the major improvement in performance occurs after the first identification in closed loop. This emphasizes the importance of plant model identification in closed loop which will be discussed in the next section.

4. Identification in Closed Loop and Robust Control

We have seen in the previous section that plant model identification in closed loop operation is motivated by the need of an improved design.

However plant model identification in closed loop is also motivated by other practical considerations, namely:

– identification of plants with integrator behavior, or unstable in open loop
– identification of a plant model when a controller already exists.

A number of methods have been developed and analyzed [21], [28]. However, in recent years, a revival of the interest in plant model identification in closed loop occurred in the context of iterative combination of identification in closed loop and control redesign, (the term "identification for control" has been coined). The result of this is, on the one hand, the development of new algorithms with very interesting features and, on the other hand, a better understanding of the subject. As a consequence, there was also a significant change of attitude. While in the past, identification in closed loop was considered difficult (if badly done, we can end up with the inverse of the controller), it is now considered as an excellent approach which offers a number of practical advantages.

Identification in closed loop cannot be dissociated from the controller and robustness considerations. To make the statements more accurate, a plant model and a controller structure will be considered (the single input single output discrete time context).

The basic scheme of a digital control loop is depicted in Fig. 21.6. The plant model is defined by the transfer operator:

$$G(q^{-1}) = \frac{q^{-d}B(q^{-1})}{A(q^{-1})} \tag{21.1}$$

where:

$$B(q^{-1}) = b_1 q^{-1} + \cdots + b_{n_B} q^{-n_B} = q^{-1}B^*(q^{-1}) \tag{21.2}$$

$$A(q^{-1}) = 1 + a_1 q^{-1} + \cdots + a_{n_A} q^{-n_A} = 1 + q^{-1}A^*(q^{-1}) \tag{21.3}$$

The plant model is operated in closed loop with a two degrees of freedom R-S-T digital controller. The output of the plant operating in closed loop is given by:

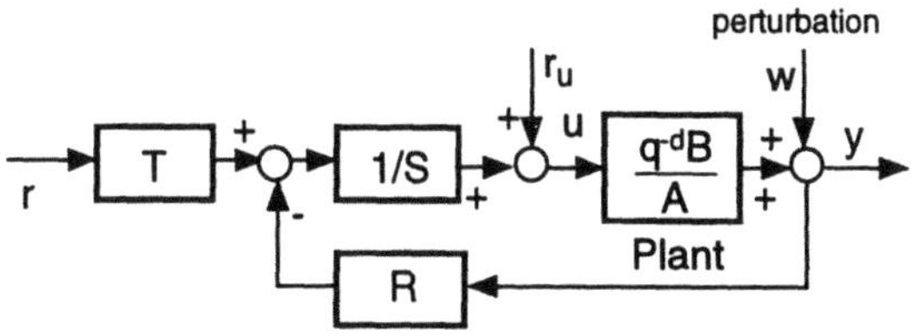

Fig. 21.6. Closed loop system with a R-S-T controller.

$$\begin{aligned} y(t+1) &= -A^* y(t) + B^* u(t-d) + Aw(t+1) \\ &= \theta^T \varphi(t) + Aw(t+1) \end{aligned} \tag{21.4}$$

where $u(t)$ is the plant input, $y(t)$ is the plant output, $w(t)$ is the output disturbance and:

$$\theta^T = [a_1 \cdots a_{n_A}, b_1 \cdots b_{n_B}] \tag{21.5}$$

$$\varphi^T(t) = [-y(t) \cdots y(t-n_A+1), u(t-d) \cdots u(t-n_B+1-d)] \tag{21.6}$$

$$u(t) = -\frac{R(q^{-1})}{S(q^{-1})} y(t) + r'_u(t) \tag{21.7}$$

where r'_u is the equivalent external excitation added to the output of the controller. In general:

$$r'_u = \frac{T(q^{-1})}{S(q^{-1})} r + r_u \tag{21.8}$$

where r is the reference signal and r_u is an external signal added to the output of the controller. For the case $T(q^{-1}) = R(q^{-1})$, the unit feedback is obtained.

A key role for control design, robustness and identification in closed loop is played by the *sensitivity* functions. Those which are of particular interest for what follows will be given next.

The *output sensitivity function* (often called the sensitivity function) is given by:

$$S_{yp}(z^{-1}) = \frac{A(z^{-1})S(z^{-1})}{P(z^{-1})} \tag{21.9}$$

where:

$$P(z^{-1}) = A(z^{-1})S(z^{-1}) + z^{-d}B(z^{-1})R(z^{-1}) \tag{21.10}$$

are the poles of the closed loop.

The *input sensitivity function* is given by:

$$S_{up}(z^{-1}) = -\frac{A(z^{-1})R(z^{-1})}{P(z^{-1})} \tag{21.11}$$

A useful remark to be made is that S_{yp} corresponds to the transfer function from w to y, but it also corresponds to the transfer function between r_u and u which are measurable quantities. As a consequence, S_{yp} can be identified from data.

The modulus of the inverse of the output sensitivity function gives at each frequency the distance of the open loop Nyquist plot with respect to the critical point $(-1, j0)$. This allows *the modulus margin*, a very important robustness margin, to be defined. The *modulus margin* (ΔM), shown in Fig. 21.7, is given by:

$$\Delta M = |S_{yp}^{-1}(j\omega)|_{\min} = (|S_{yp}(j\omega)|_{\max})^{-1} = \|S_{yp}(j\omega)\|_{\infty}^{-1} \tag{21.12}$$

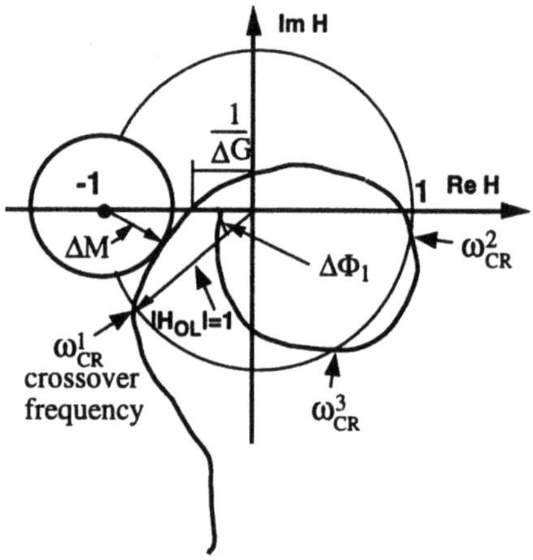

Fig. 21.7. Robustness margins.

Another important robustness margin is the *delay margin* defined as:

$$\Delta\tau = \min_i \frac{\Delta\phi_i}{\omega_{cr}^i} \tag{21.13}$$

where $\Delta\phi_i$ are the phase margins associated with the crossover frequencies ω_{cr}^i.

Sensitivity functions are related to the robust stability of the closed loop with respect to plant model uncertainties. In general for a family of plant models $\mathcal{P} \in \{\hat{G}(z^{-1}), W(z^{-1})\}$ where $\hat{G}(z^{-1})$ is the nominal model and $W(z^{-1})$ defines the size of the frequency dependent model uncertainty, the robust stability condition takes the form:

$$\|S_x(z^{-1})W_x(z^{-1})\|_{\infty} < 1 \tag{21.14}$$

or

$$|S_x(z^{-1})| < |W_x(z^{-1})|^{-1} \ ; \ \forall z = e^{j\omega} \tag{21.15}$$

where S_x and W_x are related to the type of uncertainty. Bounds on the magnitude of the frequency dependent model uncertainties convert into *upper* constraints upon the modulus of the various sensitivity functions. Relationships between the sensitivity functions may induce *lower* constraints on the various sensitivity functions. These constraints are translated into desired templates for the sensitivity functions. Conversely imposed constraints on the sensitivity functions can be translated into tolerated model uncertainties.

Typical templates for S_{yp} and S_{up} are shown in Fig. 21.8. The upper and

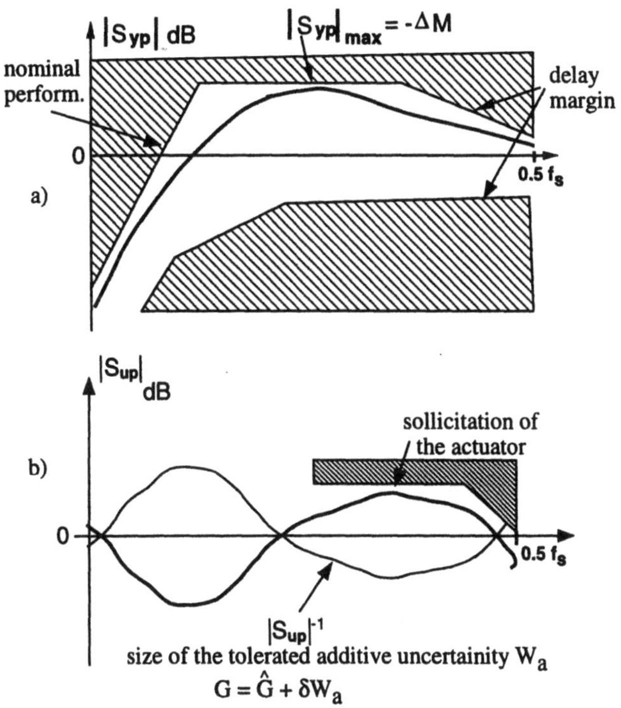

Fig. 21.8. Templates for the sensitivity functions
a) Output sensitivity function
b) Input sensitivity function.

lower bounds on the output sensitivity functions S_{yp} in the high frequencies region come from the translation of the delay margin constraints in the frequency domain [12].

The input sensitivity function reflects both the tolerance with respect to additive uncertainties and the "activity" of the input in the presence of disturbances (high values in certain frequency regions indicate low model uncertainty tolerance and important stress on the actuator).

Techniques for combining nominal performance design with the shaping of the sensitivity functions have been developed and applied in practice for the design of robust digital controllers [16], [17], [19].

Let us now consider that in the scheme of Fig. 21.6 with $r = 0$ (or constant), an excitation signal r_u is applied for identification (typically a PRBS[3]). An equivalent representation of this setting is shown in Fig. 21.9. One can see that the effective input corresponds to the external excitation filtered by the output sensitivity function, but in the meantime the noise problem becomes worse as a consequence of feedback. However, the main remark is that the spectrum of the input signal will be enhanced in the frequency region where the magnitude of the output sensitivity function is high, i.e., where the Nyquist plot is closest to the critical point $(-1, j0)$. This will allow to get a better identified model in a critical region both for stability and performance (this answers question 3.1). Therefore there is a need to develop

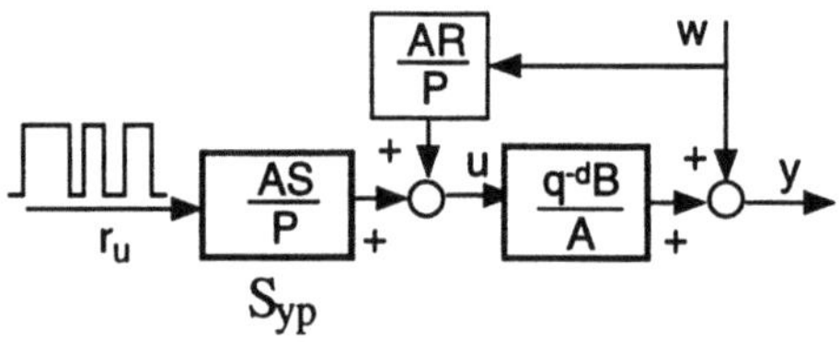

Fig. 21.9. Identification in closed loop - equivalent representation.

closed loop identification algorithms which take advantage of the "improved" input spectrum but are not affected by the noise in the context of feedback.

The answer to question 3.2 is also clear. Yes, open loop identification can be equivalent to closed loop identification (forget about the noise) provided that the excitation signal is filtered by the output sensitivity function S_{yp}. Unfortunately, this information is not available when performing an identification in open loop (except for very particular cases) since S_{yp} depends upon the plant to be identified and the controller.

In this respect identification in closed loop can be viewed as a self-adjusting redesign of the input characteristics used in open loop identification [18].

The other important issue is to see how algorithms for identification in closed loop can be generated such that the parameter estimates will not be affected (asymptotically) by the noise (question 3.3).

---

[3] Pseudo Random Binary Sequence

Let us consider the setting for controller validation and redesign of the controller shown in Fig. 21.5 (which also corresponds to one cycle of the iterative procedure of identification in closed loop and controller redesign).

The upper part represents the true closed loop system and the lower part represents the design system. The objective of the controller validation is to see the discrepancy between the two systems. One way to assess this is to examine the *closed loop output error* (the difference between the outputs of the two closed loop systems). If this error is unsatisfactory, the objective will be first to estimate a better model from data acquired in closed loop which will minimize the closed loop output error in the sense of a certain criterion and then redesign the controller based on the new model. In fact, in this context, the objective of the plant model identification in closed loop is to get a better predictor for the closed loop via a better estimation of the plant model. While this idea in an embryonic form was mentioned in [21] (pp. 393 problem 14T.2), it has not been explored till recently.

The problem of plant model identification as posed above can be viewed in two different ways which however lead to the same algorithms.

- MRAS point of view[4]
  The true closed loop system corresponds to a reference model and a parallel adjustable system (output error type) having a feedback configuration is built up.
- Identification point of view.
  Construct a re-parametrized adjustable predictor for the closed loop system in terms of a known fixed controller and an adjustable plant model.

This is illustrated in Fig. 21.10. Following this path, a family of identification algorithms called "Closed Loop Output Error" Identification Algorithms (CLOE) has been developed [14], [13] under the assumptions that:

4.1) the controller is constant,

4.2) the input-output part of the plant model belongs to the model set,

4.3) an external excitation is applied to the closed loop system.

These algorithms were further analyzed in the following contexts:

4.4) The output disturbance on the system is zero mean, finite power and independent with respect to the external excitation (no specific noise model is considered).

4.5) The plant model does not belong to the model set.

[4] Those familiar with adaptive control literature will remember that MRAS (Model Reference Adaptive Systems) and "re-parametrization" are standard concepts.

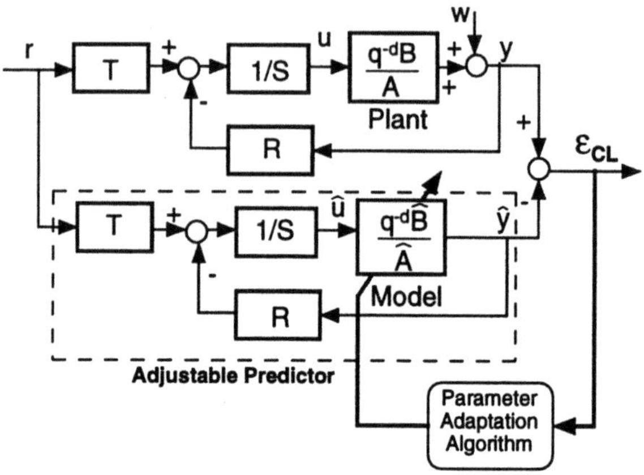

Fig. 21.10. Closed loop output error identification scheme.

Their common feature is the structure of the ajustable predictor which results from Fig. 21.10, i.e.:

$$\hat{y}^0(t+1) = \hat{\theta}^T(t)\phi(t) \tag{21.16}$$

where:

$$\hat{\theta}^T(t) = [\hat{a}_1(t)\cdots\hat{a}_{n_A}(t), \hat{b}_1(t)\cdots\hat{b}_{n_B}(t)] \tag{21.17}$$

$$\phi^T(t) = [-\hat{y}(t).. - \hat{y}(t-n_A+1), \hat{u}(t-d)..\hat{u}(t-n_B+1-d)] \tag{21.18}$$

$$\hat{y}(t) = \hat{\theta}^T(t)\phi(t-1) \tag{21.19}$$

$$\hat{u}(t) = \frac{R(q^{-1})}{S(q^{-1})}\hat{y}(t) + r'_u(t) \tag{21.20}$$

($\hat{y}^0(t)$ is the a priori predicted output and $\hat{y}(t)$ is the a posteriori predicted output). The (a posteriori) closed loop output error takes the form:

$$\epsilon(t+1) = y(t+1) - \hat{y}(t+1) = \frac{S}{P}[\theta - \hat{\theta}(t+1)]^T\phi(t) + w'(t+1) \tag{21.21}$$

where

$$w'(t+1) = \frac{AS}{P}w(t+1) = S_{yp}w(t+1) \tag{21.22}$$

They all use the following parameter adaptation algorithm:

$$\hat{\theta}(t+1) = \hat{\theta}(t) + F(t)\Phi(t)\epsilon_{CL}(t+1) \tag{21.23}$$

$$F(t+1)^{-1} = \lambda_1(t)F(t)^{-1} + \lambda_2(t)\Phi(t)\Phi(t)^T \tag{21.24}$$

$$0 < \lambda_1(t) \le 1\,;\; 0 \le \lambda_2(t) < 2$$

$$\epsilon_{CL}(t+1) = y(t+1) - \hat{y}(t+1) = \frac{y(t+1) - \hat{y}^0(t+1)}{1 + \phi^T(t)F(t)\phi(t)} \tag{21.25}$$

and what is different from one algorithm to another is the observation vector $\Phi(t)$, which is as follows:
CLOE (Closed Loop Output Error)

$$\Phi(t) = \phi(t) \tag{21.26}$$

F-CLOE (Filtered Closed Loop Output Error)

$$\Phi(t) = \frac{S(q^{-1})}{\hat{P}(q^{-1})}\phi(t) \tag{21.27}$$

where $\hat{P}$ is an estimate of the closed loop poles computed on the basis of an initial estimated plant model (obtained for example from open loop identification) and it is given by:

$$\hat{P}(q^{-1}) = \hat{A}(q^{-1})S + q^{-d}\hat{B}(q^{-1})R \tag{21.28}$$

AF-CLOE (Adaptive Filtered Closed Loop Output Error)

$$\Phi(t) = \frac{S(q^{-1})}{\hat{P}(t,q^{-1})}\phi(t) \tag{21.29}$$

with:

$$\hat{P}(t,q^{-1}) = \hat{A}(t,q^{-1})S(q^{-1}) + q^{-d}\hat{B}(t,q^{-1})R(q^{-1}) \tag{21.30}$$

where $\hat{A}(t)$ and $\hat{B}(t)$ correspond to the current plant model estimate.

Under the hypothesis 4.1, 4.2, 4.3 and 4.4, one has the following basic properties:

- global asymptotic stability for any initial parameter estimates (in the absence of noise) is assured for CLOE and F-CLOE[5],
- an asymptotical optimal predictor for the true closed loop system is obtained,
- asymptotically unbiased parameter estimation in the presence of noise is obtained (under appropriate richness conditions).

if:

$$H(z^{-1}) - \frac{\lambda}{2}\ ;\ 2 > \lambda \geq \sup_t \lambda_2(t) \tag{21.31}$$

is a strictly positive real transfer function, where:

$$H = \begin{cases} S/P & \text{for CLOE} \\ \hat{P}/P & \text{for F-CLOE} \end{cases} \tag{21.32}$$

[5] Only local stability results are available for AF-CLOE.

The asymptotic analysis of the bias distribution in the frequency domain for the case where the plant model does not belong to the model set can be done using the method introduced by Ljung [21]. For the case of Closed Loop Output Error methods, the common formula for the estimated parameter vector when N (number of data) $\to \infty$ is [10]:

$$\hat{\theta}^* = \arg\min_{\theta} \int_{-\pi}^{\pi} |S_{yp}(j\omega)|^2[|G(j\omega) - \hat{G}(j\omega)|^2(|\hat{T}_{ur}(j\omega)|^2\phi_r(\omega) + |\hat{S}_{yp}(j\omega)|^2\phi_{r_u}(\omega)) + \phi_w(\omega)]d\omega \quad (21.33)$$

where $G(\omega)$ corresponds to the true input-output plant model, $\hat{G}(\omega)$ corresponds to the estimated plant model, S_{yp} is the true sensitivity function, $\hat{T}_{u_r}$ is the transfer function between the reference r and the input of the estimated model $\hat{u}$, $\phi_r(\omega)$ is the spectral density of the reference signal, $\hat{S}_{yp}$ corresponds to the estimated output sensitivity function based on $\hat{G}$ (i.e., the sensitivity function of the closed loop predictor), $\phi_{r_u}(\omega)$ is the spectral density of the external excitation signal added to the output of the controller, and $\phi_w(\omega)$ is the spectral density of the noise.[6]

This formula clearly shows that:

- the noise does not affect the parameter estimation,
- the frequency distribution of the parameter error (bias) is not only weighted by the sensitivity function in terms of the spectrum (as expected from Fig. 21.9) but is further weighted by the estimated sensitivity function.

In other terms, using this type of algorithm, the precision of the estimation is highly improved in the critical frequency regions (where the Nyquist plot is closest to the critical point) and this is not counteracted by the effect of the noise (whatever is its model).

An alternative to the CLOE methods is the use of standard open loop identification algorithms on filtered plant input-output data acquired in closed loop (Filtered Open Loop identification algorithms - FOL).

This approach is based on the observation that the closed loop output error can also be expressed as a prediction (or output) error on filtered input-output data obtained via time-varying filters depending on the estimates [3], [22], [31].

Unfortunately, the bias distribution is strongly affected by the noise which counteracts the effect of improved excitation due to feedback (the parameter estimates are biased even in the case when the plant model is in the model set). For details see [13], [10].

Among these FOL-algorithms, the use of *open loop output error* [31] is of particular interest for the case when the controller is unknown. Eq.(21.25) can be re-written as:

6 This formula reflects in fact the properties of the predictor structure and of the adaptation error.

$$\epsilon_{CL}(t+1) = \hat{S}_{yp}(t+1, q^{-1})\epsilon_{OE}(t+1) \tag{21.34}$$

where:

$$\hat{S}_{yp} = \frac{\hat{A}S}{\hat{P}} \tag{21.35}$$

and ϵ_{OE} is the classical output error [11].

In other words, the open loop output error recursive algorithm can be used for plant model identification in closed loop provided that the input-output data are filtered through the estimated output sensitivity function. But this can be approximated by an estimation of the true sensitivity function obtained by identification (between r_u and u). This removes the need for knowledge of the controller for identification in closed loop. Unfortunately, the bias problem due to noise remains since in this case [10]:

$$\hat{\theta}^* = \arg\min_{\theta} \int_{-\pi}^{\pi} |\hat{S}_{yp}(j\omega)|^2 [|G(j\omega) - \hat{G}(j\omega)|^2 |S_{yp}(j\omega)|^2 \phi_{r_u}(\omega) + |G(j\omega) - \hat{G}(j\omega) S_{up}(j\omega) + 1|^2 \phi_w(\omega)] d\omega \tag{21.36}$$

where $\hat{S}^*_{yp}$ denotes the identified output sensitivity function.

To overcome this difficulty, the instrumental variable procedure proposed in [29] can be incorporated.

Therefore, when the plant model identification in closed loop has to be performed and the controller is unknown , one can proceed as follows:

1. Identify S_{yp} from data (it corresponds to the transfer function between r_u and u).
2. Generate an instrumental variable $\hat{u} = \hat{S}^*_{yp} r_u$, where $\hat{S}^*_{yp}$ denotes the identified output sensitivity function.
3. Filter $\hat{u}$ and y through $\hat{S}^*_{yp}$.
4. Use an open loop output error identification algorithm.

In this case, the frequency distribution of the bias is approximatively given by:

$$\hat{\theta}^* \approx \arg\min_{\theta} \int_{-\pi}^{\pi} |\hat{S}^*_{yp}(j\omega)|^2 [|G(j\omega) - \hat{G}(j\omega)|^2 |S_{yp}(j\omega)|^2 \phi_{r_u}(\omega) + |S_{yp}(j\omega)|^2 \phi_w(\omega)] d\omega \tag{21.37}$$

which is close to (21.37) obtained for CLOE algorithms. Note however that all these algorithms appear as approximation of the CLOE algorithms. They should normally use time-varying filtering of the input-output data while in general a fixed estimation of the filter is often used in practice.

The conclusions of this comparative presentation of some recent algorithms for identification in closed loop are:

- when the controller is known, CLOE algorithms are probably most suited (do not require use of time-varying filters and the noise does not affect the frequency distribution of the bias),
- when the controller is unknown, a frequency distribution of the bias similar to that for CLOE algorithms is obtained by combining an instrumental variable technique with an output error identification algorithm operating on data filtered through an estimation of the output sensitivity function.

Validation of Models Identified in Closed Loop

Like in open loop identification, validation of the models identified in closed loop is a very important step. The objective of the validation in this context will be to find to what extent the closed loop predictor formed by the model identified in closed loop and the controller used during data acquisition reproduces the behavior of the true closed loop system.

Several validation procedures can be defined [13]:

a) Statistical validation.
   Statistical tests on the closed loop output error.
b) Pole closeness validation.
   Comparison of the identified poles of the true closed loop system with those of the closed loop predictor.
c) Time domain validation.
   Comparison of the time responses for the true closed loop system with those of the closed loop predictor.

Comparative validation of various models available (in particular, comparison with the previous design model) is very useful for selecting the best model.

An important remark to be made is that independently of the criterion used to test the closeness of the two systems, the model validation will be *controller dependent.* This means that the validation will allow to decide whether the new identified model is better than the previous identified design model with respect to the actual controller but, unfortunately, can not guarantee in general that a better design model will be obtained for the re-tuning of the controller (answer to the question 3.4).

Relatively few results guaranteeing the pertinence of the identified model for an improved design are available. When the input-output part of the plant model is in the model set, under hypotheses 4.1, 4.2, 4.3 and 4.4 (presence of noise), provided that the number of data is large ($N \rightarrow \infty$), the exact model will be obtained thus ensuring a better design. For further results see [9].

In fact this problem is related to the convergence of the iterative schemes for identification and control. This will be discussed in the next section.

5. Iterative Identification in Closed Loop and Controller Redesign versus Adaptive Control

An adaptive control system will use informations collected in real time to improve tuning of the controller in real time in order to achieve or to maintain a certain level of performance.

In indirect adaptive control, the controller parameters are generally updated at each sampling instant based on the current estimates of the plant model parameters. However nothing forbids us from updating the estimates of the plant model parameters at each instant but to update the controller parameters only every N instants.

Iterative identification in closed loop and controller redesign can therefore be viewed as an indirect adaptive control scheme. If the plant to be controlled has constant parameters over a large time horizon, one can consider a large horizon N for plant model parameters estimation in closed loop followed by redesign of the controller based on these results. Of course, this procedure can be repeated. However, there is a fundamental difference between these two approaches related to the identification horizon. For small identification horizons, plant parameter estimation in indirect adaptive control is carried out in the presence of a time-varying controller and not necessarily in the presence of an external signal satisfying a persistence of excitation condition. In the iterative approach, estimation of the plant model parameters is carried out with a constant controller and normally in the presence of a persistently exciting external signal. In other words, the iterative procedure introduces a time scale separation between identification in closed loop and controller redesign. This allows a more performance oriented analysis (locally) with respect to the interaction between identification and controller design.

The most advocated concept in iterative identification and controller redesign [32], [27], [3], [30] is that the identification criterion should be the same (or similar) to the measure of the discrepancy between the desired performances and the achieved performances. This concept results from the fact that the discrepancy between the achieved performances and desired performances cannot be minimized simultaneously with respect to controller parameters and plant model parameters. For details see [30].

To obtain an identification criterion which is of the same form as the measure of the discrepancy between the achieved performances and the desired performances (which are of course achieved by the "design" system), algorithms for the identification in closed loop of the form indicated in Section 4 can be used. However, additional filtering of input-output data may be required. The effect of the noise upon the quality of the resulting model will depend upon the type of algorithm used.

As an example, consider the case when pole placement (combined with the shaping of sensitivity functions) is used as a control strategy. In this case, the closed loop output error (see Fig. 21.5) is a measure in the time domain of the discrepancy between the desired and achieved performances in tracking and has the expression:

$$\epsilon_{CL}(t+1) = S_{yp}(q^{-1})[G(q^{-1}) - \hat{G}(q^{-1})]\hat{T}_{ur}(q^{-1})r(t) + S_{yp}(q^{-1})w(t+1) \tag{21.38}$$

where S_{yp} is the achieved sensitivity function and $\hat{T}_{ur}$ is the transfer function between r and $\hat{u}$. Over an infinite time horizon and when $r(t)$ has a Fourrier transform, the discrepancy in performance takes the following expression in the frequency domain:

$$J = \int_{-\pi}^{\pi} |S_{yp}(j\omega)|^2 [G(j\omega) - \hat{G}(j\omega)|^2 |\hat{T}_{ur}(j\omega)|^2 \phi_r(\omega) + \phi_w(\omega)] d\omega \tag{21.39}$$

This is exactly the identification criterion for CLOE identification methods (for $r_u = 0$).

The particular features when using CLOE methods in combination with pole placement (or with any control design method allowing to express the discrepancy between the achieved and the desired performances in terms of the closed loop output error) are: additional data filtering is not necessary[7], the effect of the noise is removed asymptotically.

The practical conclusion is that pole placement with shaping of the sensitivity functions, which seems to be a very efficient design technique for robust digital controller (SISO case), combines nicely with CLOE identification algorithms for iterative identification and controller redesign. Other iterative strategies can also take advantage of the CLOE algorithms leading to simplification of the identification scheme.

Two questions are of interest:

1. What is the interaction between adaptive control and iterative identification and control redesign?
2. Do the iterative identification and controller redesign schemes converge (and in what sense)?

The major benefits of the interaction between these two techniques up to now are:

- the work on identification in closed loop for control design has yielded a design methodology for the data filters used in the parameter estimation part of indirect adaptive control schemes (before only qualitative guidelines were available).
- new recursive algorithms for parameter estimation in closed loop have been developed and they can be used in adaptive control schemes providing some advantages over the classical ones [15].

[7] For other methods the data should be filtered through filters which depend on the estimated model

– the "inspiration" for the development of new closed loop recursive identification algorithms came from a mixture of the two approaches.

Let us turn now to the convergence issue.

As long as the plant model is in the model set and for specific noise models and control strategies, iterative identification and controller redesign can be viewed as indirect adaptive controllers for which convergence proofs used in adaptive control will carry over [8].

When the plant model is not in the model set, under certain hypotheses the problem can be converted into the setting considered for robust adaptive control. As an example, consider the case of feedback type uncertainties on the output (Eq.21.18). The associated scheme is shown in Fig. 21.11. The

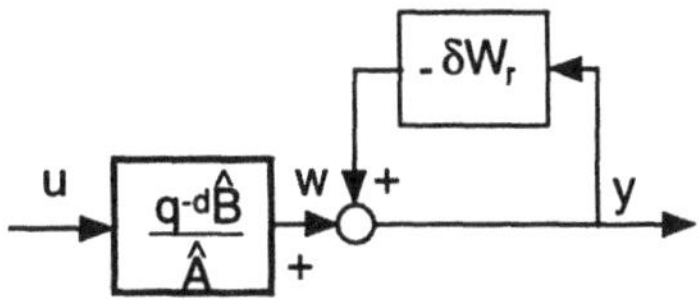

Fig. 21.11. Equivalent representation of uncertainties.

signal:

$$w(t+1) = -\delta(q^{-1})W_r(q^{-1})y(t+1) \tag{21.40}$$

can be interpreted as a disturbance and the output of the true system can be expressed as:

$$y(t+1) = \hat{G}(q^{-1})u(t) + w(t+1) = \hat{\theta}^T\varphi(t) + w(t+1) \tag{21.41}$$

Provided that $[1+\delta(z^{-1})W(z^{-1})]^{-1}$ and $\delta(z^{-1})W(z^{-1})$ are stable transfer functions, the disturbance $w(t+1)$ will satisfy:

$$\|w(t+1)\| \leq c_1 + c_2\|\varphi(t)\| \quad \text{with } 0 < c_1, c_2 < \infty \tag{21.42}$$

This allows the problem of analysis of the iterative schemes to be transformed in a robust adaptive control problem. Using standard techniques of robust adaptation (normalization, projection, deadzone) the boundedness of plant input-output signals and of the closed loop prediction error can be guaranteed under certain conditions [23], [25]. However, this type of proof does not say much about the optimality of the resulting controller (in the sense of the minimization of the discrepancy between the desired performances and the achieved ones). These proofs do not take advantage of the large horizons used for identification (when the true system is linear). Therefore, a specific theory for the convergence of the iterative identification and control redesign has yet to be developed.

Practical observations tend to lead to the idea that after a number of iterations, it is necessary to increase the order of the estimated model to further reduce the discrepancy between the desired and achieved performances.

Despite the lack of a complete theory concerning the convergence of the iterative schemes, their practical impact is remarkable. The first iteration (i.e., first identification in closed loop followed by redesign of the controller) already provides a significant improvement of achieved performances and is becoming a standard procedure for obtaining high performance controllers in practice. A number of applications have been reported in the literature, but this reflects only very partially the true impact in practice of identification in closed loop followed by redesign of the controller.

6. An example

The benefits of identification in closed loop followed by redesign of the controller will be illustrated for the case of a very flexible transmission. The flexible transmission shown in Fig. 21.12 is formed by a set of three pulleys coupled by two very elastic belts. The system is controlled by a PC via an I/O board. Data acquisition, system identification, real-time control is done using the integrated software WinPIM+TR[1]. The sampling frequency is 20Hz.

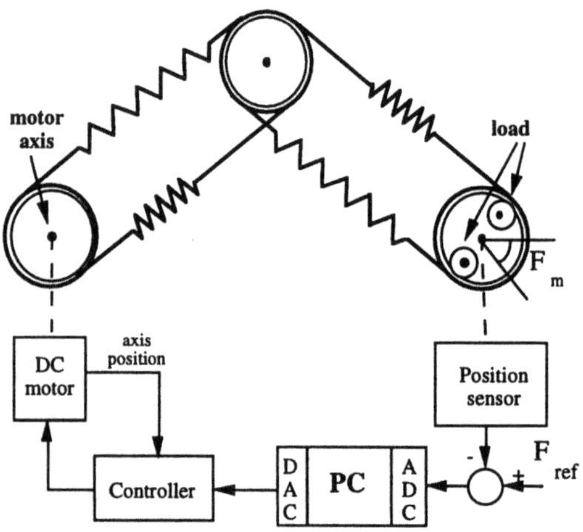

Fig. 21.12. Block diagram of the flexible transmission.

The input sequence for open loop identification is a PRBS generated by a 7-bit shift register and with a clock frequency of $f_s/2$ (the data length is $L = 255$ samples). The output error with extended prediction model algorithm provided the best results in terms of statistical validation [11]. The model obtained in open loop which passes the validation test is:

$$\begin{aligned} A(q^{-1}) &= 1 - 1.3528q^{-1} + 1.5502q^{-2} - 1.2798q^{-3} + 0.9115q^{-4} \\ B(q^{-1}) &= 0.4116q^{-1} + 0.524q^{-2} \quad ; \quad d = 2 \end{aligned}$$

The frequency characteristics of the model are shown in Fig. 21.13 (with continuous line). The main characteristics of the system are: two very oscillatory modes, an unstable zeros and a time delay of two sampling periods.

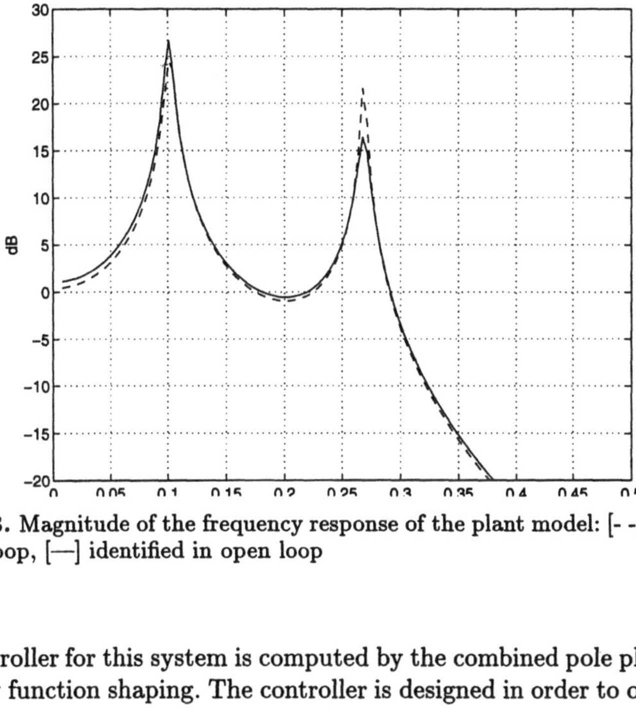

Fig. 21.13. Magnitude of the frequency response of the plant model: [- -] identified in closed loop, [—] identified in open loop

A controller for this system is computed by the combined pole placement/ sensitivity function shaping. The controller is designed in order to obtain two dominant poles with the frequency of the first mode of the open loop model but with a damping factor of 0.8. The constraints are: a modulus margin $\Delta M \geq -6dB$, a delay margin $\Delta\tau > 0.75T_s$ and a maximum for $|S_{up}| < 10dB$ beyond $0.4f_s$.

This controller (called subsequently: *open loop based controller*) was implemented in real time and is also used for system identification in closed loop. The same PRBS as in open loop identification was used as the external excitation at the reference input. The frequency response of the closed loop identified model (using CLOE method) is compared with that of the open loop model in Fig. 21.13.

In order to validate the identified model, the achieved closed loop poles (which can be obtained by identification of the whole closed loop system with standard open loop identification methods) and the computed ones (using the plant model parameters and the R-S-T controller) are given in Fig. 21.14. It is

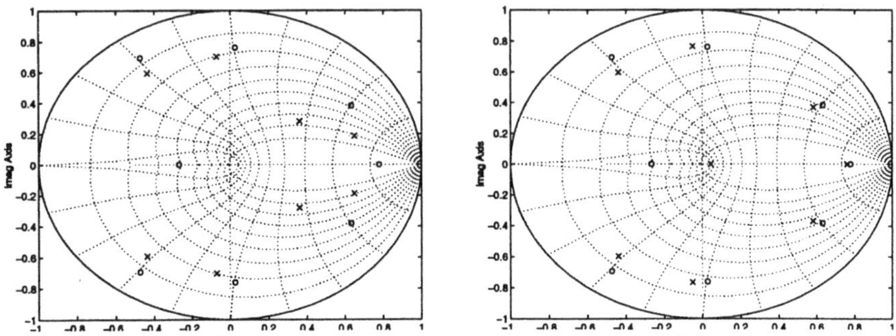

a) b)

Fig. 21.14. Closed loop pole chart.

a) [o] achieved closed loop poles,
[x] computed poles using the open loop identified model.
b) [o] achieved closed loop poles,
[x] computed poles using the closed loop identified model.

observed that the computed closed loop poles using the plant model identified in the closed loop operation are very near to the achieved ones (Fig. 21.14b) whereas this is not the case for the open loop identified model (Fig. 21.14a).

The step response of the real closed loop system also can be compared with the simulated ones using the open loop identified model (Fig.! 21.15a) and using the closed loop identified model (Fig. 21.15b). It can be clearly seen that the step response of the real closed loop system is better approximated with the closed loop identified model than with the model identified in open loop. It should be mentioned that the same controller is always used in real time and simulation experience.

We will now show that a controller design based on the model identified in the closed loop operation (which will be called: *closed loop based controller*) can significantly improve the performances of the system. Fig. 21.16 illustrates the step response of the real closed loop system and of the design system using the new controller designed on the basis of the plant model identified in the closed loop. The same specifications and the same design method as for the case of open loop identified model have been used. An almost perfect superposition of the achieved and designed step response is observed (an improvement of the achieved performances is also obtained). The response of

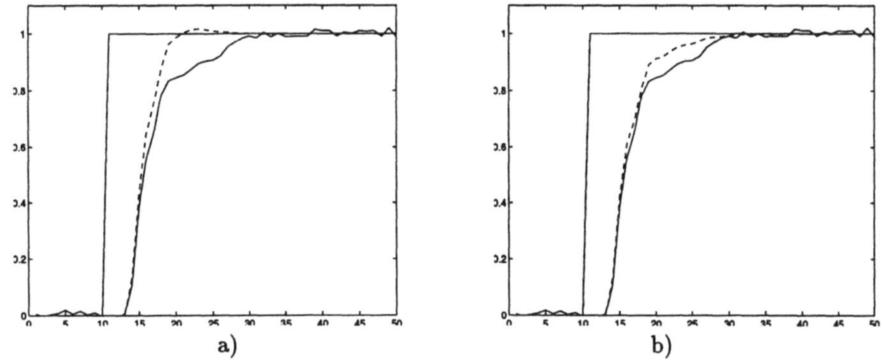

Fig. 21.15. Step response of the closed loop system

a) [—]real-time step response,
   [- -]simulated step response with the open loop identified model
b) [—]real-time step response,
   [- -]simulated step response with the closed loop identified model

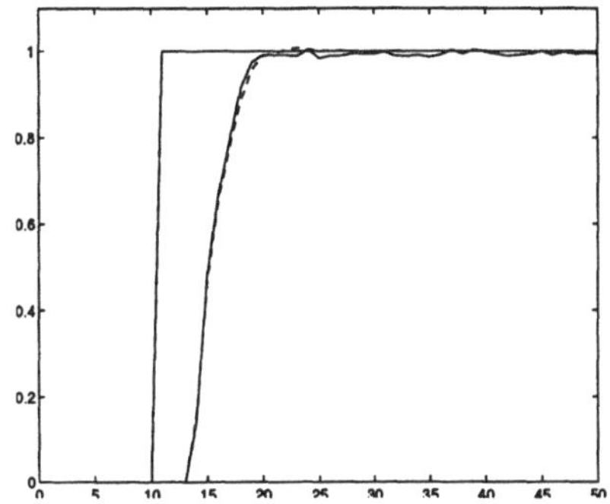

Fig. 21.16. [—] Achieved and [- -] designed step response of the closed loop system with the closed loop based controller

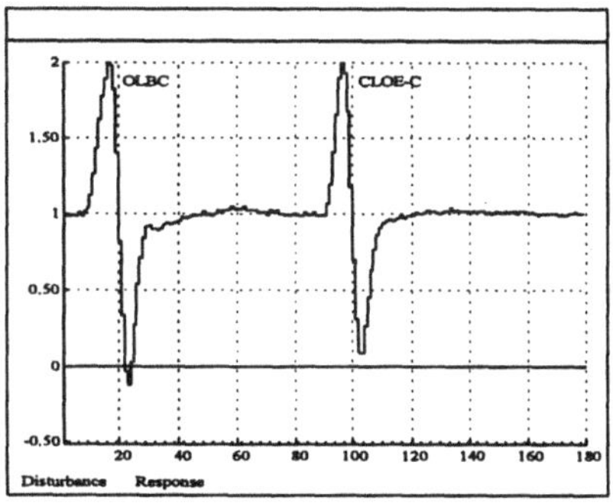

Fig. 21.17. Output disturbance rejection for the open loop based controller (OLBC) and the closed loop based controller (CLOE-C)

the system to a step disturbance at the output of the system for two different controllers is shown in Fig. 21.17. The closed loop based controller is faster than the open loop based controller in terms of the disturbance rejection.

7. Concluding Remarks

Interaction between control and other disciplines, on the one hand, and between subfields of control, on the other hand, is a need for the development of control technology and its use.

Some of the recent results on the interaction between system identification and robust control have a considerable practical impact.

The impact of these results can be explained by the fact that:

- the problems posed have their origin in pertinent questions raised by practice,
- a strong theoretical background in robust control and system identification was available,
- the problems posed at the frontier of two areas generated significant intellectual stimulus within the research community,
- simple and efficient algorithms and strategies have been developed which meet the practical demand.

Acknowledgements

The author gratefully acknowledges his colleagues A. Besançon, A. Karimi, J. Langer and C. Péchoux from Théobald S.A. for their contribution to this work.

References

1. Adaptech. *WimPIM+(includes, WinPIM, WinREG and WinTRAC) System Identification and Control Software. User's manual.* 4, rue du tour de l'eau, 38400 St. Martin-d'Hères, France, 1996.
2. B.D.O. Anderson and R.L. Kosut. Adaptive robust control: on-line learning. In *Proc. 30th IEEE-CDC*, Brighton, UK, 1991.
3. K.J. Åström. Matching criteria for control and identification. In *Proc. European Control Conference*, Groningen, The Netherlands, 1993.
4. R.R. Bitmead. Iterative control design approaches. In *Prepr. 12th IFAC World Congress*, volume 9, pages 381–384, Sydney, Australia, 1993.
5. J.C. Doyle, B.A. Francis, and A.R. Tannenbaum. *Feedback Control Theory.* Mac Millan, N.Y., 1992.
6. M. Gevers. Towards a joint design of identification and control? In *Essays on Control: Perspectives in the Theory and its Applications*, pages 111–152. Birkhäuser, Boston, U.S.A., 1993.
7. M. Gevers. Identification for control. In *Prepr. IFAC Symposium ACASP 95*, Budapest, Hungary, June 1995.
8. G.C. Goodwin and K.S. Sin. *Adaptive Filtering Prediction and Control.* Prentice Hall, N. J., 1984.
9. H. Hjalmarsson, M. Gevers, and F. De Bruyne. For model-based control design, closed loop identification gives better performance. *Automatica*, 32(12):1659–1673, 1996.
10. A. Karimi and I.D. Landau. Comparison of the closed loop identification methods in terms of the bias distribution. Technical report, LAG, January 1997. To appear in *Systems and Control Letters*, 1998.
11. I.D. Landau. *System identification and control design.* Prentice Hall, Englewood Cliffs, N.J., USA, 1990.
12. I.D. Landau. Robust digital control of systems with time delay. *Int. J. of Control*, 62(2):325–347, 1995.
13. I.D. Landau and A. Karimi. Recursive algorithms for identification in closed loop - a unified approach and evaluation. In *Proc. 36 IEEE CDC*, Kobe, Japan (extended version: to appear in Automatica 1997), December 1996.
14. I.D. Landau and A. Karimi. An output error recursive algorithm for unbiased identification in closed loop. *Automatica*, 33(4), 1997.
15. I.D. Landau, M. M'Saad and R. Lozano. *Adaptive Control* Springer Verlag, London, 1997.
16. I.D. Landau, A. Karimi, A. Voda-Besançon, and D. Rey. Robust digital control of flexible transmissions using the combined pole placement/sensitivity function shaping method. *European Journal of Control*, 1(2):122–133, 1995.
17. I.D. Landau, J. Langer, D. Rey, and J. Barnier. Robust control of a 360° flexible arm using the combined pole/placement sensitivity function shaping method. *IEEE T-CST*, 4(4):369–383, 1996.
18. J. Langer and I.D. Landau. Improvement of robust digital control by identification in closed loop, application to a 360° flexible arm. *Control Eng. Practice*, 4(8):1079–1088, 1996.
19. J. Langer and I.D. Landau. Combined pole placement/sensitivity function shaping method using convex optimization criteria. In *Proc. ECC 97*, Bruxelles, Belgium, July 1997.
20. S. Lee, B.D.O. Anderson, R.L. Kosut, and I.M.Y. Mareels. A new approach to adaptive robust control. *Int. J. ACASP*, 7:183–211, 1993.

21. L. Ljung. *System Identification: Theory for the User*. Prentice-Hall, Englewood Cliffs, N.J.,USA, 1987.
22. L. Ljung. Information contents in identification data from closed-loop operation. In *Proc. 32th IEEE-CDC*, San Antonio, Texas, USA, December 1993.
23. R. Lozano and X.H. Zhao. Adaptive pole placement without excitation probing signals. *IEEE Tran.Automatic Control*, 39(1):47–58, 1994.
24. A.G.J. Mac Farlane. Information knowledge and control in essays on control. In *Essays on Control: Perspectives in the Theory and its Applications*, pages 1–28. Birkhäuser, Boston, U.S.A., 1993.
25. R. Ortega and Y. Tang. Robustness of adaptive controllers - a survey. *Automatica*, 25(5):651–677, 1989.
26. C. Péchoux and Y. Bourande. Régulation active du rapport air/gaz d'un brûleur. In *113ème Congrès du Gaz*, Paris, France, September 1996.
27. R.J.P. Schrama. Accurate models for control design: the necessity of an iterative scheme. *IEEE T-AC*, 37:991–994, 1992.
28. T. Söderström and P. Stoica. *System Identification*. Prentice-Hall, U.K., 1989.
29. P. van den Hof and R. Schrama. An indirect method for transfer function estimation from closed loop data. *Automatica*, 29(6), December 1993.
30. P. van den Hof and R. Schrama. Identification and control — closed-loop issues. *Automatica*, 31(12), 1995.
31. A. Voda-Besançon and I.D. Landau. An iterative method for the autocalibration of the digital controllers. application. In *Proc. ECC95*, Rome, Italy, 1995.
32. Z. Zang, R.R. Bitmead, and M. Gevers. Iterative weighted model refinement and control robustness enhancement. In *Proc. 30th IEEE-CDC*, Brighton, UK, 1991.
33. Z. Zang, R.R. Bitmead, and M. Gevers. Iterative weighted least-squares identification and weighted lqg control design. *Automatica*, 31(11), 1995.

Publications (since 1968)

Books

1. I.D. Landau. *Adaptive Control : the Model Reference Approach.* Dekker, New York, 1979. Translated in Chinese (1985).
2. I.D. LANDAU and M. TOMIZUKA. *Theory and Practice of Adaptive Systems.* Ohm Shia, Tokyo, 1981. (in japanese).
3. I.D. LANDAU. *Adaptive Control.* University of California, Continuing Education, Berkeley, 1983. Lecture notes.
4. I.D. LANDAU. *Identification et Commande de Systèmes.* Hermès, 1988. 2nd edition in 1993. Translated in Romanian in 1997.
5. I.D. LANDAU. *System Identification and Control Design.* Prentice Hall, Englewood Cliffs, N.J., U.S.A., 1990.
6. I.D. LANDAU, R. LOZANO, and M. M'SAAD. *Adaptive Control.* Springer, London, 1997.
7. I.D. LANDAU. *Identification des Systèmes.* Collection pédagogique d'automatique. Hermès, Paris, 1998.

Edited Books

1. I.D. LANDAU (Editor). *Outils et Modèles Mathématiques pour l'Automatique l'Analyse de Systèmes et le Traitement du Signal.* Vol.1, 2 et 3, Editions du C.N.R.S., 1981, 1982, 1983.
2. I.D. LANDAU, M. TOMIZUKA, and D. AUSLANDER (Editors). *Adaptive Systems in Control and Signal Processing.* Pergamon Press, 1984.
3. I.D. LANDAU and L. DUGARD (Editors). *La Commande Adaptative - Aspects Pratiques et Théoriques.* Masson, 1986.
4. L. DUGARD and I.D. LANDAU (Editors). *Commande Adaptative - Méthodologie et Applications.* Hermès, 1988.
5. R. DAVID, J.M. DION, and I.D. LANDAU (Editors). *Résultats et Perspectives en Automatique.* Hermès, 1988.

Contributions in Books

1. I.D. LANDAU. Digital model reference adaptive control. In S.G. Tzafestas, editor, *Applied Digital Control.* North Holland, 1985.
2. I.D. LANDAU, F. ROLLAND, C. CYROT, and A. VODA. Régulation numérique robuste. Le placement de poles avec calibrage de la fonction de sensibilité. In A. Oustaloup, editor, *La Robustesse.* Hermès, Paris, 1994.
3. J.M. DEVESNE, I.D. LANDAU, and J. LANGER. Régulation numérique robuste d'un bras manipulateur souple - Banc d'essais comparatif de différentes méthodes de synthèse. In J. Bernussou, editor, *La Commande Robuste.* Hermès, Paris, 1995.

4. I.D. LANDAU, C. CYROT, D. REY, and J. BARNIER. Régulation numérique robuste d'un bras manipulateur souple 360° utilisant le placement de poles avec calibrage de la fonction de sensibilité. In J. Bernussou, editor, *La Commande Robuste*. Hermès, 1995.
5. I.D. LANDAU. Identification and digital control of very flexible mechanical structures. In C. Shaper, editor, *Progresses in Control and Communications*. Birkhauser, N.Y., 1995.
6. I.D. LANDAU. Régulation numérique robuste. In A. Rachid, editor, *Systèmes de régulation*. Masson, 1996.

Journal Papers

1. I.D. LANDAU, A. GROSSU, and St-GAVAT. L'utilisation de la commande adaptive à l'aide d'un modèle dans la régulation des moteurs électriques. *Automatisme*, (4):146–152, 1968.
2. I.D. LANDAU. Wide range speed control of three phases squirrel cage induction motors using static frequency converters. *IEEE Trans. on Industry and General Applications*, IGA-5(1):53–60, 1969.
3. I.D. LANDAU. A hyperstability criterion for model reference adaptive control systems. *IEEE Trans. on Automatic Control*, AC-14(5):552–555, 1969.
4. I.D. LANDAU. Analyse et synthèse des commandes adaptatives à l'aide d'un modèle par des méthodes d'hyperstabilité. *Automatisme*, (7, 8):301–309, 1969.
5. I.D. LANDAU. Les systèmes adaptatifs avec modèle (théorie, mise en oeuvre, application). *Automatisme*, (5):272-291, 1971.
6. I.D LANDAU. Synthesis of discrete model reference adaptive systems. *IEEE, Trans. on Automatic Control (corresp.)*, AC-16(5):507–508, 1971.
7. I.D. LANDAU. A generalization of the hyperstability conditions for model reference adaptive systems. *IEEE Trans. on Automatic Control (corresp.)*, AC-17(2):246–247, 1972.
8. I.D. LANDAU. Model reference adaptive systems - A survey (MRAS - What is possible and why ?). *ASME Trans. Journal of Dynamic Systems, Measurement and Control*, 94 series G(2):119–132, 1972.
9. I.D. LANDAU and B. COURTIOL. Adaptive model following systems for flight control and simulation. *A.I.A.A. Journal of Aircraft*, 9(9):668–674, September 1972.
10. I.D. LANDAU. Hyperstabilité des systèmes adaptatifs avec modèle de référence parallèle. *C.R. de l'Académie des Sciences*, 275 série A:675–678, 1972.
11. I.D. LANDAU. Algorithmes d'adaptation hyperstables pour une classe de systèmes adaptatifs échantillonnés. *C.R. de l'Académie des Sciences*, 275 série A:1391–1394, 1972.
12. I.D. LANDAU. Algorithme pour l'identification à l'aide d'un modèle ajustable parallèle. *C.R. de l'Académie des Sciences*, 277 série A:197–200, 1973.
13. I.D. LANDAU. A survey of model reference adaptive techniques (theory and applications). *AUTOMATICA*, 10(4):II–32, July 1974.
14. I.D. LANDAU and B. COURTIOL. Design of multivariable adaptive model following control systems. *AUTOMATICA*, 10(5):II–34, September 1974.
15. I.D. LANDAU and H. UNBEHAUEN. The development of adaptive systems in Germany and France. *Trans. A.S.M.E. D.S.M.C. Journ.*, 96 séries G(4):405–413, December 1974.

16. I.D. LANDAU. Algorithme d'identification récursive utilisant le concept de positivité. *RAIRO*, série jaune J-2:69–87, May 1975.
17. M. ESPANA and I.D. LANDAU. Approximation bilinéaire des procédés de distillation. *Richerche di Automatica*, December 1975.
18. B. COURTIOL and I.D. LANDAU. High speed adaptation system for controlled electrical drives. *AUTOMATICA (IFAC)*, 11(2):119–127, March 1975.
19. I.D. LANDAU. Unbiased recursive identification using model reference adaptive techniques. *IEEE Trans. on Automatic Control*, AC-20(2), 1976.
20. J.M. DION and I.D. LANDAU. Aperçu de méthodes algébriques utilisées pour l'étude de la commande des systèmes linéaires. *RAIRO série Automatique*, 11(1 et 2):125–160 et 239–256, 1977.
21. I.D. LANDAU. An addendum to unbiased recursive identification using model reference adaptive techniques. *IEEE Trans. on Automatic Control*, AC-22(1), 1978.
22. M. ESPANA and I.D. LANDAU. Reduced order bilinear models for distillation columns. *AUTOMATICA (IFAC)*, 14(4), July 1978.
23. A. GAUTHIER and I.D. LANDAU. On the recursive identification of multi-input, multi-output systems. *AUTOMATICA (IFAC)*, 14(6), November 1978.
24. I.D. LANDAU. Elimination of the real positivity condition in the design of parallel MRAS. *IEEE Trans. on Automatic Control*, AC-23(6), December 1978.
25. M. ESPANA and I.D. LANDAU. Distillation columns - a class of dynamic systems with multiplicative inputs. *Trans. of ASME, D.S.M.C., Journ.*, 101, March 1979.
26. I.D. LANDAU and H.M. SILVEIRA. A stability theorem with applications to adaptive control. *IEEE Trans. on Automatic Control*, AC-24(2), April 1979.
27. H.M. SILVEIRA and I.D. LANDAU. Une nouvelle méthode pour la synthèse des systèmes adaptatifs avec modèle de référence. *RAIRO Systems Analysis and Control*, 13(1), 1979.
28. I.D. LANDAU. Dualité asymptotique entre les systèmes de commande adaptive avec modèle et les régulateurs à variance minimale auto-ajustables. *Ibid*, 14(2), 1980.
29. L. DUGARD and I.D. LANDAU. Recursive output error identification algorithms - Theory and evaluation. *AUTOMATICA (IFAC)*, 16(4), September 1980.
30. L. DUGARD and I.D. LANDAU. Adaptive state estimation using MRAS techniques - Convergence analysis and evaluation. *IEEE Trans. on Automatic Control*, AC-25(5), October 1980.
31. I.D. LANDAU. An extension of a stability theorem applicable to adaptive control. *Ibid*, AC-25(4), August 1980.
32. I.D. LANDAU and M. ESPANA. Identification de modèles bilinéaires pour les colonnes à distiller. *Astérisque 75-76*, 1980.
33. I.D. LANDAU and R. LOZANO. Unification and evaluation of discrete time explicit model reference adaptive control design. *AUTOMATICA (IFAC)*, 17, July 1981.
34. R. LOZANO and I.D. LANDAU. Redesign of explicit and implicit discrete time model reference adaptive control. *Int. Journal of Control*, February 1981.
35. I.D. LANDAU. Model reference adaptive controllers and stochastic self-tuning regulators - A unified approach. *Trans. ASME, J. of Dyn. Syst. Meas. and Contr.*, 103:404–414, December 1981.
36. L. DUGARD and I.D. LANDAU. Stabilité des algorithmes récursifs d'adaptation. *Annales de Télécommunications*, December 1981.
37. I.D. LANDAU. Combining model reference adaptive controllers and stochastic self-tuning regulators. *AUTOMATICA*, January 1982.

38. I.D. LANDAU. Near supermartingales for convergence analysis of recursive identification and adaptive control schemes. *Int. J. of Control*, 35(2):197–226, 1982.
39. L. DUGARD, B. EGARDT, and I.D. LANDAU. Design and convergence analysis of stochastic model reference adaptive controllers. *Int. J. of Control*, 35(5):755–773, 1982.
40. I.D. LANDAU and L. DUGARD. Importance de l'aspect stochastique dans les problèmes de régulation (linéaire et adaptative). *Bulletin de liaison de la Recherche en Informatique et Automatique*, 1982.
41. I.D. LANDAU. Martingale convergence analysis of adaptive schemes - A feedback approach. *IEEE Trans. on A.C.*, June 1982.
42. W.W. LOY and I.D. LANDAU. An on-line procedure for recognition of handprinted alphanumeric characters. *IEEE Trans. PAMI*, July 1982.
43. R. LOZANO and I.D. LANDAU. Quasi-direct adaptive control of nominimum phase systems. *ASME Trans. J. of Dyn. Meas and Control*, 104:311–316, December 1982.
44. I.D. LANDAU. Deterministic and stochastic adaptive poles zeros placement for minimum phase systems. *Ricerche di Automatica*, November 1982.
45. F. ALIX, J.M. DION, L. DUGARD, and I.D. LANDAU. Adaptive control of non-minimum phase systems comparison of several algorithms and improvement. *Ricerche di Automatica*, November 1982.
46. I.D. LANDAU. The links between model reference adaptive controllers and stochastic self-tuning regulators. *Journal of Japanese Assoc. of Instruments and Control Eng.*, 23(5):460–472, 1984.
47. I.D. LANDAU. A feedback system approach to adaptive filtering. *IEEE Trans. on Information Theory*, April 1984.
48. M. M'SAAD, R. ORTEGA, and I.D. LANDAU. Adaptive controllers for discrete-time systems with arbitrary zeros : an overview. *AUTOMATICA*, 21(4):413–423, 1985.
49. I.D. LANDAU, M. M'SAAD, M'SIRDI, and C. CANUDAS. Identification d'un échangeur de chaleur à l'aide d'un apple II. *Revue Générale de Thermique*, (28O):387, April 1985.
50. R. ORTEGA, L. PRALY, and I.D. LANDAU. Robustness of discrete-time direct adaptive controllers. *IEEE Trans. on Automatic Control*, AC-30(6):1179–1187, November 1985.
51. I.D. LANDAU. Logiciels de modélisation et commande - régulation de procédés industriels. *Minis et Micros*, (235):57–61, 1985.
52. I.D. LANDAU. Utilisation des séries temporelles en automatique. *S.M.A.I.*, 1985.
53. I.D. LANDAU, N. M'SIRDI, and M. M'SAAD. Techniques de modélisation récursive pour l'analyse spectrale paramétrique adaptive. *Revue de Traitement du Signal*, 1986.
54. I.D. LANDAU. Model reference adaptive control. *Pergamon Press*, 1988. "Systems and Control Encyclopedia".
55. I.D. LANDAU. Hyperstability and positivity concepts for adaptive control. *Pergamon Press*, 1988. "Systems and Control Encyclopedia".
56. I.D. LANDAU. Parameter adaptation algorithms - A feedback system approach. *Pergamon Press*, 1988. "Systems and Control Encyclopedia".
57. I.D. LANDAU. A unified approach to the design of adaptive controllers for minimum phase systems. *Pergamon Press*, 1988. "Systems and Control Encyclopedia".
58. I.D. LANDAU. Control, adaptive systems. *Encyclopedia of Physical Sciences and Technology*, 1988. Academic Press.

59. I.D. LANDAU and R. HOROWITZ. Application of the passive system approach to the stability analysis of adaptive controllers for robot manipulators. *Int. J. of ACASP*, 3(1):23–38, 1989.
60. M. M'SAAD, I.D. LANDAU, and M. DUQUE. Example applications of the partial state model reference adaptive design technique. *Int. J. of ACASP*, 3(1):155 - 165, 1989.
61. I.D. LANDAU and M. M'SAAD. Evoluzione del controllo adattativo : robustezza e miglioramento delle prestazioni. *Automazione e Strumentazione*, October 1990.
62. M. SAMAAN, I.D. LANDAU, M. M'SAAD, and B. MANGUIN. Expert-ad : un software per il controllo adattativo :descrizione ed application. *Automazione e Strumentazione*, December 1990.
63. M. M'SAAD, I.D. LANDAU, and M. SAMAAN. Further evaluation for partial state model reference adaptive design. *Int. J. Adaptive Control and Signal Processing*, 4(2):133–148, 1990.
64. R. LOZANO-LEAL, B. BROGLIATO, and I.D. LANDAU. Passivity and global stabilization of cascaded nonlinear systems. *IEEE Trans. on Automatic Control*, 37(9):1386–1388, 1992.
65. B. BROGLIATO, I.D. LANDAU, and R. LOZANO-LEAL. Adaptive motion control of robot manipulators : a unified approach based on passivity. *Int. J. on Robust and Nonlinear Control*, 1992.
66. I.D. LANDAU and F. ROLLAND. Identification and digital control of electrical drives. *Control Engineering Practice*, (3), 1993.
67. C. FENOT, F. ROLLAND, G. VIGNERON, and I.D. LANDAU. Open loop adaptive feedback control of deposited zinc in hot dip galvanizing. *Control Engineering Practice*, (5), 1993.
68. I.D. LANDAU. Evolution of adaptive control. *Trans. ASME - Journ. of Dyn. Syst. Meas. Contr.*, 115:381–391, June 1993.
69. B. BROGLIATO, R. LOZANO-LEAL, and I.D. LANDAU. New relationships between Lyapunov functions and the passivity theorem. *Int. J. of Adaptive Control and Signal Processing*, 7:353–365, November 1993.
70. I.D. LANDAU. Recherche en contrôle avancé au CNRS. *Revue Générale d'-Electricité*, December 1993.
71. H.N. DUONG and I.D. LANDAU. On test horizon for model validation by output error. *IEEE Trans. on Automatic Control*, 39(1):102–107, 1994.
72. H.N. DUONG and I.D. LANDAU. On statistical properties of a test for model structure selection using the extended instrumental variable approach. *IEEE Trans. on Automatic Control*, 39(1):211–215, 1994.
73. B.D. ANDERSON and I.D. LANDAU. Least squares identification and the robust strict positive real property. *IEEE Transactions on Circuits and Systems*, 41(9), September 1994.
74. A. VODA and I.D. LANDAU. A method for the auto-calibration of PID controllers. *AUTOMATICA*, 31(1):41–53, 1995.
75. I.D. LANDAU. Robust digital control of systems with time delay. *Int. J. of Control*, June 1995.
76. A. VODA and I.D. LANDAU. The auto-calibration of PI controllers based on two frequency measurements. *Int. Journal of ACASP*, 9(5):395–422, 1995.
77. I.D. LANDAU, A. KARIMI, A. VODA, and D. REY. Robust digital control of a flexible transmission usign the combined pole placement/sensitivity function shaping method. *European J. of Control*, 1(2), 1995.
78. I.D. LANDAU, D. REY, A. KARIMI, A. VODA, and A. FRANCO. A flexible transmission system as a benchmark for robust digital control. *European J. of Control*, 1(2), 1995.

79. N.H. DUONG and I.D. LANDAU. An IV based criterion for model order selection. *AUTOMATICA*, 32:909–914, June 1996.
80. I.D. LANDAU, J. LANGER, D. REY, and J. BARNIER. Robust control of a 360° flexible arm using the combined pole placement/sensitivity function shaping method. *IEEE Trans. on Control Systems Technology*, 4(4):369–383, July 1996.
81. J. LANGER and I.D. LANDAU. Improvement of robust digital control by identification in the closed loop. Application to a 360° flexible arm. *Control Engineering Practice*, 4(12):1637–1646, 1996.
82. I.D. LANDAU and A. KARIMI. An output error recursive algorithm for unbiased identification in closed loop. *AUTOMATICA*, 33(5):933–938, 1997.
83. I.D. LANDAU and A. KARIMI. Recursive algorithms for identification in closed loop - A unified approach and evaluation. *AUTOMATICA*, 33(8):1499–1523, 1997.
84. I.D. LANDAU and A. KARIMI. Robust digital control using pole placement with sensitivity function shaping method. *Int. J. Robust and Nonlinear Control*, 8:191–210, 1998.
85. I.D. LANDAU. The R-S-T digital controller. Design and applications. *Control Engineering Practice*, 6, 1998.
86. I.D. LANDAU and A. KARIMI. A recursive algorithm for ARMAX model identification in closed loop. *IEEE Trans. on Automatic Control*, 43, 1998.

MIX
Papier aus verantwortungsvollen Quellen
Paper from responsible sources
FSC® C105338

If you have any concerns about our products,
you can contact us on
ProductSafety@springernature.com

In case Publisher is established outside the EU,
the EU authorized representative is:
Springer Nature Customer Service Center GmbH
Europaplatz 3, 69115 Heidelberg, Germany

Printed by Libri Plureos GmbH
in Hamburg, Germany